DESIGN AND ANALYSIS OF EXPERIMENTS

Board of Advisors, Engineering

Design and Analysis of Experiments

Douglas C. Montgomery

Georgia Institute of Technology

JOHN WILEY & SONS

New York • Chichester • Brisbane • Toronto • Singapore

Library of Congress Cataloging in Publication Data:

Montgomery, Douglas C
   Design and analysis of experiments.

   Bibliography: p.
   Includes index.
   1. Experimental design.   I. Title.
QA279.M66   001.4'24   76-21075
ISBN 0-471-61421-1

Printed in the United States of America

10 9 8 7 6 5

To Martha, Meredith, and Neil.

Preface

This is an introductory textbook dealing with the statistical design and analysis of experiments. It is an outgrowth of lecture notes from a course in design of experiments that I have taught at the Georgia Institute of Technology over the past six years. The book is intended for readers who have completed a first course in statistical methods. The mathematical maturity that results from calculus would be helpful, but not essential, and some familiarity with matrix algebra is required in the latter part of Chapter 13 and Chapter 14.

Because of the relatively modest mathematical prerequisites, the book can be used in a second statistics course for undergraduate engineering, physical science, mathematics, biology, and social science students. The material can be covered more rapidly, with more emphasis on the mathematical aspects of the subject, in an experimental design course for first-year graduate students. There are numerical examples illustrating most of the design techniques, and this makes the book useful as a reference work for experimenters in various disciplines.

The book contains 15 chapters. Chapter 1 presents the basic philosophy of the statistical approach to experimental design. Chapter 2 reviews elementary statistical methods, and introduces terminology and notation used in subsequent chapters.

In Chapter 3, we begin the study of designs for experiments with a single factor. The analysis of variance is introduced as the appropriate method of statistical analysis. Chapters 4 and 5 continue the development of single-factor experiments, with randomized blocks, Latin squares, and related designs discussed in Chapter 4. Chapter 5 introduces incomplete block designs. Factorial designs are introduced in Chapter 6. Chapter 7 presents a set of rules for deriving computing formulas for sums of squares and expected mean squares for any balanced multifactor design.

The 2^k and 3^k factorial designs are introduced in Chapter 8. The 2^k and 3^k factorial designs may be run in incomplete blocks by sacrificing information on certain interactions. Procedures for constructing and analyzing these designs are given in Chapter 9. The relatively high cost of industrial experimentation has led to the extensive use of fractional 2^k and 3^k factorial designs, which are discussed in Chapter 10. The basic presentation of multifactor designs is continued in Chapter 11, which discusses nested arrangements. These are nonfactorial designs in which the levels of one factor are contained entirely within the levels of another factor. Chapter 12 illustrates how randomization restrictions are employed in multifactor experiments. An example of such a design would be running a factorial experiment in a randomized block.

Regression analysis is introduced in Chapter 13 as a methodology for the analysis of unplanned experiments. In Chapter 14 we discuss response surface methodology, a collection of mathematical and statistical techniques for determining the optimum operating conditions for industrial processes. This chapter concludes with a section on evolutionary operation, a process control method developed initially for chemical plants. The final chapter treats the analysis of covariance, which, like blocking, is a methodology for improving the precision of comparisons between treatments.

The book contains more material than can usually be covered comfortably in a first course. Hopefully, the instructor will be able to vary his course content from one offering to another, or perhaps discuss certain topics in greater depth, depending on class interest. There are problem sets at the end of each substantive chapter. These problems vary in scope from computational exercises designed to reinforce the fundamentals of the analysis of variance to extensions or elaborations of basic principles.

Many individuals have contributed toward the completion of the book. I thank Dr. R. N. Lehrer for his support and for his providing resources to develop the manuscript. I have benefited from reviews of the manuscript by Professors H. M. Wadsworth, K. S. Stephens, R. L. Rardin, R. G. Heikes, L. A. Johnson, and R. V. Fuller of the Georgia Institute of Technology; Professor A. L. Dorris of the University of Oklahoma; Professor J. J. Moder of the University of Miami; and Professor C. J. Tompkins of the University of Virginia. Their constructive criticism and suggestions have substantially improved the book. Various drafts of the manuscript have been used in my classes over the past two years, and I am grateful to the many students who corrected numerical and typographical errors, and provided critical evaluation. J. R. Black, G. N. Gollobin, R. E. Rosenthal, and R. D. Stewart were particularly helpful in this respect. I am indebted to Professor E. S. Pearson and the *Biometrika* Trustees, Wiley, Prentice-Hall, The Ronald Press, the editor of *Biometrics*, the editor of the *Annals of Statistics*, the editor of *Technometrics*, the Institute of Mathematical Statistics, and the American Statistical Association for permission to use copyrighted material. Finally, I thank Ms. Jeanie Hagen, Ms. Kaye Watkins, and Ms. Amelia L. N. Williams for typing the several drafts of the manuscript.

Douglas C. Montgomery
Atlanta, Georgia

Contents

Chapter One
Introduction

1-1 THE NEED FOR DESIGNED EXPERIMENTS

Experiments are carried out by investigators in all fields of study either to discover something about a particular process or to compare the effect of several conditions on some phenomena. For example, suppose a metallurgical engineer is interested in studying the effect of two different hardening processes, oil quenching and salt water quenching, on an aluminum alloy. Here the objective of the experimenter is to determine the quenching solution that produces the maximum hardness for this particular alloy. The engineer decides to subject a number of alloy specimens to each quenching medium and measure the hardness of the specimens after quenching. The average hardness of the specimens treated in each quenching solution will be used to determine which solution is best.

As we think about this experiment, a number of important questions come to mind.

1. Are these two solutions the only quenching media of potential interest?
2. Are there any other factors that might affect hardness that should be investigated or controlled in this experiment?
3. How many specimens of alloy should be tested in each quenching solution?

4. How should the specimens be assigned to the quenching solutions, and in what order should the data be collected?
5. What method of data analysis should be used?
6. What difference in average observed hardness between the two quenching media will be considered important?

All of these questions, and perhaps many others, will have to be satisfactorily answered before the experiment is performed.

In any experiment, the results and conclusions that can be drawn depend to a large extent on the manner in which the data were collected. To illustrate this point, suppose that the metallurgical engineer in the above experiment used specimens from one heat in the oil quench and specimens from a second heat in the salt water quench. Now when the mean hardness is compared, the engineer is unable to say how much of the observed difference is due to the quenching media and how much is due to inherent differences between the heats.[1] Thus the method of data collection has adversely affected the conclusions that can be drawn from the experiment.

1-2 BASIC PRINCIPLES OF EXPERIMENTAL DESIGN

If an experiment is to be performed most efficiently, then a scientific approach to planning the experiment must be considered. By the *statistical design of experiments*, we refer to the process of planning the experiment so that appropriate data will be collected, which may be analyzed by statistical methods resulting in valid and objective conclusions. The statistical approach to experimental design is necessary if we wish to draw meaningful conclusions from the data. When the problem involves data that are subject to experimental errors, statistical methodology is the only objective approach to analysis. Thus, there are two aspects to any experimental problem: the design of the experiment and the statistical analysis of the data. These two subjects are closely related, since the method of analysis depends directly on the design employed.

The two basic principles of experimental design are *replication* and *randomization*. By replication we mean a repetition of the basic experiment. In the metallurgical experiment above, a replication would consist of a specimen treated by oil quenching and a specimen tested by salt water quenching. Thus, if five specimens are treated in each quenching medium, we say that five *replicates* have been obtained. Replication has two important properties. First, it allows the experimenter to obtain an estimate of the experimental error. This estimate of error becomes a basic unit of measurement for determining whether observed differences in the data are really *statistically* different. Second, if the sample mean (e.g., $\bar{y}$) is used to estimate the effect of a factor in the experiment, then replication permits the experimenter to obtain a more precise es-

---

[1] A statistician would say that the effects of quenching media and heat were *confounded*; that is, the two effects cannot be separated.

timate of this effect; for if σ^2 is the variance of the data, and there are n replicates, then the variance of the sample mean is

$$\sigma_{\bar{y}}^2 = \frac{\sigma^2}{n}$$

The practical implication of this is that if we had $n = 1$ replicate, and observed $y_1 = 145$ (oil quench) and $y_2 = 147$ (salt water quench) we would probably be unable to make satisfactory inferences about the effect of the quenching medium. That is, the observed difference could be due to experimental error. On the other hand, if n was reasonably large, and the experimental error was sufficiently small, then if we observed $\bar{y}_1 < \bar{y}_2$, we would be reasonably safe in concluding that salt water quenching produces a higher hardness in this particular aluminum alloy than does oil quenching.

Randomization is the cornerstone underlying the use of statistical methods in experimental design. By randomization we mean that both the allocation of the experimental material and the order in which the individual runs or trials of the experiment are to be performed are randomly determined. Statistical methods require that the observations (or errors) are independently distributed random variables. Randomization usually makes this assumption valid. By properly randomizing the experiment, we will also assist in "averaging out" the effects of extraneous factors that may be present. For example, suppose that the specimens in the above experiment are of slightly different thicknesses, and the effectiveness of the quenching medium may be affected by specimen thickness. If all the specimens subjected to the oil quench are thicker than those subjected to the salt water quench, then we may be continually handicapping one quenching medium over the other. Randomly assigning the specimens to the quenching media will alleviate this problem.

In order to use the statistical approach to designing and analyzing an experiment, it is necessary that everyone involved in the experiment have a clear idea in advance of exactly what is to be studied, how the data is to be collected, and at least a qualitative understanding of how this data is to be analyzed. An outline of the recommended procedure is as follows:

1. **Recognition of and statement of the problem.** This may seem to be a rather obvious point, but in practice it is often not simple to realize that a problem requiring experimentation exists, and to develop a clear and generally accepted statement of this problem. It is necessary to develop all ideas about the objectives of the experiment. A clear statement of the problem often contributes substantially to a better understanding of the phenomena and the final solution of the problem.

2. **Choice of factors and levels.** The experimenter must select the independent variables or factors to be investigated in the experiment. For example, in the hardness testing experiment described previously, the single factor is quenching media. The factors in an experiment may be either quantitative or qualitative. If they are quantitative, thought should be given as to how these factors are to be controlled at the desired values and measured. We must also select the values or levels of the factors to be used in the experiment. These levels may be chosen specifically, or selected at random from the set of all possible factor levels.

3. **Selection of a response variable.** In choosing a response or dependent variable, the experimenter must be certain that the response to be measured really provides information about the problem under study. Thought must also be given to how the response will be measured, and the probable accuracy of those measurements.

4. **Choice of experimental design.** This step is of primary importance in the experimental process. The experimenter must determine the difference in true response he wishes to detect and the magnitude of the risks he is willing to tolerate so that an appropriate sample size (number of replicates) may be chosen. He must also determine the order in which the data will be collected and the method of randomization to be employed. It is always necessary to maintain a balance between statistical accuracy and cost. Most recommended experimental designs are both statistically efficient and economical, so that the experimenter's efforts to obtain statistical accuracy usually result in economic efficiency. A mathematical model for the experiment must also be proposed, so that a statistical analysis of the data may be performed.

5. **Performing the experiment.** This is the actual data collection process. The experimenter should carefully monitor the progress of the experiment to insure that it is proceeding according to the plan. Particular attention should be paid to randomization, measurement accuracy, and maintaining as uniform an experimental environment as possible.

6. **Data analysis.** Statistical methods should be employed in analyzing the data from the experiment. Numerical accuracy is an important concern here, although present-day computers have largely relieved the experimenter from this problem, and simultaneously reduced the computational burden. Graphical methods are also frequently useful in the data analysis process.

7. **Conclusions and recommendations.** Once the data has been analyzed, the experimenter may draw conclusions or inferences about his results. The statistical inferences must be physically interpreted, and the practical significance of these findings evaluated. Then recommendations concerning these findings must be made. These recommendations may include a further round of experiments, as experimentation is usually an *iterative* process, with one experiment answering some questions and simultaneously posing others. In presenting his results and conclusions to others, the experimenter should be careful to minimize the use of unnecessary statistical terminology, and phrase his information as simply as possible. The use of charts and graphs is a very effective way to present important experimental results to management.

In this book we will concentrate primarily on step 4, the choice of experimental design, and step 6, the statistical analysis of the data. However, throughout the book we will emphasize the importance of the entire seven-step process.

1-3 HISTORICAL PERSPECTIVE

The late Sir Ronald A. Fisher was the innovator in the use of statistical methods in experimental design. For several years he was responsible for statistics and data analysis at the Rothamsted Agricultural Experiment Station in London, England. Fisher developed and first used the analysis of variance as the primary method of statistical analysis in experimental design. Frank Yates worked with Fisher at the Rotham-sted station, and the two collaborated on many projects. Yates also became a primary contributor to the literature of experimental design. In 1933 Fisher took a professor-ship at the University of London. He later was on the faculty of Cambridge Univer-sity, and held visiting professorships at several universities throughout the world.

Many of the early applications of experimental design methodology were in the agricultural and biological sciences. As a result, much of the terminology of the dis-cipline is derived from this agricultural background. For example, an agricultural scientist may plant a variety of a crop in several plots, then apply different fertilizers or *treatments* to the plots, and observe the effect of the fertilizers on crop yield. Each plot will produce one observation on yield. An engineer or physical scientist, how-ever, will think of the independent variable or *factor* that affects his response (rather than a treatment) and use the term *run* to characterize one observation. However, much experimental design terminology, such as "treatment," "plot," and "block" have lost their strictly agricultural connotation, and are in wide use in many fields of application. We will use the phrases "levels of the factor" and "treatment" inter-changeably. Sometimes "treatment combination" will be used to denote a particular combination of factor levels to be used in one run of the experiment.

Modern-day experimental design methods are widely employed in all fields of in-quiry. Agricultural science, biology, medicine, the engineering sciences, the physical sciences, and the social sciences are disciplines where the statistical approach to the design and analysis of experiments is an accepted practice.

1-4 AN EXAMPLE OF A DESIGNED EXPERIMENT

To illustrate some of the preceding ideas, we present an example of a designed experi-ment. A hardness testing machine presses a rod with a pointed tip into a metal spec-imen with a known force. By measuring the depth of the depression caused by the tip, the hardness of the specimen is determined. Two different tips are available for this machine, and although the precision (variability) of the two tips seems to be the same, it is suspected that one tip produces different hardness readings than the other.

An experiment could be performed as follows. A number of metal specimens (e.g., 10) could be randomly selected. Half of these specimens could be tested by tip 1 and the other half by tip 2. The exact assignment of specimens to tips would be randomly determined. After the hardness data have been collected, the average hardness of the two samples could be compared using the t test. That is, if n_1 specimens are tested with tip 1 and n_2 specimens are tested with tip 2, and $\bar{y}_1$ and $\bar{y}_2$ are the resulting aver-age hardness from tips 1 and 2, respectively, then assuming that the hardness data is normally distributed, the test statistic is

$$t_0 = \frac{\bar{y}_1 - \bar{y}_2}{S_p \sqrt{\dfrac{1}{n_1} + \dfrac{1}{n_2}}} \qquad (1\text{-}1)$$

where

$$S_p = \left[\frac{\sum_{j=1}^{n_1} (y_{1j} - \bar{y}_1)^2 + \sum_{j=1}^{n_2} (y_{2j} - \bar{y}_2)^2}{n_1 + n_2 - 2} \right]^{1/2} \qquad (1\text{-}2)$$

is the pooled or combined estimate of the variability (experimental error). Formally stated, the hypothesis we are testing is

$$H_0: \mu_1 = \mu_2$$

$$H_1: \mu_1 \neq \mu_2$$

where μ_1 and μ_2 are the true means of the distributions of hardness values produced by tips 1 and 2, respectively. We will reject the null hypothesis $H_0: \mu_1 = \mu_2$ if

$$|t_0| > t_{\alpha/2, n_1 + n_2 - 2}$$

where $t_{\alpha/2, n_1 + n_2 - 2}$ is the upper $\alpha/2$ percentage point of the t distribution with $n_1 + n_2 - 2$ degrees of freedom.

A little reflection will reveal a serious disadvantage of this design. Suppose the metal specimens were cut from different bar stock that were produced in different heats, or were not exactly homogeneous in some other way that might affect the hardness. This lack of homogeneity between specimens will contribute to the variability of the hardness measurements and tend to inflate the experimental error, thus making a true difference between tips harder to detect.

To protect against this possibility, consider an alternate experimental design. Assume that each specimen is large enough so that *two* hardness determinations may be made on it. This alternative design would consist of dividing each specimen into two parts, then randomly assigning one tip to one half of each specimen and the other tip to the remaining half. The order in which the tips are tested for a particular specimen would also be randomly selected. The experiment, when performed according to this design with 10 specimens, produced the data shown in Table 1-1.

Table 1-1 Data for the Hardness Testing Experiment

Specimen	Tip 1	Tip 2
1	7	6
2	3	3
3	3	5
4	4	3
5	8	8
6	3	2
7	2	4
8	9	9
9	5	4
10	4	5

We may write a statistical model that describes the data from this experiment as

$$y_{ij} = \mu_i + \beta_j + \epsilon_{ij} \quad \begin{cases} i = 1, 2 \\ j = 1, 2, \ldots, 10, \end{cases} \tag{1-3}$$

where y_{ij} is the observation for tip i on specimen j, μ_i is the true mean hardness of the ith tip, β_j is an effect on hardness due to the jth specimen, and ϵ_{ij} is a random experimental error. Note that if we compute the jth paired difference, for example,

$$d_j = y_{1j} - y_{2j} \quad j = 1, 2, \ldots, 10 \tag{1-4}$$

the expected value of this difference is

$$\mu_d = E(d_j)$$
$$= E(y_{1j} - y_{2j})$$
$$= E(y_{1j}) - E(y_{2j})$$
$$= \mu_1 + \beta_j - (\mu_2 + \beta_j)$$
$$= \mu_1 - \mu_2$$

That is, we may make inferences about the difference in means of the two tips $\mu_1 - \mu_2$ by making inferences about the mean of the differences μ_d. Notice that the additive effect of the specimens β_j cancels out when the observations are paired in this manner.

Testing $H_0: \mu_1 = \mu_2$ is equivalent to testing

$$H_0: \mu_d = 0$$
$$H_1: \mu_d \neq 0$$

The test statistic for this hypothesis is

$$t_0 = \frac{\bar{d}}{S_d/\sqrt{n}} \tag{1-5}$$

where

$$\bar{d} = \frac{1}{n} \sum_{j=1}^{n} d_j \tag{1-6}$$

is the sample mean of the differences and

$$S_d = \left[\frac{\sum_{j=1}^{n} (d_j - \bar{d})^2}{n-1} \right]^{1/2} = \left[\frac{\sum_{j=1}^{n} d_j^2 - \frac{1}{n} \left(\sum_{j=1}^{n} d_j \right)^2}{n-1} \right]^{1/2} \tag{1-7}$$

is the sample standard deviation of the differences. $H_0: \mu_d = 0$ would be rejected if $|t_0| > t_{\alpha/2, n-1}$. For the data in Table 1-1, we find

$$d_1 = 7 - 6 = 1 \qquad d_6 = 3 - 2 = 1$$

$$d_2 = 3 - 3 = 0 \qquad\qquad d_7 = 2 - 4 = -2$$

$$d_3 = 3 - 5 = -2 \qquad\qquad d_8 = 9 - 9 = 0$$

$$d_4 = 4 - 3 = 1 \qquad\qquad d_9 = 5 - 4 = 1$$

$$d_5 = 8 - 8 = 0 \qquad\qquad d_{10} = 4 - 5 = -1$$

Thus,

$$\bar{d} = \frac{1}{n} \sum_{j=1}^{n} d_j = \frac{1}{10}(-1) = -0.10$$

$$S_d = \left[\frac{\sum_{j=1}^{n} d_j^2 - \frac{1}{n} \left(\sum_{j=1}^{n} d_j \right)^2}{n-1} \right]^{1/2} = \left[\frac{13 - \frac{1}{10}(-1)^2}{10-1} \right]^{1/2} = 1.21$$

The test statistic is

$$t_0 = \frac{\bar{d}}{S_d/\sqrt{n}}$$

$$= \frac{-0.10}{1.21/\sqrt{10}}$$

$$= -0.26$$

and since $t_{.05,9} = 1.833$, we cannot reject the hypothesis $H_0: \mu_d = 0$. That is, there is no evidence to indicate that the two tips produce different hardness readings.

The initial design proposed for this experiment is called the *pooled* or *independent t* test, because the *experimental units* (the metal specimens) are assigned randomly to the two tips. It is an example of a *completely randomized* design for a single *factor* (the tips). In Chapter 3 we will investigate this type of design in more detail, indicating situations where it is appropriate, and offer a method of analysis that is a generalization of the pooled t test.

The design actually used for this experiment is called the *paired t* test. Actually, it is a special case of a more general type of design called the *randomized block* design. The term "block" refers to a relatively homogeneous experimental unit (in our case, the metal specimens are the blocks), and they represent a restriction on complete randomization because the treatment combinations are only randomized within the block. We shall study designs of this type in Chapters 4 and 5. In those chapters the mathematical model for the design, Equation 1-3, will be written in a slightly different form.

Before leaving this experiment, several points should be made. Note that although $2n = 2(10) = 20$ observations have been taken, only $n - 1 = 9$ degrees of freedom are available for the t statistic (1-5) (we know that as the degrees of freedom for t increase the test becomes more sensitive). By pairing, we have effectively "lost" $n - 1$ degrees of freedom, but hopefully we have gained a better knowledge of the situation by eliminating an additional source of variability (the difference between specimens). We may obtain an indication of the quality of information produced from the paired design by comparing the standard deviation of the differences S_d with the pooled standard deviation S_p that would have resulted, had the experiment been conducted in a completely randomized manner and the data of Table 1-1 obtained. Using the data in Table 1-1 as two independent samples, we may compute the pooled standard deviation from (1-2) as $S_p = 2.32$. Comparing this value to $S_d = 1.21$, we see that blocking or pairing has reduced the estimate of variability by nearly 50 percent. We may also express this information in terms of a confidence interval on $\mu_1 - \mu_2$. Using the paired data, a 95 percent confidence interval on $\mu_1 - \mu_2$ is

$$\bar{d} \pm t_{.025,9} \, S_d/\sqrt{n}$$

$$-0.10 \pm (2.262)\,(1.21)/\sqrt{10}$$

$$-0.10 \pm 0.866,$$

while using the pooled or independent analysis, a 95 percent confidence interval on $\mu_1 - \mu_2$ is

$$\bar{y}_1 - \bar{y}_2 \pm t_{.025,18} \, S_p \sqrt{\frac{1}{n_1} + \frac{1}{n_2}}$$

$$4.80 - 4.90 \pm (2.101)\,(2.32) \sqrt{\frac{1}{10} + \frac{1}{10}}$$

$$-0.10 \pm 2.18$$

The confidence interval based on the paired analysis is much narrower than the confidence interval from the independent analysis.

Blocking is not always the best design strategy. If the within-block variability is the same as the between-block variability, then the variance of $\bar{y}_1 - \bar{y}_2$ will be the same regardless of which design is used. Actually, blocking in this situation would be a poor choice of design, since blocking results in the loss of $n - 1$ degress of freedom, and will actually lead to a wider confidence interval on $\mu_1 - \mu_2$. A further discussion of blocking will be given in Chapter 4.

Chapter Two
Fundamental Statistical Concepts

2-1 INTRODUCTION

In this chapter we will review some elementary statistical methods from the point of view of designing simple comparative experiments. These are experiments in which the parameter of a distribution (either the mean or the variance) is compared against either a standard value or the corresponding parameter of a second distribution.

The mathematical basis of statistical inference is probability theory. Let y be the random variable of interest in an experiment, and suppose that the probability structure of y is described by a probability distribution. For example, in an experiment designed to study the burning rate of a particular type of solid-fuel rocket propellant, the random variable of interest would be the observed burning rate. Random variables may be either discrete or continuous. If the set of all possible values of the random variable y is either *finite* or *countably infinite*, then the random variable is *discrete*, and its probability structure is described by the probability function $p(y)$. However, if the set of all possible values of y is an *interval*, then the random variable is continuous, and its probability structure is characterized by the probability density function $f(y)$. Frequently, $p(y)$ and $f(y)$ are called probability distributions.

The mean of a probability distribution is a measure of its central tendency or location. Mathematically, we define the mean (e.g., μ) as

$$\mu = \begin{cases} \displaystyle\int_{-\infty}^{\infty} yf(y)\,dy, \ y \text{ continuous} \\[2ex] \displaystyle\sum_{\text{all } y} yp(y), \quad y \text{ discrete} \end{cases} \tag{2-1}$$

We may also express the mean in terms of the *expected value* or long-run average value of the random variable y as

$$\mu = E(y) = \begin{cases} \displaystyle\int_{-\infty}^{\infty} yf(y)\,dy, \ y \text{ continuous} \\[2ex] \displaystyle\sum_{\text{all } y} yp(y), \quad y \text{ discrete} \end{cases} \tag{2-2}$$

where E denotes the expected value *operator*.

The spread or dispersion of a probability distribution can be measured by the *variance*, defined as

$$\sigma^2 = \begin{cases} \displaystyle\int_{-\infty}^{\infty} (y - \mu)^2 f(y)\,dy, \ y \text{ continuous} \\[2ex] \displaystyle\sum_{\text{all } y} (y - \mu)^2 p(y), \quad y \text{ discrete} \end{cases} \tag{2-3}$$

Note that the variance can be expressed entirely in terms of expectation, since

$$\sigma^2 = E[(y - \mu)^2] \tag{2-4}$$

Finally, the variance is used so extensively that it is convenient to define a variance operator V such that

$$V(y) \equiv E[(y - \mu)^2] = \sigma^2 \tag{2-5}$$

The concepts of expected value and variance are used extensively throughout this book, and it may be helpful to review several elementary results concerning these operators. If y is a random variable and c is a constant, then

(i) $E(c) = c$
(ii) $E(y) = \mu$
(iii) $E(cy) = cE(y) = c\mu$
(iv) $V(c) = 0$
(v) $V(y) = \sigma^2$
(vi) $V(cy) = c^2 V(y) = c^2 \sigma^2$

If there are two random variables, for example, y_1 with $E(y_1) = \mu_1$ and $V(y_1) = \sigma_1^2$, and y_2 with $E(y_2) = \mu_2$ and $V(y_2) = \sigma_2^2$, then we have

(vii) $E(y_1 + y_2) = E(y_1) + E(y_2) = \mu_1 + \mu_2$

It is possible to show that

(viii) $V(y_1 + y_2) = V(y_1) + V(y_2) + 2\text{Cov}(y_1, y_2)$

where

$$\text{Cov}(y_1, y_2) = E[(y_1 - \mu_1)(y_2 - \mu_2)] \qquad (2\text{-}6)$$

is the *covariance* of the random variables y_1 and y_2. The covariance is a measure of *independence*; that is, if y_1 and y_2 are independent,[1] then $\text{Cov}(y_1, y_2) = 0$. We may also show that

(ix) $V(y_1 - y_2) = V(y_1) + V(y_2) - 2\text{Cov}(y_1, y_2)$

If y_1 and y_2 are *independent*, then we have

(x) $V(y_1 \pm y_2) = V(y_1) + V(y_2) = \sigma_1^2 + \sigma_2^2$

and

(xi) $E(y_1 \cdot y_2) = E(y_1) \cdot E(y_2) = \mu_1 \cdot \mu_2$

However, note that, in general,

(xii) $E\left(\dfrac{y_1}{y_2}\right) \neq \dfrac{E(y_1)}{E(y_2)}$

regardless of whether or not y_1 and y_2 are independent.

2-2 SAMPLING AND SAMPLING DISTRIBUTIONS

The objective of statistical inference is to draw conclusions about a population using a sample from that population. Most of the methods that we will study assume that random samples are used. That is, if the population contains N elements, and a sample of n of them is to be selected, then if each of the $N!/(N - n)!n!$ possible samples has an equal probability of being chosen, the procedure employed is called random sampling. In practice, it is sometimes difficult to obtain random samples, and tables of random numbers, such as Table IX in the Appendix, may be helpful.

Statistical inference makes considerable use of quantities computed from the observations in the sample. We define a *statistic* as any function of the observations in a sample that does not contain unknown parameters. For example, suppose that $y_1, y_2, \ldots, y_n$ represents a sample. Then the *sample mean*

$$\overline{y} = \frac{\displaystyle\sum_{i=1}^{n} y_i}{n} \qquad (2\text{-}7)$$

[1]Note that the converse of this is not necessarily so; that is, we may have $\text{Cov}(y_1, y_2) = 0$ and yet this does not imply independence. For an example, see Hines and Montgomery (1972, pp. 103–104).

and the *sample variance* *is a biased estimator*

the unbiased estimator of population variance →

$$S^2 = \frac{\sum_{i=1}^{n}(y_i - \bar{y})^2}{n-1} = \frac{\sum(y-\bar{y})^2}{n} \quad (2\text{-}8)$$

are both statistics. These quantities are measures of the central tendency and dispersion of the sample, respectively. Sometimes $S = \sqrt{S^2}$, called the *sample standard deviation*, is used as a measure of dispersion.

Often we will be able to determine the probability distribution of a particular statistic if we know the probability distribution of the population from which the sample was drawn. The probability distribution of a statistic is called a *sampling distribution*. We will now briefly discuss several useful sampling distributions.

One of the most important sampling distributions is the *normal* distribution. If y is a normal random variable, then the probability distribution of y is

$$f(y) = \frac{1}{\sigma\sqrt{2\pi}} e^{-(1/2)[(y-\mu)/\sigma]^2}, \quad -\infty < y < \infty \quad (2\text{-}9)$$

where $-\infty < \mu < \infty$ is the mean of the distribution and $\sigma^2 > 0$ is the variance. The normal distribution is shown in Figure 2-1.

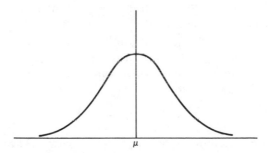

Figure 2-1. The normal distribution.

We will often use the shorthand notation $y \sim N(\mu, \sigma^2)$ to denote that y is distributed normally with mean μ and variance σ^2. The normal distribution plays a central role in statistical methodology. Also, many important sampling distributions may be defined in terms of normal random variables.

An important special case of the normal distribution is the *standard normal distribution*, that is, $\mu = 0$ and $\sigma^2 = 1$. We see that if $y \sim N(\mu, \sigma^2)$ then the random variable

$$z = \frac{y - \mu}{\sigma} \quad (2\text{-}10)$$

follows the standard normal distribution, denoted $z \sim N(0, 1)$. The operation demonstrated in Equation 2-10 is often called *standardizing* the normal random variable y. A table of the cumulative standard normal distribution is given in Appendix Table I.

Many statistical techniques assume that the random variable is normally distributed. The central limit theorem is often a justification of approximate normality.

THEOREM 2-1. THE CENTRAL LIMIT THEOREM. If $y_1, y_2, \ldots, y_n$ is a sequence of n independent random variables with $E(y_i) = \mu_i$ and $V(y_i) = \sigma_i^2$ (both finite), and $x = y_1 + y_2 + \cdots + y_n$, then

$$z_n = \frac{x - \sum_{i=1}^{n} \mu_i}{\sqrt{\sum_{i=1}^{n} \sigma_i^2}}$$

has an approximate $N(0, 1)$ distribution in the sense that if $F_n(z)$ is the distribution function of z_n, and $\Phi(z)$ is the distribution function of the $N(0, 1)$ random variable, then $\lim_{n \to \infty} [F_n(z)/\Phi(z)] = 1$.

This result states essentially that the sum of n independent random variables is approximately normally distributed. In many cases this approximation is good for very small n, say $n < 10$, while in other cases large n is required, say $n > 100$. Frequently we think of the error in an experiment as arising in an additive manner from several independent sources, and consequently, the normal distribution becomes a plausible model for the combined experimental error.

An important sampling distribution that can be defined in terms of normal random variables is the chi-square or χ^2 distribution. If $z_1, z_2, \ldots, z_k$ are normally and independently distributed random variables with mean zero and variance one, abbreviated NID(0, 1), then the random variable

$$\chi_k^2 = z_1^2 + z_2^2 + \cdots + z_k^2$$

follows the chi-square distribution with k degrees of freedom. The density function of chi-square is

$$f(\chi^2) = \frac{1}{2^{k/2}\, \Gamma\left(\dfrac{k}{2}\right)} (\chi^2)^{(k/2)-1}\, e^{-\chi^2/2}, \chi^2 > 0 \qquad (2\text{-}11)$$

Several chi-square distributions are shown in Figure 2-2. The distribution is asymmetric or *skewed*, with mean and variance

is positively skewed

$$\mu = k$$
$$\sigma^2 = 2k$$

respectively. A table of percentage points of the chi-square distribution is given in Table III of the Appendix.

As an example of a random variable that follows the chi-square distribution, suppose

k = 1

$k = 2$

mode is at $x^2 = y - 2$ (= *k-2*)

$k = 5$

$k = 15$

Figure 2-2.   Several chi-square distributions.

that $y_1, y_2, \ldots, y_n$ is a random sample from a $N(\mu, \sigma^2)$ distribution. Then

$$\frac{SS}{\sigma^2} = \frac{\sum\limits_{i=1}^{n} (y_i - \overline{y})^2}{\sigma^2} \sim \chi^2_{n-1} \tag{2-12}$$

That is, SS/σ^2 is distributed as chi-square with $n - 1$ degrees of freedom.

The quantity $SS = \sum_{i=1}^{n} (y_i - \overline{y})^2$ in the numerator of (2-12) is called a *corrected sum of squares*. Many of the techniques used in this book involve the computation and manipulation of sums of squares. The result given in (2-12) is extremely important and occurs repeatedly; a sum of squares in normal random variables when divided by σ^2 follows the chi-square distribution.

Examining Equation 2-8, we see that the sample variance can be written as

$$S^2 = \frac{SS}{n-1} \tag{2-13}$$

If the observations in the sample are $NID(\mu, \sigma^2)$, then the distribution of S^2 is $[\sigma^2/(n-1)] \chi^2_{n-1}$. Thus the sampling distribution of the sample variance is a constant times the chi-square distribution, if the population is normally distributed.

If z and χ^2_k are independent standard normal and chi-square random variables, respectively, then the random variable

$$t_k = \frac{z}{\sqrt{\chi^2_k/k}} \tag{2-14}$$

follows the t distribution with k degrees of freedom, denoted t_k. The density function of t is

$$f(t) = \frac{\Gamma[(k+1)/2]}{\sqrt{k\pi}\,\Gamma(k/2)} \frac{1}{[(t^2/k)+1]^{(k+1)/2}}, \quad -\infty < t < \infty \tag{2-15}$$

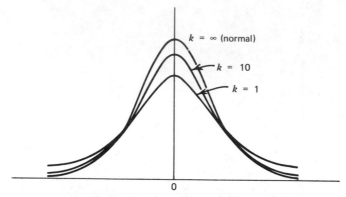

Figure 2-3.   Several t distributions.

and the mean and variance of t are $\mu = 0$ and $\sigma^2 = k/(k - 2)$ for $k > 2$, respectively. Several t distributions are shown in Figure 2-3. Note that if $k = \infty$ the t distribution becomes the standard normal distribution. A table of percentage points of the t distribution is given in Table II of the Appendix. If $y_1, y_2, \ldots, y_n$ is a random sample from the $N(\mu, \sigma^2)$ distribution, then the quantity

$$t = \frac{\bar{y} - \mu}{S/\sqrt{n}} \tag{2-16}$$

is distributed as t with $n - 1$ degrees of freedom.

The final sampling distribution that we will consider is the F distribution. If χ_u^2 and χ_v^2 are two independent chi-square random variables with u and v degrees of freedom, respectively, then the ratio

$$F_{u,v} = \frac{\chi_u^2/u}{\chi_v^2/v} \tag{2-17}$$

follows the F distribution with u *numerator* degrees of freedom and v *denominator* degrees of freedom. The probability distribution of F is

$$f(F) = \frac{\Gamma\left(\dfrac{u + v}{2}\right)\left(\dfrac{u}{v}\right)^{u/2} F^{(u/2)-1}}{\Gamma\left(\dfrac{u}{2}\right)\Gamma\left(\dfrac{v}{2}\right)\left[\left(\dfrac{u}{v}\right)F + 1\right]^{(u+v)/2}}, \quad 0 < F < \infty \tag{2-18}$$

Several F distributions are shown in Figure 2-4. This distribution is very important in the statistical analysis of designed experiments. A table of percentage points of the F distribution is given in Appendix IV.

As an example of a statistic that is distributed as F, suppose we have two normal populations with common variance σ^2. If $y_{11}, y_{12}, \ldots, y_{1n_1}$ is a random sample of n_1 observations from the first population, and if $y_{21}, y_{22}, \ldots, y_{2n_2}$ is a random sam-

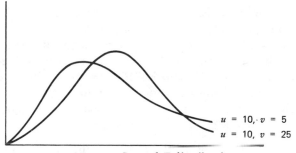

$u = 10, v = 5$
$u = 10, v = 25$

Figure 2-4.   Several F distributions.

ple of n_2 observations from the second, then

$$\frac{S_1^2}{S_2^2} \sim F_{n_1-1,\,n_2-1} \tag{2-19}$$

where S_1^2 and S_2^2 are the two sample variances. This result follows directly from (2-12) and (2-17).

2-3 ESTIMATION

A random variable is characterized or described by its probability distribution. This distribution is described by one or more *parameters*. For example, the mean μ and variance σ^2 are the parameters of the normal distribution (Equation 2-9), and the number of degrees of freedom k is the parameter of the chi-square distribution (Equation 2-11). Since the parameters characterize the distribution, we are usually interested in making inferences about them. One way to do this is to estimate the parameters of the distribution.

An *estimator* of an unknown parameter is simply a statistic that corresponds to that parameter. A particular numerical value of an estimator, computed from sample data, is called an *estimate*. We may distinguish between a point estimator and an interval estimator. A *point estimator* is a procedure leading to a single numerical value for the estimate of the unknown parameter. An *interval estimator* is a random interval in which the true value of the parameter lies with some probability. These random intervals are usually called *confidence intervals*.

2-3.1 Point Estimation

Consider a random variable y with probability distribution $f(y)$. Suppose that the mean μ and variance σ^2 of this distribution are both unknown. If $y_1, y_2, \ldots, y_n$ is a random sample of n observations on the random variable y, then the sample mean $\overline{y}$ is a point estimator of the population mean μ, and the sample variance S^2 is a point estimator of the population variance σ^2. For example, suppose we wished to estimate the mean and variance of the breaking strength of a particular type of textile fiber. A random sample of $n = 25$ fiber specimens were tested and the breaking strength recorded for each. The sample mean and variance were computed according to Equations 2-7 and 2-8, respectively, and were $\overline{y} = 18.6$ and $S^2 = 1.20$. Therefore, the estimate of μ is $\overline{y} = 18.6$ and the estimate of σ^2 is $S^2 = 1.20$.

There are a number of properties required of good point estimators. Two of the most important are the following:

1. The point estimator should be *unbiased*. That is, the long-run average or expected value of the point estimator should be the parameter that is being estimated.
2. The point estimator should have *minimum variance*. Since the point estimator is a statistic, it is a random variable. This property states that the minimum variance point estimator has a variance that is smaller than the variance of any other estimator of that parameter.

We may easily show that $\overline{y}$ and S^2 are unbiased estimators of μ and σ^2, respectively. First consider $\overline{y}$. Using the properties of expectation, we have

$$E(\overline{y}) = E\left(\frac{\sum\limits_{i=1}^{n} y_i}{n}\right)$$

$$= \frac{1}{n} E\left(\sum\limits_{i=1}^{n} y_i\right)$$

$$= \frac{1}{n} \sum\limits_{i=1}^{n} E(y_i)$$

$$= \frac{1}{n} \sum\limits_{i=1}^{n} \mu$$

$$= \mu$$

since the expected value of each observation y_i is μ. Thus, $\overline{y}$ is an unbiased estimator of μ.

Now consider the sample variance S^2. We have

$$E(S^2) = E\left[\frac{\sum\limits_{i=1}^{n} (y_i - \overline{y})^2}{n-1}\right]$$

$$= \frac{1}{n-1} E\left[\sum\limits_{i=1}^{n} (y_i - \overline{y})^2\right]$$

$$= \frac{1}{n-1} E(SS)$$

where SS is the corrected sum of squares of the observations y_i. Now

$$E(SS) = E\left[\sum_{i=1}^{n} (y_i - \overline{y})^2\right] \tag{2-20}$$

$$= E\left[\sum_{i=1}^{n} y_i^2 - n\overline{y}^2\right]$$

$$= \sum_{i=1}^{n} (\mu^2 + \sigma^2) - n(\mu^2 + \sigma^2/n)$$

$$= (n - 1)\sigma^2 \tag{2-21}$$

Therefore,

$$E(S^2) = \frac{1}{n-1} E(SS)$$

$$= \sigma^2$$

The quantity $n - 1$ in Equation 2-21 is called the *number of degrees of freedom* of the sum of squares SS. This is a very general result; that is, if y is a random variable with variance σ^2 and $SS = \Sigma(y_i - \overline{y})^2$ has ν degrees of freedom, then

$$E\left(\frac{SS}{\nu}\right) = \sigma^2 \tag{2-22}$$

The number of degrees of freedom of a sum of squares is equal to the number of independent elements in that sum of squares. For example, $SS = \Sigma_{i=1}^{n} (y_i - \overline{y})^2$ in (2-20) consists of the sum of squares of the n elements $y_1 - \overline{y}, y_2 - \overline{y}, \ldots, y_n - \overline{y}$. These elements are not all independent since $\Sigma_{i=1}^{n} (y_i - \overline{y}) = 0$; and in fact, only $n - 1$ of them are independent, implying that SS has $n - 1$ degrees of freedom.

2-3.2 Interval Estimation

An interval estimate of a parameter is the distance between two statistics that includes the true value of the parameter with some probability. Thus, to obtain an interval estimate of a parameter (e.g., θ), we need to find two statistics L and U such that the probability statement

$$P(L \leqslant \theta \leqslant U) = 1 - \alpha \tag{2-23}$$

is true. The interval

$$L \leqslant \theta \leqslant U \tag{2-24}$$

is called a $100(1 - \alpha)$ percent *confidence interval* for the parameter θ. The interpretation of this interval is that if, in repeated random samplings, a large number of such intervals are constructed, $100(1 - \alpha)$ percent of them will contain the true value of θ. The statistics L and U are called the lower and upper *confidence limits*, respectively, and $1 - \alpha$ is called the *confidence coefficient*. If $\alpha = .05$, then (2-24) is called a 95 percent confidence interval for θ.

As an example of a confidence interval, suppose that y is normally distributed with unknown mean μ and known variance σ^2. If a random sample of n observations is taken and $\overline{y}$ computed, then

$$\frac{\overline{y} - \mu}{\sigma/\sqrt{n}} \sim N(0, 1)$$

Expressing this as a probability statement, we have

$$P\left(-Z_{\alpha/2} \leqslant \frac{\overline{y} - \mu}{\sigma/\sqrt{n}} \leqslant Z_{\alpha/2}\right) = 1 - \alpha \tag{2-25}$$

where $Z_{\alpha/2}$ is the upper-tail percentage point of the standard normal distribution such that the probability to the right of $Z_{\alpha/2}$ is $\alpha/2$. Equation 2-25 may be rewritten as

$$P(\overline{y} - Z_{\alpha/2} \, \sigma/\sqrt{n} \leqslant \mu \leqslant \overline{y} + Z_{\alpha/2} \, \sigma/\sqrt{n}) = 1 - \alpha \tag{2-26}$$

Comparing (2-26) with (2-23), we see that

$$\overline{y} \pm Z_{\alpha/2} \, \sigma/\sqrt{n} \tag{2-27}$$

is a $100(1 - \alpha)$ percent confidence interval for μ.

If the variance of the distribution is unknown then the sample variance S^2 may be used to estimate σ^2, and then

$$\frac{\overline{y} - \mu}{S/\sqrt{n}}$$

is distributed as t with $n - 1$ degrees of freedom. The $100(1 - \alpha)$ percent confidence interval on μ becomes

$$\overline{y} \pm t_{\alpha/2, \, n-1} \, S/\sqrt{n}, \tag{2-28}$$

where $t_{\alpha/2, \, n-1}$ denotes the upper-tail percentage point of the t distribution with $n - 1$ degrees of freedom such that the probability to the right of this point is $\alpha/2$.

2-4   HYPOTHESIS TESTING

A hypothesis is a statement about the values of the parameters of a probability distribution. For example, suppose we think that the mean yield of a chemical process is 94.5 percent. This statement is expressed formally as

$$H_0: \ \mu = 94.5$$
$$H_1: \ \mu \neq 94.5 \tag{2-29}$$

The statement H_0: $\mu = 94.5$ in (2-29) is called the *null* hypothesis, and H_1: $\mu \neq 94.5$ is called the *alternative* hypothesis. Since H_1 specifies values of the mean μ that are either greater than 94.5 or less than 94.5, it is called a *two-sided* alternative hypothesis. The value of the mean specified in the null hypothesis is determined in one of three ways. It may result from past evidence, knowledge, or experimentation. It may be the result of some theory or model regarding the process under study. Finally, it may be the result of contractual specifications.

To test a hypothesis we devise a procedure for taking a random sample, computing an appropriate test statistic, and then rejecting or failing to reject the null hypothesis H_0. Part of this procedure is specifying the set of values for the test statistic which lead to rejection of H_0. This set of values is called the critical region or rejection region for the test.

Two kinds of errors may be committed when testing hypotheses. If the null hypothesis is rejected when it is true, then a type I error has occurred. If the null hypothesis is *not* rejected when it is false, then a type II error has been made. The probabilities of these two errors are given special symbols:

$$\alpha = P(\text{type I error}) = P \text{ (reject } H_0 \,|\, H_0 \text{ is true)}$$

$$\beta = P(\text{type II error}) = P \text{ (fail to reject } H_0 \,|\, H_0 \text{ is false)}$$

Sometimes it is more convenient to work with the *power* of the test, where

$$\text{Power} = 1 - \beta = P \text{ (reject } H_0 \,|\, H_0 \text{ is false)}$$

The general procedure in hypothesis testing is to specify a value of the probability of type I error α, often called the *significance level* of the test, and then design the test procedure so that the probability of type II error β has a suitably small value.

We will now briefly review several frequently encountered hypothesis testing problems. The following situations will be considered:

1. Comparison of means when the variance is known.
2. Comparison of means when the variance is unknown.
3. Comparison of variances.

For a complete development of these and other hypothesis testing procedures, refer to Hines and Montgomery (1972) or Bowker and Lieberman (1972).

2-4.1 Tests on Means

Suppose that y is a random variable with unknown mean μ and known variance σ^2. We wish to test the hypotheses

$$H_0: \mu = \mu_0$$
$$H_1: \mu \neq \mu_0 \tag{2-30}$$

where μ_0 is a specified constant. The procedure for testing this hypothesis is to take a random sample of n observations on y, compute the test statistic

$$Z_0 = \frac{\bar{y} - \mu_0}{\sigma/\sqrt{n}} \qquad (2\text{-}31)$$

and reject H_0 if $|Z_0| > Z_{\alpha/2}$, where $Z_{\alpha/2}$ is the upper $\alpha/2$ percentage point of the standard normal distribution. A justification of this test procedure follows. From the central limit theorem, we know that the sample mean $\bar{y} \sim N(\mu, \sigma^2/n)$, approximately. Now if H_0 is true, Z_0 in (2-31) is distributed as $N(0, 1)$ and, consequently, we would expect $100(1 - \alpha)$ percent of the values of Z_0 to fall between $-Z_{\alpha/2}$ and $Z_{\alpha/2}$. A sample producing a value of Z_0 outside these limits would be unusual if the null hypothesis were true, and is evidence that H_0 should be rejected. Note that α is the probability of type I error for the test.

In some problems the experimenter may wish to reject H_0 only if the true mean is larger than μ_0. Thus, he would specify a *one-sided alternative* hypothesis $H_1 : \mu > \mu_0$, and would reject H_0 only if $Z_0 > Z_{\alpha}$. If he wants to reject H_0 only if μ is less than μ_0, then the alternative hypothesis is $H_1 : \mu < \mu_0$, and he would reject H_0 if $Z_0 < -Z_{\alpha}$.

Now suppose there are two populations with known variances σ_1^2 and σ_2^2. We wish to test the hypothesis that the means of these two populations, μ_1 and μ_2, differ by a constant amount γ. Expressed mathematically, we wish to test

$$\begin{aligned} H_0 &: \mu_1 - \mu_2 = \gamma \\ H_1 &: \mu_1 - \mu_2 \neq \gamma \end{aligned} \qquad (2\text{-}32)$$

The appropriate test procedure is to obtain a random sample of n_1 observations from population 1 and n_2 observations from population 2, compute the test statistic

$$Z_0 = \frac{\bar{y}_1 - \bar{y}_2 - \gamma}{\sqrt{\dfrac{\sigma_1^2}{n_1} + \dfrac{\sigma_2^2}{n_2}}} \qquad (2\text{-}33)$$

and reject H_0 if $|Z_0| > Z_{\alpha/2}$. If the one-sided alternative $H_1 : \mu_1 - \mu_2 > \gamma$ is used in (2-32), then H_0 would be rejected if $Z_0 > Z_{\alpha}$. The other one-sided alternative is $H_1 : \mu_1 - \mu_2 < \gamma$, and in this case H_0 would be rejected if $Z_0 < -Z_{\alpha}$.

A summary of these test procedures is given in Table 2-1.

■**EXAMPLE 2-1.** A vendor submits lots of fabric to a textile manufacturer. The manufacturer wants to know if the lot average-breaking strength exceeds 200 psi. Past experience indicates that the variance of breaking strength is $100\,(\text{psi})^2$. The hypothesis to be tested is

$$H_0 : \mu = 200$$

$$H_1 : \mu > 200$$

Table 2-1 Tests on Means with Variance Known

Hypothesis	Test Statistic	Criteria for Rejection
$H_0: \mu = \mu_0$ $H_1: \mu \neq \mu_0$		$\lvert Z_0 \rvert > Z_{\alpha/2}$
$H_0: \mu = \mu_0$ $H_1: \mu < \mu_0$	$Z_0 = \dfrac{\overline{y} - \mu_0}{\sigma/\sqrt{n}}$	$Z_0 < -Z_\alpha$
$H_0: \mu = \mu_0$ $H_1: \mu > \mu_0$		$Z_0 > Z_\alpha$
$H_0: \mu_1 - \mu_2 = \gamma$ $H_1: \mu_1 - \mu_2 \neq \gamma$		$\lvert Z_0 \rvert > Z_{\alpha/2}$
$H_0: \mu_1 - \mu_2 = \gamma$ $H_1: \mu_1 - \mu_2 < \gamma$	$Z_0 = \dfrac{(\overline{y}_1 - \overline{y}_2) - \gamma}{\sqrt{\dfrac{\sigma_1^2}{n_1} + \dfrac{\sigma_2^2}{n_2}}}$	$Z_0 < -Z_\alpha$
$H_0: \mu_1 - \mu_2 = \gamma$ $H_1: \mu_1 - \mu_2 > \gamma$		$Z_0 > Z_\alpha$

Four specimens are randomly selected, and the sample average breaking strength is $\overline{y} = 214$ psi. The value of the test statistic is

$$Z_0 = \frac{\overline{y} - \mu_0}{\sigma/\sqrt{n}} = \frac{214 - 200}{10/\sqrt{4}} = 2.80$$

If a type I error of $\alpha = .05$ is specified, we find $Z_\alpha = Z_{.05} = 1.645$ from Appendix Table I. Thus H_0 is rejected, and we conclude that the lot average-breaking strength exceeds 200 psi. ∎

If the variance of the population is unknown, we must make the additional assumption that the population is normally distributed. The normality assumption is required to formally develop the test procedure; however, moderate departures from normality will not seriously affect the results.

To test (2-30) in the variance unknown case, the sample variance S^2 is used to estimate σ^2. Replacing σ by S in (2-31), we have the test statistic

$$t_0 = \frac{\overline{y} - \mu_0}{S/\sqrt{n}} \tag{2-34}$$

The null hypothesis $H_0: \mu = \mu_0$ would be rejected if $\lvert t_0 \rvert > t_{\alpha/2,n-1}$, where $t_{\alpha/2,n-1}$ denotes the upper $\alpha/2$ percentage point of the t distribution with $n - 1$ degrees of freedom. The critical regions for the one-sided alternative hypothesis are shown in Table 2-2.

Table 2-2   Tests on Means of Normal Distributions, Variance Unknown

Hypothesis	Test Statistic	Criteria for Rejection
$H_0: \mu = \mu_0$ $H_1: \mu = \mu_0$		$\lvert t_0 \rvert > t_{\alpha/2, n-1}$
$H_0: \mu = \mu_0$ $H_1: \mu < \mu_0$	$t_0 = \dfrac{\bar{y} - \mu_0}{S/\sqrt{n}}$	$t_0 < -t_{\alpha, n-1}$
$H_0: \mu = \mu_0$ $H_1: \mu > \mu_0$		$t_0 > t_{\alpha, n-1}$
$H_0: \mu_1 - \mu_2 = \gamma$ $H_1: \mu_1 - \mu_2 \neq \gamma$	$t_0 = \dfrac{\bar{y}_1 - \bar{y}_2 - \gamma}{S_p \sqrt{\dfrac{1}{n_1} + \dfrac{1}{n_2}}}$ $\nu = n_1 + n_2 - 2$, or	$\lvert t_0 \rvert > t_{\alpha/2, \nu}$
$H_0: \mu_1 - \mu_2 = \gamma$ $H_1: \mu_1 - \mu_2 < \gamma$	$t_0 = \dfrac{\bar{y}_1 - \bar{y}_2 - \gamma}{\sqrt{\dfrac{S_1^2}{n_1} + \dfrac{S_2^2}{n_2}}}$	$t_0 < -t_{\alpha, \nu}$
$H_0: \mu_1 - \mu_2 = \gamma$ $H_1: \mu_1 - \mu_2 > \gamma$	$\nu = \dfrac{\left(\dfrac{S_1^2}{n_1} + \dfrac{S_2^2}{n_2}\right)^2}{\dfrac{(S_1^2/n_1)^2}{n_1 + 1} + \dfrac{(S_2^2/n_2)^2}{n_2 + 1}} - 2$	$t_0 > t_{\alpha, \nu}$

Now suppose there are two normal populations with unknown means μ_1 and μ_2 and unknown variances σ_1^2 and σ_2^2. The test procedure for (2-32) depends on whether $\sigma_1^2 = \sigma_2^2$. First consider the case where this assumption is reasonable. To test H_0, two random samples of n_1 and n_2 observations are taken from populations 1 and 2, respectively, and an estimate of the common variance computed as

$$S_p^2 = \frac{(n_1 - 1) S_1^2 + (n_2 - 1) S_2^2}{n_1 + n_2 - 2} \tag{2-35}$$

where S_1^2 and S_2^2 are the two sample variances. Then the test statistic

$$t_0 = \frac{\bar{y}_1 - \bar{y}_2 - \gamma}{S_p \sqrt{\dfrac{1}{n_1} + \dfrac{1}{n_2}}} \tag{2-36}$$

is calculated, and H_0 is rejected if $\lvert t_0 \rvert > t_{\alpha/2, n_1 + n_2 - 2}$. This procedure is often called the pooled t test, because the two samples are combined or "pooled" to estimate the common variance.

If we cannot reasonably assume that $\sigma_1^2 = \sigma_2^2$, then a slightly different procedure is followed. The test statistic for (2-32) becomes

$$t_0 = \frac{\overline{y}_1 - \overline{y}_2 - \gamma}{\sqrt{\dfrac{S_1^2}{n_1} + \dfrac{S_2^2}{n_2}}} \tag{2-37}$$

and the number of degrees of freedom for t are

$$\nu = \frac{\left(\dfrac{S_1^2}{n_1} + \dfrac{S_2^2}{n_2}\right)^2}{\dfrac{(S_1^2/n_1)^2}{n_1 + 1} + \dfrac{(S_2^2/n_2)^2}{n_2 + 1}} - 2 \tag{2-38}$$

This is only an approximate test procedure. The one-sided alternative hypotheses are summarized in Table 2-2.

Occasionally, when comparing two means, we find that the observations should be taken in pairs, because the experimental units employed in the test are not homogeneous. This paired t test is illustrated in Section 1-4.

2-4.2 Tests on Variances

We will now review testing hypothesis about the variance of a normal population. Unlike the tests on means, the procedures for tests on variances are rather sensitive to the normality assumption. A good discussion of this is in Appendix 2A of Davies (1956).

Suppose we wish to test the hypothesis that the variance of a normal population equals a constant, for example, σ_0^2. Stated formally, we wish to test

$$\begin{aligned} H_0 &: \sigma^2 = \sigma_0^2 \\ H_1 &: \sigma^2 \neq \sigma_0^2 \end{aligned} \tag{2-39}$$

The test statistic for (2-39) is

$$\chi_0^2 = \frac{SS}{\sigma_0^2} \tag{2-40}$$

where $SS = \sum_{i=1}^{n} (y_i - \overline{y})^2$ is the corrected sum of squares of the sample data. The null hypothesis is rejected if $\chi_0^2 > \chi_{\alpha/2, n-1}^2$ or if $\chi_0^2 < \chi_{1-(\alpha/2), n-1}^2$ where $\chi_{\alpha/2, n-1}^2$ and $\chi_{1-(\alpha/2), n-1}^2$ are the upper $\alpha/2$ and lower $1 - (\alpha/2)$ percentage points of the chi-square distribution with $n - 1$ degrees of freedom, respectively. Table 2-3 gives the critical regions for the one-sided alternative hypotheses.

Now consider testing the equality of the variances of two normal populations. If random samples of size n_1 and n_2 are taken from populations 1 and 2, respectively, then the test statistic for

$$\begin{aligned} H_0 &: \sigma_1^2 = \sigma_2^2 \\ H_1 &: \sigma_1^2 \neq \sigma_2^2 \end{aligned} \tag{2-41}$$

Table 2-3   Tests on Variances of Normal Distributions

Hypothesis	Test Statistic	Criteria for Rejection
$H_0: \sigma^2 = \sigma_0^2$ $H_1: \sigma^2 \neq \sigma_0^2$		$\chi_0^2 > \chi_{\alpha/2,n-1}^2$ or $\chi_0^2 < \chi_{1-\alpha/2,n-1}^2$
$H_0: \sigma^2 = \sigma_0^2$ $H_1: \sigma^2 < \sigma_0^2$	$\chi_0^2 = \dfrac{SS}{\sigma_0^2}$	$\chi_0^2 < \chi_{1-\alpha,n-1}^2$
$H_0: \sigma^2 = \sigma_0^2$ $H_1: \sigma^2 > \sigma_0^2$		$\chi_0^2 > \chi_{\alpha,n-1}^2$
$H_0: \sigma_1^2 = \sigma_2^2$ $H_1: \sigma_1^2 \neq \sigma_2^2$	$F_0 = \dfrac{S_1^2}{S_2^2}$	$F_0 > F_{\alpha/2,n_1-1,n_2-1}$ or $F_0 < F_{1-\alpha/2,n_1-1,n_2-1}$
$H_0: \sigma_1^2 = \sigma_2^2$ $H_1: \sigma_1^2 < \sigma_2^2$	$F_0 = \dfrac{S_2^2}{S_1^2}$	$F_0 > F_{\alpha,n_2-1,n_1-1}$
$H_0: \sigma_1^2 = \sigma_2^2$ $H_1: \sigma_1^2 > \sigma_2^2$	$F_0 = \dfrac{S_1^2}{S_2^2}$	$F_0 > F_{\alpha,n_1-1,n_2-1}$

is the ratio of the sample variances,

$$F_0 = \frac{S_1^2}{S_2^2} \tag{2-42}$$

The null hypothesis would be rejected if $F_0 > F_{\alpha/2,n_1-1,n_2-1}$ or if $F_0 < F_{1-(\alpha/2),n_1-1,n_2-1}$, where $F_{\alpha/2,n_1-1,n_2-1}$ and $F_{1-(\alpha/2),n_1-1,n_2-1}$ denote the upper $\alpha/2$ and lower $1-(\alpha/2)$ percentage points of the F distribution with $n_1 - 1$ and $n_2 - 1$ degrees of freedom. Table IV of the Appendix gives only upper-tail percentage points of F; however, the upper and lower tail points are related by

$$F_{1-\alpha,\nu_1,\nu_2} = \frac{1}{F_{\alpha,\nu_2,\nu_1}} \tag{2-43}$$

Test procedures for more than two variances are discussed in Section 3-9.

■**EXAMPLE 2-2.**   Suppose we wish to test the hypotheses

$$H_0: \sigma_1^2 = \sigma_2^2$$

$$H_1: \sigma_1^2 > \sigma_2^2$$

Two random samples of $n_1 = 12$ and $n_2 = 10$ observations are taken and the sample variances are $S_1^2 = 14.5$ and $S_2^2 = 10.8$. The test statistic is

$$F_0 = \frac{S_1^2}{S_2^2} = \frac{14.5}{10.8} = 1.34.$$

From Appendix Table IV we find that $F_{.05,11,9} = 3.10$, so the null hypothesis cannot be rejected. That is, we have found insufficient statistical evidence to conclude that $\sigma_1^2 > \sigma_2^2$. ∎

2-4.3 The Probability of Type II Error

In any hypothesis testing situation it is important to determine the probability of type II error or, equivalently, the power of the test. As an example, we will determine the probability of type II error for testing the hypothesis

$$H_0: \mu = \mu_0$$

$$H_1: \mu \neq \mu_0$$

with the variance σ^2 known. The test statistic for this hypothesis is given in Equation 2-31. To find the probability of type II error, we must assume that the null hypothesis $H_0: \mu = \mu_0$ is false. Suppose that the true mean of the distribution is $\mu_1 = \mu_0 + \delta$, where $\delta > 0$. Now this implies that the *alternative* hypothesis $H_1: \mu \neq \mu_0$ is true, and under this assumption the distribution of the test statistic Z_0 is

$$Z_0 \sim N\left(\frac{\delta\sqrt{n}}{\sigma}, 1\right) \tag{2-44}$$

The distribution of the test statistic under both hypotheses H_0 and H_1 is shown in Figure 2-5. From examining Figure 2-5, we see that the probability of type II error is the probability of Z_0 falling between $-Z_{\alpha/2}$ and $Z_{\alpha/2}$ *given that H_1 is true*. The probability is shown as the shaded portion of the figure. Expressed mathematically, this probability is just

$$\beta = \Phi\left(Z_{\alpha/2} - \frac{\delta\sqrt{n}}{\sigma}\right) - \Phi\left(-Z_{\alpha/2} - \frac{\delta\sqrt{n}}{\sigma}\right) \tag{2-45}$$

where $\Phi(z)$ denotes the probability to the *left* of z on the standard normal distribution. Note that Equation 2-45 was obtained by evaluating the probability between $-Z_{\alpha/2}$ and $Z_{\alpha/2}$ on the distribution of Z_0 when H_1 is true. These two points were

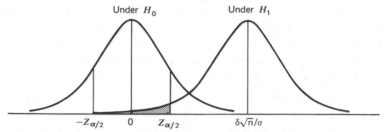

Figure 2-5. The distribution of Z_0 under H_0 and H_1.

standardized so that the table of the cumulative standard normal distribution (Appendix Table I) could be used to evaluate β.

■EXAMPLE 2-3. The burning rate of a rocket propellant is being studied. Standards specify that the mean burning rate must be 40 inches per second, and from past experience it is known that the variance of burning rate is 4.0. The appropriate hypotheses are

$$H_0: \mu = 40$$

$$H_1: \mu \neq 40$$

The experimenter decides to use a sample of $n = 25$ observations and $\alpha = .05$. Thus the test statistic becomes

$$Z_0 = \frac{\overline{y} - 40}{2/\sqrt{25}}$$

and H_0 would be rejected if $|Z_0| > Z_{.025} = 1.96$. Now suppose that the experimenter would like to know the probability of type II error if the true mean burning rate is $\mu_1 = 41$ inches per second. Since $\delta = \mu_1 - \mu_0 = 41 - 40 = 1$, we find that

$$\beta = \Phi \left(Z_{\alpha/2} - \frac{\delta\sqrt{n}}{\sigma} \right) - \Phi \left(-Z_{\alpha/2} - \frac{\delta\sqrt{n}}{\sigma} \right)$$

$$= \Phi \left(1.96 - \frac{(1)(5)}{2} \right) - \Phi \left(-1.96 - \frac{(1)(5)}{2} \right)$$

$$= \Phi(4.46) - \Phi(0.54)$$

$$= 1.0000 - .7054$$

$$= .2946.$$

That is, the probability of incorrectly failing to reject H_0 if the true burning rate is 41 inches per second is .2946. Equivalently, we could state that the power of the test is $1 - \beta = 1 - .2946 = .7054$.   ■

From examining Equation 2-45 and Figure 2-5, we find that:

1. The further the true mean μ_1 is from the hypothesized value μ_0, the smaller the probability of type II error for a given n and α. That is, for a specified sample size and α, the test will detect large differences more easily than small ones.
2. As the sample size n gets larger the probability of type II error gets smaller for a given δ and α. That is, to detect a specified difference δ we may make the test more powerful by increasing the sample size.

It is useful to plot a graph of the probability of type II error for various values of n. Such a graph is shown in Figure 2-6 for a level of significance of $\alpha = .05$. These graphs are usually called *operating characteristic* (O.C.) curves. The parameter on the horizontal axis is $d = |\delta|/\sigma$.

β

d

Figure 2-6. Operating characteristic curves for the two-sided normal test with $\alpha = 0.05$. (Reproduced with permission from "Operating Characteristics for the Common Statistical Tests of Significance," C. L. Ferris, F. E. Grubbs, and C. L. Weaver, *Annals of Mathematical Statistics*, June, 1946.)

The operating characteristic curves can be used to determine how large a sample is required to detect a specified difference in means with a given probability. As an illustration, suppose that in Example 2-3 if the true burning rate was as large as 42 inches per second we wish to reject H_0: $\mu = 40$ with probability .90. That is, we want the *power* of the test to be .90, or $\beta = .10$. Since $\delta = 42 - 40 = 2$, we have $d = |\delta|/\sigma = |2|/2 = 1.0$, and from Figure 2-6 with $\beta = .10$ and $d = 1.0$ we find that the required sample size is $n = 10$, approximately.

The use of operating characteristic curves for the statistical tests in Tables 2-1, 2-2, and 2-3 are discussed in Hines and Montgomery (1972) and Bowker and Lieberman (1972). Operating characteristic curves play an important role in the design of experiments, and their use in that respect will be discussed later in the book.

2-5 PROBLEMS

2-1. The breaking strength of a fiber is required to be at least 150 psi. Past experience has indicated that the standard deviation of breaking strength is $\sigma = 3$ psi. A random sample of four specimens are tested and the average breaking strength is observed to be 148 psi. Should the fiber be judged acceptable with $\alpha = .05$? Construct a 95 percent confidence interval on the mean breaking strength.

2-2. The diameter of steel shafts produced by a certain manufacturing process are known to have a standard deviation of $\sigma = 0.0001$ inch. A random sample of 10 shafts has an average diameter of 0.2545 inches. Test the hypothesis that the true mean diameter equals 0.255 inches, using $\alpha = .05$. Construct a 90 percent confidence interval on the mean diameter.

2-3. A normally distributed random variable has an unknown mean μ and a known variance $\sigma^2 = 9$. Find the sample size required to construct a 95 percent confidence interval on the mean which has total width of 1.0.

2-4. The shelf life of a carbonated beverage is of interest. Ten bottles are randomly selected and tested, and the following results obtained.

108 days	138 days
124	163
124	159
106	134
115	139

Assume that the alternative hypothesis is that the mean shelf life is greater than 125 days. Can the null hypothesis $H_0: \mu = 125$ be rejected? Construct a 95 percent confidence interval on the true mean shelf life.

2-5. The time to repair an electronic instrument is a normally distributed random variable measured in hours. The repair times for 16 such instruments chosen at random are as follows:

Hours			
159	280	101	212
224	379	179	264
222	362	168	250
149	260	485	170

Does it seem reasonable that the true mean repair time is greater than 225 hours? Determine a confidence interval on the true mean repair time.

2-6. Two machines are used for filling plastic bottles with a net volume of 16.0 ounces. The filling processes can be assumed normal, with standard deviations $\sigma_1 = 0.015$ and $\sigma_2 = 0.018$. The quality control department suspects that both machines fill to the same net volume, whether or not this volume is 16.0 ounces. A random sample is taken from the output of each machine:

Machine 1		Machine 2	
16.03	16.01	16.02	16.03
16.04	15.96	15.97	16.04
16.05	15.98	15.96	16.02
16.05	16.02	16.01	16.01
16.02	15.99	15.99	16.00

Do you think the quality control department is correct?

2-7. Two types of plastic are suitable for use by an electronic calculator manufacturer. The breaking strength of this plastic is important. It is known that $\sigma_1 = \sigma_2 = 1.0$ psi. From a random sample of size $n_1 = 10$ and $n_2 = 12$ we obtain $\bar{y}_1 = 162.5$ and $\bar{y}_2 = 155.0$. The company will not adopt plastic 1 unless its breaking strength exceeds that of plastic 2 by at least 10 psi. Based on the sample information, should they use plastic 1? Construct a 99 percent confidence interval on the true mean difference in breaking strength.

2-8. The following are the burning times of flares of two different types:

Type 1	Type 2
65 82	64 56
81 67	71 69
57 59	83 74
66 75	59 82
82 70	65 79

(a) Test the hypothesis that the two variances are equal. Use $\alpha = .05$.

(b) Using the results of (a), test the hypothesis that the mean burning times are equal.

2-9. A new filtering device is installed in a chemical unit. Before its installation, a random sample yielded the following information about the percentage of impurity: $\bar{y}_1 = 12.5$, $S_1^2 = 101.17$, and $n_1 = 8$. After installation, a random sample yielded $\bar{y}_2 = 10.2$, $S_2^2 = 94.73$, $n_2 = 9$. Can you conclude that the two variances are equal? Has the filtering device reduced the percentage of impurity significantly?

2-10. Suppose that the random variable $y \sim N(\mu, \sigma^2)$ and the following random sample is available.

$$5.34 \quad 6.65 \quad 4.76 \quad 5.98 \quad 7.25$$
$$6.00 \quad 7.55 \quad 5.54 \quad 5.62 \quad 6.21$$
$$5.97 \quad 7.35 \quad 5.44 \quad 4.39 \quad 4.98$$
$$5.25 \quad 6.35 \quad 4.61 \quad 6.00 \quad 5.32$$

(a) Compute a 95 percent confidence interval estimate of σ^2.

(b) Test the hypothesis that $\sigma^2 = 1.0$.

2-11. The diameter of a ball bearing was measured by 12 individuals, each using two different kinds of calipers. The results are shown below:

Caliper 1	Caliper 2
0.265	0.264
0.265	0.265
0.266	0.264
0.267	0.266
0.267	0.267
0.265	0.268
0.267	0.264
0.267	0.265
0.265	0.265
0.268	0.267
0.268	0.268
0.265	0.269

Is there a significant difference between the means of the population of measurements represented by the two samples? Use $\alpha = .05$.

2-12. Two popular pain medications are being compared on the basis of the speed of absorption by the body. Specifically, tablet 1 is claimed to be absorbed twice as fast as tablet 2. Assume that σ_1^2 and σ_2^2 are known. Develop a test statistic for

$$H_0: 2\mu_1 = \mu_2$$
$$H_1: 2\mu_1 \neq \mu_2$$

2-13. Suppose we are testing

$$H_0: \mu_1 = \mu_2$$
$$H_1: \mu_1 \neq \mu_2$$

where σ_1^2 and σ_2^2 are known. Our sampling resources are constrained such that $n_1 + n_2 = N$. How should we allocate the N observations between the two populations to obtain the most powerful test?

2-14. Develop an expression for a $100(1 - \alpha)$ percent confidence interval for the variance of a normal distribution.

2-15. Develop an expression for a $100(1 - \alpha)$ percent confidence interval for the ratio σ_1^2/σ_2^2, where σ_1^2 and σ_2^2 are the variances of two normal distributions.

Chapter Three
Experiments With a Single Factor

3-1 INTRODUCTION

In this chapter we will study experiments with a single factor. For example, an experimenter might be interested in comparing the strength of fiber produced by several different production methods. Each method produces fiber with a particular mean strength, so the experimenter is interested in testing the equality of several means. It might seem that this problem could be solved by performing a t test on all possible pairs of means. However, such a solution would be incorrect, since it would lead to considerable distortion in the type I error. For example, suppose we wish to test the equality of 5 means, using pairwise comparisons. There are 10 possible pairs, and if the probability of correctly accepting the null hypothesis for each individual test is $1 - \alpha = .95$, then the probability of correctly accepting the null hypothesis for all 10 tests is $(.95)^{10} = .60$, if the tests are independent. Thus, a substantial increase in the type I error has occurred.

The appropriate procedure for testing the equality of several means is the analysis of variance. However, the analysis of variance has a much wider application than the problem above. It is probably the most useful technique in the field of statistical inference.

3-2 THE ONE-WAY CLASSIFICATION ANALYSIS OF VARIANCE

Suppose we have a different levels of a single factor that we wish to compare. The different levels of the factor are often called *treatments*. The observed response from each of the a treatments is a random variable. The data would appear as in Table 3-1. An entry in Table 3-1 (e.g., y_{ij}) represents the jth observation taken under treatment i. There will be, in general, n observations under the ith treatment.

Table 3-1 Typical Data for One-Way
Classification Analysis of Variance

		Observation			
	1	y_{11}	y_{12}	$\cdots$	y_{1n}
	2	y_{21}	y_{22}	$\cdots$	y_{2n}
Treatment	.	.	.	$\cdots$	.
	.	.	.	$\cdots$	.
	.	.	.	$\cdots$	.
	a	y_{a1}	y_{a2}	$\cdots$	y_{an}

We will find it useful to describe the observations by the *linear statistical model*

$$y_{ij} = \mu + \tau_i + \epsilon_{ij} \begin{cases} i = 1, 2, \ldots, a \\ j = 1, 2, \ldots, n \end{cases} \tag{3-1}$$

where y_{ij} is the (ij)th observation, μ is a parameter common to all treatments called the *overall* mean, τ_i is a parameter peculiar to the ith treatment called the ith *treatment effect*, and ϵ_{ij} is a random error component. Our objectives will be to test appropriate hypotheses about the treatment effects and to estimate them. For hypothesis testing, the model errors are assumed to be normally and independently distributed random variables with mean zero and variance σ^2. The variance σ^2 is assumed constant for all levels of the factor.

This model is called the *one-way classification* analysis of variance, because only one factor is investigated. Furthermore we will require that the experiment be performed in a random order so that the environment in which the treatments are used (often called the experimental units) is as uniform as possible. Such an experimental design is called a *completely randomized* design.

Actually, we can see that the statistical model, Equation 3-1, describes two different situations with respect to the treatment effects. First, the a treatments could have been specifically chosen by the experimenter. In this situation we wish to test hypotheses about the τ_i, and conclusions will apply only to the factor levels considered in the analysis. The conclusions cannot be extended to similar treatments that were not considered. Also, we may wish to estimate the τ_i. This is called the *fixed effects* model.

Alternatively the a treatments could be a *random sample* from a larger population of treatments. In this situation we should like to be able to extend the conclusions (which are based on the sample of treatments) to all treatments in the population, whether they were explicitly considered in the analysis or not. Here the τ_i are random variables, and knowledge about the particular ones investigated is relatively useless. Instead, we test hypotheses about the variability of the τ_i and try to estimate this variability. This is called the *random effects*, or *components of variance*, model.

3-3 THE FIXED EFFECTS MODEL

3-3.1 Statistical Analysis

In this section we will develop the analysis of variance for the fixed effects model, one-way classification. In the fixed effects model, the treatment effects τ_i are usually defined as deviations from the overall mean, so that

$$\sum_{i=1}^{a} \tau_i = 0 \tag{3-2}$$

Let $y_{i.}$ represent the total of the observations under the ith treatment and $\bar{y}_{i.}$ represent the average of the observations under the ith treatment. Similarly, let $y_{..}$ represent the grand total of all observations and $\bar{y}_{..}$ represent the grand mean of all observations. Expressed symbolically,

$$y_{i.} = \sum_{j=1}^{n} y_{ij}, \quad \bar{y}_{i.} = y_{i.}/n, \quad i = 1, 2, \ldots, a$$

$$\tag{3-3}$$

$$y_{..} = \sum_{i=1}^{a}\sum_{j=1}^{n} y_{ij}, \quad \bar{y}_{..} = y_{..}/N$$

where $N = an$ is the total number of observations. We see that the "dot" subscript notation implies summation over the subscript that it replaces.

We are interested in testing the equality of the a treatment effects. In light of (3-2), the appropriate hypotheses are

$$H_0: \tau_1 = \tau_2 = \tau_3 = \cdots = \tau_a = 0$$

$$H_1: \tau_i \neq 0 \text{ for at least one } i$$

That is, if the null hypothesis is true then each observation is made up of the overall mean μ plus a realization or value of the random error ϵ_{ij}. The appropriate test procedure is the analysis of variance.

The name "analysis of variance" is derived from a partitioning of total variability into its component parts. The total corrected sum of squares may be written as

$$\sum_{i=1}^{a}\sum_{j=1}^{n} (y_{ij} - \bar{y}_{..})^2 = \sum_{i=1}^{a}\sum_{j=1}^{n} [(\bar{y}_{i.} - \bar{y}_{..}) + (y_{ij} - \bar{y}_{i.})]^2 \tag{3-4}$$

or,

$$\sum_{i=1}^{a}\sum_{j=1}^{n}(y_{ij} - \bar{y}_{..})^2 = n\sum_{i=1}^{a}(\bar{y}_{i.} - \bar{y}_{..})^2 + \sum_{i=1}^{a}\sum_{j=1}^{n}(y_{ij} - \bar{y}_{i.})^2$$

$$+ 2\sum_{i=1}^{a}\sum_{j=1}^{n}(\bar{y}_{i.} - \bar{y}_{..})(y_{ij} - \bar{y}_{i.}) \quad (3\text{-}5)$$

However, the cross-product term in (3-5) is zero, since

$$\sum_{j=1}^{n}(y_{ij} - \bar{y}_{i.}) = y_{i.} - n\bar{y}_{i.} = y_{i.} - n(y_{i.}/n) = 0$$

Therefore, we have

$$\sum_{i=1}^{a}\sum_{j=1}^{n}(y_{ij} - \bar{y}_{..})^2 = n\sum_{i=1}^{a}(\bar{y}_{i.} - \bar{y}_{..})^2 + \sum_{i=1}^{a}\sum_{j=1}^{n}(y_{ij} - \bar{y}_{i.})^2 \quad (3\text{-}6)$$

Equation 3-6 states that the total variability in the data, as measured by the total corrected sum of squares, can be partitioned into a sum of squares of differences *between* treatment means and the grand mean, plus a sum of squares of differences of observations *within* treatments from the treatment mean. Now differences between observed treatment means and the grand mean is a measure of differences between treatments, while differences of observations within a treatment from the treatment mean can be due only to random error. Thus we may write (3-6) symbolically as

$$SS_T = SS_{\text{Treatments}} + SS_E$$

where SS_T is the total sum of squares, $SS_{\text{Treatments}}$ is called the sum of squares due to treatments (i.e., between treatments), and SS_E is called the sum of squares due to error (i.e., within treatments). There are $an = N$ total observations, thus SS_T has $N - 1$ degrees of freedom. There are a levels of the factor (and a treatment means), so $SS_{\text{Treatments}}$ has $a - 1$ degrees of freedom. Finally, within any treatment there are n replicates providing $n - 1$ degrees of freedom with which to estimate the experimental error. Since there are a treatments, we have $a(n - 1) = an - a = N - a$ degrees of freedom for error.

Now consider the distributional properties of these sums of squares. Since we have assumed that ϵ_{ij} is NID$(0, \sigma^2)$ the observations y_{ij} are NID$(\mu + \tau_i, \sigma^2)$. Thus SS_T/σ^2 is distributed as chi-square with $N - 1$ degrees of freedom, since SS_T is a sum of squares in normal random variables. We may also show that $SS_{\text{Treatments}}/\sigma^2$ is chi-square with $a - 1$ degrees of freedom if H_0 is true, and SS_E/σ^2 is chi-square with $N - a$ degrees of freedom. However, all three sums of squares are not independent since $SS_{\text{Treatments}}$ and SS_E add up to SS_T. The following theorem, which is a special form of one due to Cochran, is useful.

THEOREM 3-1. (COCHRAN). Let Z_i be NID$(0, 1)$ for $i = 1, 2, \ldots, \nu$ and

$$\sum_{i=1}^{\nu} Z_i^2 = Q_1 + Q_2 + \cdots + Q_s$$

where $s \leqslant \nu$, and Q_i has ν_i degrees of freedom $(i = 1, 2, \ldots, s)$. Then the $Q_1, Q_2, \ldots, Q_s$ are independent chi-square random variables with $\nu_1, \nu_2, \ldots, \nu_s$ degrees of freedom, respectively, if and only if

$$\nu = \nu_1 + \nu_2 + \cdots + \nu_s$$

Applying the theorem, we see that the degrees of freedom for $SS_{\text{Treatments}}$ and SS_E add up to $N - 1$, so that $SS_{\text{Treatments}}/\sigma^2$ and SS_E/σ^2 are independently distributed chi-square random variables. Therefore, under the null hypothesis, the statistic

$$F_0 = \frac{SS_{\text{Treatments}}/(a - 1)}{SS_E/(N - a)} = \frac{MS_{\text{Treatments}}}{MS_E} \tag{3-7}$$

follows the $F_{a-1, N-a}$ distribution. The quantities $MS_{\text{Treatments}}$ and MS_E are called *mean squares*.

We may use the expected values of the mean squares to show that F_0 in Equation 3-7 is an appropriate test statistic for $H_0: \tau_i = 0$. Consider

$$E(MS_E) = E\left(\frac{SS_E}{N - a}\right) = \frac{1}{N - a} E\left[\sum_{i=1}^{a} \sum_{j=1}^{n} (y_{ij} - \bar{y}_{i.})^2\right]$$

$$- \frac{1}{N - a} E\left[\sum_{i=1}^{a} \sum_{j=1}^{n} (y_{ij}^2 - 2y_{ij}\bar{y}_{i.} + \bar{y}_{i.}^2)\right]$$

$$= \frac{1}{N - a} E\left[\sum_{i=1}^{a} \sum_{j=1}^{n} y_{ij}^2 - 2n \sum_{i=1}^{a} \bar{y}_{i.}^2 + n \sum_{i=1}^{a} \bar{y}_{i.}^2\right]$$

$$= \frac{1}{N - a} E\left[\sum_{i=1}^{a} \sum_{j=1}^{n} y_{ij}^2 - \frac{1}{n} \sum_{i=1}^{a} y_{i.}^2\right]$$

Substituting the model, Equation 3-1, into this equation we obtain

$$E(MS_E) = \frac{1}{N - a} E\left[\sum_{i=1}^{a} \sum_{j=1}^{n} (\mu + \tau_i + \epsilon_{ij})^2 - \frac{1}{n} \sum_{i=1}^{a} \left(\sum_{j=1}^{n} \mu + \tau_i + \epsilon_{ij}\right)^2\right]$$

Now upon squaring and taking expectation of the quantities within brackets, we see that terms involving ϵ_{ij}^2 and $\epsilon_{i.}^2$ are replaced by σ^2 and $n\sigma^2$, respectively, because $E(\epsilon_{ij}) = 0$. Furthermore, all cross-products involving ϵ_{ij} have zero expectation. Therefore, after squaring and taking expectation, the last equation becomes

$$E(MS_E) = \frac{1}{N - a}\left[N\mu^2 + n \sum_{i=1}^{a} \tau_i^2 + N\sigma^2 - N\mu^2 - n \sum_{i=1}^{a} \tau_i^2 - a\sigma^2\right]$$

or

$$E(MS_E) = \sigma^2$$

By a similar approach, we may also show that

$$E(MS_{Treatments}) = \sigma^2 + \frac{n \sum_{i=1}^{a} \tau_i^2}{a - 1}$$

From the expected mean square we see that, in general, MS_E is an unbiased estimator of σ^2. Also, under the null hypothesis, $MS_{Treatments}$ is an unbiased estimator of σ^2. However, if the null hypothesis is false, then the expected value of $MS_{Treatments}$ is greater than σ^2. Therefore, under the alternative hypothesis the expected value of the numerator of the test statistic (Equation 3-7) is greater than the expected value of the denominator, and we should reject H_0 on values of the test statistic which are too large. This implies an upper-tail, one-tail critical region. We would reject H_0 if

$$F_0 > F_{\alpha, a-1, N-a}$$

where F_0 is computed from Equation 3-7.

Computing formulas for the sums of squares may be obtained by rewriting and simplifying the definitions of $SS_{Treatments}$ and SS_T in Equation 3-6. This yields

$$SS_T = \sum_{i=1}^{a} \sum_{j=1}^{n} y_{ij}^2 - \frac{y_{..}^2}{N} \tag{3-8}$$

and

$$SS_{Treatments} = \sum_{i=1}^{a} \frac{y_{i.}^2}{n} - \frac{y_{..}^2}{N} = n \sum_{i=1}^{a} (\bar{y}_{i.} - \bar{y}_{..})^2 = n(a-1) S_{\bar{y}_{i.}}^2 \tag{3-9}$$

The error sum of squares is obtained by subtraction as

$$SS_E = SS_T - SS_{Treatments} = n-1 \sum_{i=1}^{a} s_i^2 \tag{3-10}$$

The test procedure is summarized in Table 3-2. This is called an analysis of variance table.

Table 3-2 The Analysis of Variance Table for the One-Way Classification Fixed-Effects Model

Source of Variation	Sum of Squares	Degrees of Freedom	Mean Square	F_0
Between treatments	$SS_{Treatments}$	$a - 1$	$MS_{Treatments}$	$F_0 = \dfrac{MS_{Treatments}}{MS_E}$
Error (within treatments)	SS_E	$N - a$	MS_E	
Total	SS_T	$N - 1$		

■**EXAMPLE 3-1.** The tensile strength of synthetic fiber used to make cloth for men's shirts is of interest to the manufacturer. It is suspected that strength is affected by the percentage of cotton in the fiber. Five levels of cotton percentage are of interest, 15 percent, 20 percent, 25 percent, 30 percent, and 35 percent. Five observations are to be taken at each level of cotton percentage, and the 25 total observations are to be run in random order.

To illustrate how the order of the runs is randomized, suppose we number the runs as follows:

Percentage of Cotton

15	20	25	30	35
1	6	11	16	21
2	7	12	17	22
3	8	13	18	23
4	9	14	19	24
5	10	15	20	25

Now select a random number between 1 and 25. Suppose this number is 8. Then the number 8 observation (20 percent cotton) is run first. This process would be repeated until all 25 observations have been assigned a position in the test sequence.[1] Suppose that the test sequence obtained is

Test Sequence	Run Number	Percentage of Cotton
1	8	20
2	18	30
3	10	20
4	23	35
5	17	30
6	5	15
7	14	25
8	6	20
9	15	25
10	20	30
11	9	20
12	4	15
13	12	25
14	7	20
15	1	15
16	24	35

---

[1] The only restriction on randomization here is that if the same number (e.g., 8) is drawn again it is discarded. This is a minor restriction and is ignored.

Test Sequence	Run Number	Percentage of Cotton
17	21	35
18	11	25
19	2	15
20	13	25
21	22	35
22	16	30
23	25	35
24	19	30
25	3	15

The 25 observations are taken in this test sequence and the data obtained is shown in Table 3-3.

Table 3-3   Tensile Strength of Synthetic Fiber (lb/in.^{2})

Percentage of Cotton	Observations					$y_i.$
	1	2	3	4	5	
15	7	7	15	11	9	49
20	12	17	12	18	18	77
25	14	18	18	19	19	88
30	19	25	22	19	23	108
35	7	10	11	15	11	54

$$376 = y_{..}$$

The sums of squares are computed as follows:

$$SS_T = \sum_{i=1}^{5} \sum_{j=1}^{5} y_{ij}^2 - \frac{y_{..}^2}{N}$$

$$= (7)^2 + (7)^2 + (15)^2 + \cdots + (15)^2 + (11)^2 - \frac{(376)^2}{25} = 636.96$$

$$SS_{\text{Treatments}} = \sum_{i=1}^{5} \frac{y_{i.}^2}{n} - \frac{y_{..}^2}{N}$$

$$= \frac{(49)^2 + \cdots + (54)^2}{5} - \frac{(376)^2}{25} = 475.76$$

$$SS_E = SS_T - SS_{\text{Treatments}}$$

$$= 636.96 - 475.76 = 161.20$$

The analysis of variance is summarized in Table 3-4. Since $F_{.01, 4, 20} = 4.43$, we reject H_0 and conclude that the percentage of cotton in the fiber significantly affects the strength.

Table 3-4 Analysis of Variance for the Tensile Strength Data

Source of Variation	Sum of Squares	Degrees of Freedom	Mean Square	F_0
Treatments	175.76	4	118.94	$F_0 = 14.76$
Error	161.20	20	8.06	
Total	636.96	24		∎

∎ **EXAMPLE 3-2. Coding the Observations.** The calculations in the analysis of variance may often be made more accurate or simplified by coding the observations. For example, consider the tensile strength data in Example 3-1. Suppose we subtract 15 from each observation. The coded data are shown in Table 3-5.

Table 3-5 Coded Tensile Strength Data from Example 3-2

Percentage of Cotton	Observations					
	1	2	3	4	5	$y_{i.}$
15	−8	−8	0	−4	−6	−26
20	−3	2	−3	3	3	2
25	−1	3	3	4	4	13
30	4	10	7	4	8	33
35	−8	−5	−4	0	−4	−21
						$1 = y_{..}$

It is easy to verify that

$$SS_T = (-8)^2 + (-8)^2 + \cdots + (-4)^2 - \frac{(1)^2}{25} = 636.96$$

$$SS_{\text{Treatments}} = \frac{(-26)^2 + (2)^2 + \cdots + (-21)^2}{5} - \frac{(1)^2}{25} = 475.76$$

and

$$SS_E = 161.20$$

Comparing these sums of squares to those obtained in Example 3-1, we see that subtracting a constant from the original data does not change the sums of squares. When the data consists of very large numbers, this is often a useful computational simplification.

Now suppose that we multiply each observation in Example 3-1 by 2. It is easy to verify that the sums of squares for the transformed data are $SS_T = 2547.84$, $SS_{Treatments} = 1903.04$, and $SS_E = 644.80$. These sums of squares appear to differ considerably from those obtained in Example 3-1. However, if they are divided by 4 (i.e., 2^2), then the results are identical. For example, for the treatment sum of squares $1903.04/4 = 475.76$. Also, for the coded data, the F ratio is $F = (1903.04/4)/(644.80/20) = 14.76$, which is identical to the F ratio for the original data. Thus the analyses of variance are equivalent. ∎

3-3.2 Estimation of the Model Parameters

We will now develop estimators for the parameters in the one-way classification model

$$y_{ij} = \mu + \tau_i + \epsilon_{ij}$$

using the method of least squares. In estimating μ and τ_i by least squares the normality assumption on the errors ϵ_{ij} is not needed. To find the least squares estimators of μ and τ_i we form the sum of squares of the errors

$$L = \sum_{i=1}^{a} \sum_{j=1}^{n} \epsilon_{ij}^2 = \sum_{i=1}^{a} \sum_{j=1}^{n} (y_{ij} - \mu - \tau_i)^2 \tag{3-11}$$

and choose values of μ and τ_i, say $\hat{\mu}$ and $\hat{\tau}_i$, which minimize L. The appropriate values would be, of course, the solutions to the $a + 1$ simultaneous equations

$$\left. \frac{\partial L}{\partial \mu} \right|_{\hat{\mu}, \hat{\tau}_i} = 0$$

$$\left. \frac{\partial L}{\partial \tau_i} \right|_{\hat{\mu}, \hat{\tau}_i} = 0, \qquad i = 1, 2, \ldots, a$$

Upon differentiating (3-11) with respect to μ and τ_i and equating to zero, we obtain

$$-2 \sum_{i=1}^{a} \sum_{j=1}^{n} (y_{ij} - \hat{\mu} - \hat{\tau}_i) = 0$$

and

$$-2 \sum_{j=1}^{n} (y_{ij} - \hat{\mu} - \hat{\tau}_i) = 0, \qquad i = 1, 2, \ldots, a$$

which, after simplification, yield

$$N\hat{\mu} + n\hat{\tau}_1 + n\hat{\tau}_2 + \cdots + n\hat{\tau}_a = y_{..}$$

$$n\hat{\mu} + n\hat{\tau}_1 \qquad\qquad = y_{1.}$$

$$n\hat{\mu} \qquad + n\hat{\tau}_2 \qquad\qquad = y_{2.}$$

$$\vdots \qquad\qquad\qquad \vdots$$

$$n\hat{\mu} \qquad\qquad\qquad + n\hat{\tau}_a = y_{a.}$$

(3-12)

The $a + 1$ equations (3-12) in $a + 1$ unknowns are called the *least squares normal equations*. Notice that if we add the last a normal equations we obtain the first normal equation. Therefore, the normal equations are not linearly independent and no unique solution for $\mu, \tau_1, \ldots, \tau_a$ exists. This difficulty can be overcome by several methods. Since we have defined the treatment effects as deviations from the overall mean, it seems reasonable to apply the constraint

$$\sum_{i=1}^{a} \hat{\tau}_i = 0 \tag{3-13}$$

Using this constraint, we obtain as the solution to the normal equations

$$\hat{\mu} = \bar{y}_{..}$$

$$\hat{\tau}_i = \bar{y}_{i.} - \bar{y}_{..}, \qquad i = 1, 2, \ldots, a \tag{3-14}$$

This solution has considerable intuitive appeal, for we see that the overall mean is estimated by the grand average of the observations and any treatment effect is just the difference between the treatment average and the grand average.

This solution is obviously not unique, and depends on the constraint (Equation 3-13) that we have chosen. At first this may seem unfortunate, because two different experimenters could analyze the same data and obtain different results if they apply different constraints. However, certain *functions* of the model parameter *are* uniquely estimated, regardless of the constraint. Some examples are $\tau_i - \tau_j$, which would be estimated by $\hat{\tau}_i - \hat{\tau}_j = \bar{y}_{i.} - \bar{y}_{j.}$, and $\mu + \tau_i$, which would be estimated by $\hat{\mu} + \hat{\tau}_i = \bar{y}_{i.}$ from (3-12). Since we are usually interested in differences among the treatment effects rather than their actual values, it causes no concern that the τ_i cannot be uniquely estimated. In general, any function of the model parameters that is a linear combination of the left-hand side of the normal equations (3-12) can be uniquely estimated. Functions that are uniquely estimated regardless of which constraint is used are called *estimable* functions.

A confidence interval estimate of the ith treatment mean may be easily determined. The mean of the ith treatment is

$$\mu_i = \mu + \tau_i$$

A point estimator of μ_i would be $\hat{\mu}_i = \hat{\mu} + \hat{\tau}_i = \bar{y}_{i.}$. Now, if we assume that the errors are normally distributed, each $\bar{y}_{i.}$ is $NID(\mu_i, \sigma^2/n)$. Thus, if σ^2 were known, we could use the normal distribution to define the confidence interval. Using the MS_E as an estimator of σ^2, we would base the confidence interval on the t distribution. Therefore, a $100(1 - \alpha)$ percent confidence interval on the ith treatment mean μ_i is

$$[\bar{y}_{i.} \pm t_{\alpha/2, N-a}\sqrt{MS_E/n}] \tag{3-15}$$

A $100(1 - \alpha)$ percent confidence interval on the difference in any two treatments means, say $\mu_i - \mu_j$, would be

$$[\bar{y}_{i.} - \bar{y}_{j.} \pm t_{\alpha/2, N-a}\sqrt{2MS_E/n}] \tag{3-16}$$

■**EXAMPLE 3-3.** Using the data in Example 3-1, we may find the estimates of the overall mean and the treatment effects as $\hat{\mu} = 376/25 = 15.04$ and

$$\hat{\tau}_1 = \bar{y}_{1.} - \bar{y}_{..} = 9.80 - 15.04 = -5.24$$

$$\hat{\tau}_2 = \bar{y}_{2.} - \bar{y}_{..} = 15.40 - 15.04 = +0.36$$

$$\hat{\tau}_3 = \bar{y}_{3.} - \bar{y}_{..} = 17.60 - 15.04 = +2.56$$

$$\hat{\tau}_4 = \bar{y}_{4.} - \bar{y}_{..} = 21.60 - 15.04 = +6.56$$

$$\hat{\tau}_5 = \bar{y}_{5.} - \bar{y}_{..} = 10.80 - 15.04 = -4.24$$

A 95 percent confidence interval on the mean of treatment 4 is computed from Equation 3-15 as

$$[21.60 \pm (2.086)\sqrt{8.06/5}]$$

or

$$[21.60 \pm 2.65]$$

Thus the desired confidence interval is $18.95 \leqslant \mu_4 \leqslant 24.25$.

3-3.3   The Unbalanced Case

In some single-factor experiments the number of observations taken under each treatment may be different. We then say that the design is *unbalanced*. The analysis of variance described above may still be used, but slight modifications must be made in the sum of squares formulas. Let n_i observations be taken under treatment i ($i = 1, 2, \ldots, a$) and define $N = \sum_{i=1}^{a} n_i$. The computational formulas for SS_T and $SS_{\text{Treatments}}$ become

$$SS_T = \sum_{i=1}^{a} \sum_{j=1}^{n_i} y_{ij}^2 - \frac{y_{..}^2}{N}$$

and

$$SS_{\text{Treatments}} = \sum_{i=1}^{a} \frac{y_{i.}^2}{n_i} - \frac{y_{..}^2}{N}$$

In solving the normal equations, the constraint $\sum_{i=1}^{a} n_i \hat{\tau}_i = 0$ is used. No other changes are required in the analysis of variance.

However, there are two advantages in choosing a balanced design. First, the test statistic is relatively insensitive to small departures from the assumption of equal variances for the a treatments if the sample sizes are equal. This is not the case for unequal sample sizes. Second, the power of the test is maximized if the samples are of equal size.

3-4 TESTS ON INDIVIDUAL TREATMENT MEANS

Suppose in conducting an analysis of variance for the fixed effects model the null hypothesis is rejected. Thus, there are differences between the treatments, but exactly *which* treatments differ is not specified. In this situation, further comparisons between groups of treatments means may be useful. The ith treatment mean is defined as $\mu_i = \mu + \tau_i$, and μ_i is estimated by $\bar{y}_{i.}$. Comparisons between treatment means are usually made in terms of the treatment totals $\{y_{i.}\}$.

Consider the synthetic fiber testing problem of Example 3-1. Since the hypothesis $H_0: \tau_i = 0$ was rejected, we know that some cotton percentages produce different tensile strengths than others, but which ones actually cause this difference? We might suspect at the outset of the experiment that cotton percentages 4 and 5 produce the same tensile strength, implying that we would like to test the hypothesis

$$H_0: \mu_4 = \mu_5$$
$$H_1: \mu_4 \neq \mu_5$$

This hypothesis could be tested by investigating an appropriate linear combination of treatment totals, say

$$y_{4.} - y_{5.} = 0$$

If we had suspected that the *average* of cotton percentages 1 and 3 did not differ from the *average* of cotton percentages 4 and 5, then the hypothesis would have been

$$H_0: \mu_1 + \mu_3 = \mu_4 + \mu_5$$
$$H_1: \mu_1 + \mu_3 \neq \mu_4 + \mu_5$$

which implies the linear combination

$$y_{1.} + y_{3.} - y_{4.} - y_{5.} = 0$$

In general, the comparison of treatment means of interest will imply a linear combination of treatment totals such as

$$C = \sum_{i=1}^{a} c_i y_{i.}$$

with the restriction that $\sum_{i=1}^{a} c_i = 0$. Such linear combinations are called *contrasts*. The sum of squares for any contrast is

$$SS_C = \frac{\left(\sum_{i=1}^{a} c_i y_{i.} \right)^2}{n \sum_{i=1}^{a} c_i^2} \tag{3-17}$$

and has a single degree of freedom. If the design is unbalanced, then the comparison of treatment means requires that $\sum_{i=1}^{a} n_i c_i = 0$, and (3-17) becomes

$$SS_C = \frac{\left(\sum_{i=1}^{a} c_i y_{i.} \right)^2}{\sum_{i=1}^{a} n_i c_i^2} \tag{3-18}$$

A contrast is tested by comparing its sum of squares to the mean square error. The resulting statistic would be distributed as F with 1 and $N - a$ degrees of freedom.

A very important special case of the above procedure is that of *orthogonal* contrasts. Two contrasts with coefficients $\{c_i\}$ and $\{d_i\}$ are orthogonal if

$$\sum_{i=1}^{a} c_i d_i = 0$$

or for an unbalanced design, if

$$\sum_{i=1}^{a} n_i c_i d_i = 0$$

For a treatments a set of $a - 1$ orthogonal contrasts will partition the sum of squares due to treatments into $a - 1$ independent single-degree-of-freedom components. Thus, tests performed on orthogonal contrasts are independent.

There are many ways to choose the orthogonal contrast coefficients for a set of treatments. Usually, something in the nature of the experiment should suggest which comparisons will be of interest. For example if there are $a = 3$ treatments with treatment 1 a control and treatments 2 and 3 actual levels of the factor of interest to the experimenter, then appropriate orthogonal contrasts might be as follows:

Treatment	Coefficients for Orthogonal Contrasts	
1 (control)	-2	0
2 (level 1)	1	-1
3 (level 2)	1	1

Note that contrast 1 with $c_i = -2, 1, 1$ compares the average effect of the factor with the control, while contrast 2 with $d_i = 0, -1, 1$ compares the two levels of the factor of interest.

Contrast coefficients must be chosen prior to running the experiment and examining the data. The reason for this is that if comparisons are selected after examining the data, most experimenters would construct tests that correspond to large observed differences in means. These large differences could be due to the presence of real effects or they could be due to random error. If the experimenter consistently picks the largest differences to compare, he will inflate the type I error of the test, since it is likely that in an unusually high percentage of the comparisons selected the observed differences will be due to error.

■**EXAMPLE 3-4.** Consider the data in Example 3-1. There are five treatment means, and four degrees of freedom between these treatments. One set of comparisons between these means, and the associated orthogonal contrasts are

Hypothesis	*Contrast*
$H_0: \mu_4 = \mu_5$	$C_1 = \qquad\qquad - y_{4.} + y_{5.}$
$H_0: \mu_1 + \mu_3 = \mu_4 + \mu_5$	$C_2 = y_{1.} \qquad + y_{3.} - y_{4.} - y_{5.}$
$H_0: \mu_1 = \mu_3$	$C_3 = y_{1.} \qquad - y_{3.}$
$H_0: 4\mu_2 = \mu_1 + \mu_3 + \mu_4 + \mu_5$	$C_4 = - y_{1.} + 4y_{2.} - y_{3.} - y_{4.} - y_{5.}$

Notice that the contrast coefficients are orthogonal. Using the data in Table 3-3, we find the numerical values of the contrasts and the sums of squares as follows:

$$C_1 = \qquad\qquad -1(108) + 1(54) = -54, \quad SS_{C_1} = \frac{(-54)^2}{5(2)} = 291.60$$

$$C_2 = +1(49) \qquad +1(88) - 1(108) - 1(54) = -25, \quad SS_{C_2} = \frac{(-25)^2}{5(4)} = 31.25$$

$$C_3 = +1(49) \qquad -1(88) \qquad\qquad = -39, \quad SS_{C_3} = \frac{(-39)^2}{5(2)} = 152.10$$

$$C_4 = -1(49) + 4(77) - 1(88) - 1(108) - 1(54) = \quad 9, \quad SS_{C_4} = \frac{(9)^2}{5(20)} = 0.81$$

Table 3-6 Analysis of Variance for the Tensile Strength Data

Source of Variation	Sum of Squares	Degrees of Freedom	Mean Square	F_0
Cotton percentage	475.76	4	118.94	14.76[a]
orthogonal contrasts				
C_1: $\mu_4 = \mu_5$	(291.60)	1	291.60	36.18[a]
C_2: $\mu_1 + \mu_3 = \mu_4 + \mu_5$	(31.25)	1	31.25	3.88
C_3: $\mu_1 = \mu_3$	(152.10)	1	152.10	18.87[a]
C_4: $4\mu_2 = \mu_1 + \mu_3 + \mu_4 + \mu_5$	(0.81)	1	0.81	0.10
Error	161.20	20	8.06	
Total	636.96	24		

[a]Significant at 1 percent.

These contrast sums of squares completely partition the treatment sum of squares. The tests on the contrasts are usually incorporated in the analysis of variance, such as shown in Table 3-6. From this display, we conclude that there are significant differences between cotton percentages 4 and 5 and 1 and 3; but that the *average* of 1 and 3 does not differ from the average of 4 and 5, nor does 2 differ from the average of the other four percentages. ∎

In many experiments the investigator may not know in advance how to construct appropriate orthogonal contrasts or he may wish to test more than $a - 1$ comparisons using the same data. For example, the experimenter may wish to test all possible pairs of means. The null hypotheses would then be H_0: $\mu_i = \mu_j$, for all $i \neq j$, and, if we employ the t test, the statistic $t_0 = (\bar{y}_{i.} - \bar{y}_{j.})/(2MS_E/n)^{1/2}$ is distributed as t_{N-a} under H_0. Thus assuming a two-sided alternative, any pair of means would be declared significantly different if $|\bar{y}_{i.} - \bar{y}_{j.}| > t_{\alpha/2, N-a}(2MS_E/n)^{1/2}$. The quantity $t_{\alpha/2, N-a}(2MS_E/n)^{1/2}$ is called the least significant difference.

When all possible pairs of means are tested using this approach, the probability of type I error may be considerably increased, as discussed in Section 3-1. There are several tests in wide use that avoid this criticism. Among the more popular of these tests are those of Newman (1939), Tukey (1953), and Duncan (1955). New interest in Newman's test was provided by Keuls (1952), and the procedure is usually called the Newman-Keuls test. Tukey's test is conservative, in that the probability of type I error per experiment is really smaller than α. The Newman-Keuls and Duncan tests are somewhat more liberal. Carmer and Swanson (1973), in an extensive comparison of these and other test procedures, show that Duncan's test is superior to the Newman-Keuls test in detecting true differences between pairs of means. We will describe Duncan's method, usually called the *Duncan multiple range test*.

To apply Duncan's multiple range test for equal sample sizes, the a treatment means are arranged in ascending order, and the standard error of each mean is determined as

$$S_{\bar{y}_{i.}} = \sqrt{\frac{MS_E}{n}} \tag{3-19}$$

From Duncan's table of significant ranges (Appendix Table VII), obtain the values $r_\alpha(p, f)$, for $p = 2, 3, \ldots, a$, where α is the significance level and f is the number of degrees of freedom for error. Convert these ranges into a set of $a - 1$ least significant ranges (e.g., R_p) for $p = 2, 3, \ldots, a$, by calculating

$$R_p = r_\alpha(p, f) S_{\bar{y}_{i.}}, \text{ for } p = 2, 3, \ldots, a$$

Then the observed ranges between means are tested, beginning with largest versus smallest, which would be compared with the least significant range R_a. Then the range of the largest and the second smallest is computed and compared with the least significant range R_{a-1}. These comparisons are continued until all means have been compared with the largest mean. Then the range of the second largest mean and the smallest is computed and compared against the least significant range R_{a-1}. This process is continued until the ranges of all possible $a(a-1)/2$ pairs of means have been considered. If an observed range is greater than the corresponding least significant range, then we conclude that the pair of means in question are significantly different. To prevent contradictions, no differences between a pair of means is considered significant if the two means involved fall between two other means that do not differ significantly.

■**EXAMPLE 3-5.** We will apply Duncan's multiple range test to the data of Example 3-1. Recall that $MS_E = 8.06$, $N = 25$, $n = 5$, and there are 20 error degrees of freedom. Ranking the treatment means in ascending order, we have

$$\bar{y}_{1.} - 9.8$$
$$\bar{y}_{5.} = 10.8$$
$$\bar{y}_{2.} = 15.4$$
$$\bar{y}_{3.} - 17.6$$
$$\bar{y}_{4.} = 21.6$$

The standard error of each mean is $S_{\bar{y}_{i.}} = \sqrt{8.06/5} = 1.27$. From the table of significant ranges in Appendix Table VII for 20 degrees of freedom and $\alpha = .05$ we obtain $r_{.05}(2, 20) = 2.95$, $r_{.05}(3, 20) = 3.10$, $r_{.05}(4, 20) = 3.18$, and $r_{.05}(5, 20) = 3.25$. Thus the least significant ranges are

$$R_2 = r_{.05}(2, 20) S_{\bar{y}_{i.}} = (2.95)(1.27) = 3.75$$
$$R_3 = r_{.05}(3, 20) S_{\bar{y}_{i.}} = (3.10)(1.27) = 3.94$$
$$R_4 = r_{.05}(4, 20) S_{\bar{y}_{i.}} = (3.18)(1.27) = 4.04$$
$$R_5 = r_{.05}(5, 20) S_{\bar{y}_{i.}} = (3.25)(1.27) = 4.13$$

The comparisons would yield

$$4 \text{ vs. } 1 = 21.6 - 9.8 = 11.8 > 4.13 \; (R_5)$$
$$4 \text{ vs. } 5 = 21.6 - 10.8 = 10.8 > 4.04 \; (R_4)$$

$$4 \text{ vs. } 2 = 21.6 - 15.4 = 6.2 > 3.94 \ (R_3)$$

$$4 \text{ vs. } 3 = 21.6 - 17.6 = 4.0 > 3.75 \ (R_2)$$

$$3 \text{ vs. } 1 = 17.6 - 9.8 = 7.8 > 4.04 \ (R_4)$$

$$3 \text{ vs. } 5 = 17.6 - 10.8 = 6.8 > 3.95 \ (R_3)$$

$$3 \text{ vs. } 2 = 17.6 - 15.4 = 2.2 < 3.75 \ (R_2)$$

$$2 \text{ vs. } 1 = 15.4 - 9.8 = 5.6 > 3.94 \ (R_3)$$

$$2 \text{ vs. } 5 = 15.4 - 10.8 = 4.6 > 3.75 \ (R_2)$$

$$5 \text{ vs. } 1 = 10.8 - 9.8 = 1.0 < 3.75 \ (R_2)$$

From the analysis we see that there are significant differences between all pairs of means except 3 and 2 and 5 and 1. It is helpful to draw a graph underlining those means which are not significantly different as in Fig. 3-1. From this graph it is immediately obvious that treatment 4 produces a significantly greater tensile strength than the other treatments. ∎

$\bar{y}_{1.}$	$\bar{y}_{5.}$	$\bar{y}_{2.}$	$\bar{y}_{3.}$	$\bar{y}_{4.}$
9.8	10.8	15.4	17.6	21.6

Figure 3-1. Results of Duncan's multiple range test

Carmer and Swanson (1973) also show that the least significant difference method described above is a very effective test for detecting true differences in means if applied *only after* the F statistic in the analysis of variance is significant at 5 percent. Although this procedure will result in more type I errors than Duncan's multiple range test, some experimenters may prefer it as the critical value depends on the observed analysis of variance F value rather than on the number of means in the experiment.

The difference between two means is, of course, a contrast. In some experiments after examining the data, we may wish to test more elaborate contrasts involving *several* means. Scheffé (1953) has proposed a method for testing such comparisons. His test gives the experimenter freedom to examine the data and test any estimable function that seems interesting. However, Carmer and Swanson (1973) point out that Scheffé's test does not perform very well with respect to detecting true differences between pairs of means, and should not be used in the pairwise multiple comparison problem.

3-5   THE RANDOM EFFECTS MODEL

An experimenter is frequently interested in a factor that has a large number of possible levels. If the experimenter randomly selects a of these levels from the population of factor levels, then we say that the factor is *random*. Because the levels of the factor actually used in the experiment were chosen randomly, inferences will be made about the entire population of factor levels. We will assume that the population of factor levels is either of infinite size, or is large enough to be considered infinite. Situations in which the population of factor levels is small enough to employ a finite population approach are not encountered frequently, and one may refer to Bennett and Franklin (1954) and Searle and Fawcett (1970) for discussion of the finite population case.

The linear statistical model is

$$y_{ij} = \mu + \tau_i + \epsilon_{ij} \begin{cases} i = 1, 2, \ldots, a \\ j = 1, 2, \ldots, n \end{cases} \tag{3-20}$$

where *both* τ_i *and* ϵ_{ij} are random variables. If τ_i has variance σ_τ^2 and is independent of ϵ_{ij}, the variance of any observation is

$$V(y_{ij}) = \sigma_\tau^2 + \sigma^2$$

The variances σ_τ^2 and σ^2 are called *variance components*, and the model (3-20) is called the *components of variance* or *random effects* model. To test hypotheses in this model we require that the $\{\epsilon_{ij}\}$ are NID($0, \sigma^2$), that the $\{\tau_i\}$ are NID($0, \sigma_\tau^2$), and that τ_i and ϵ_{ij} are independent.[1]

The sum of squares identity

$$SS_T = SS_{\text{Treatments}} + SS_E \tag{3-21}$$

is still valid. That is, we partition the total variability in the observations into a component that measures variation between treatments ($SS_{\text{Treatments}}$) and a component that measures variation within treatments (SS_E). Testing hypotheses about individual treatment effects is meaningless, so instead we test the hypotheses

$$H_0: \sigma_\tau^2 = 0$$

$$H_1: \sigma_\tau^2 > 0$$

If $\sigma_\tau^2 = 0$, all treatments are identical; but if $\sigma_\tau^2 > 0$, variation exists between treatments. As before, SS_E/σ^2 is distributed as chi-square with $N - a$ degrees of freedom, and under the null hypothesis $SS_{\text{Treatments}}/\sigma^2$ is distributed as chi-square with $a - 1$ degrees of freedom. Both random variables are independent. Thus under the null hypothesis, the ratio

$$F_0 = \frac{\dfrac{SS_{\text{Treatments}}}{a - 1}}{\dfrac{SS_E}{N - a}} = \frac{MS_{\text{Treatments}}}{MS_E} \tag{3-22}$$

---

[1] The assumption that the $\{\tau_i\}$ are independent random variables implies that the usual assumption of $\sum_{i=1}^{a} \tau_i = 0$ from the fixed effects model does not apply to the random effects model.

is distributed as F with $a - 1$ and $N - a$ degrees of freedom. However, we need to examine the expected mean squares to fully describe the test procedure.

Consider

$$E(MS_{Treatments}) = \frac{1}{a - 1} E(SS_{Treatments}) = \frac{1}{a - 1} E \left[\sum_{i=1}^{a} \frac{y_{i.}^2}{n} - \frac{y_{..}^2}{N} \right]$$

$$= \frac{1}{a - 1} E \left[\frac{1}{n} \sum_{i=1}^{a} \left(\sum_{j=1}^{n} \mu + \tau_i + \epsilon_{ij} \right)^2 - \frac{1}{N} \left(\sum_{i=1}^{a} \sum_{j=1}^{n} \mu + \tau_i + \epsilon_{ij} \right)^2 \right]$$

Upon squaring and taking expectation of the quantities in brackets, we see that terms involving τ_i^2 are replaced by σ_τ^2, as $E(\tau_i) = 0$. Also, terms involving ϵ_{ij}^2, $\epsilon_{..}^2$, and $\sum_{i=1}^{a} \sum_{j=1}^{n} \tau^2$ are replaced by $n\sigma^2$, $an\sigma^2$, and $an^2\sigma_\tau^2$, respectively. Furthermore, all cross-product terms involving τ_i and ϵ_{ij} have zero expectation. This leads to

$$E(MS_{Treatments}) = \frac{1}{a - 1} [N\mu^2 + N\sigma_\tau^2 + a\sigma^2 - N\mu^2 - n\sigma_\tau^2 - \sigma^2]$$

or

$$E(MS_{Treatments}) = \sigma^2 + n\sigma_\tau^2 \tag{3-23}$$

Similarly, we may show that

$$E(MS_E) = \sigma^2 \tag{3-24}$$

From the expected mean squares, we see that under H_0 both numerator and denominator of the test statistic (Equation 3-22) are unbiased estimators of σ^2, while under H_1 the expected value of the numerator is greater than the expected value of the denominator. Therefore we should reject H_0 for values of F_0 which are too large. This implies an upper-tail, one-tail critical region, so we reject H_0 if $F_0 > F_{\alpha, a-1, N-a}$.

The computational procedure and analysis of variance table for the random effects model are identical to the fixed effects case. The conclusions, however, are quite different because they apply to the entire population of treatments.

We may be interested in estimating the variance components (σ^2 and σ_τ^2) in the model. The procedure that we will use to estimate σ^2 and σ_τ^2 is called the "analysis of variance method," because it makes use of the lines in the analysis of variance table. The procedure consists of equating the expected mean squares to their observed value in the analysis of variance table and solving for the variance components. Upon equating observed and expected mean squares in the one-way classification random effects model we obtain

$$MS_{Treatments} = \sigma^2 + n\sigma_\tau^2$$

and

$$MS_E = \sigma^2$$

Therefore, the estimators of the variance components are

$$\hat{\sigma}^2 = MS_E \tag{3-25}$$

and

$$\hat{\sigma}_\tau^2 = \frac{MS_{\text{Treatments}} - MS_E}{n} \tag{3-26}$$

For unequal sample sizes replace n in Equation (3-26) by

$$n_0 = \frac{1}{a-1} \left[\sum_{i=1}^{a} n_i - \frac{\sum_{i=1}^{a} n_i^2}{\sum_{i=1}^{a} n_i} \right]$$

The analysis of variance method of variance component estimation does not require the normality assumption. It does yield estimators of σ^2 and σ_τ^2 which are best quadratic unbiased (i.e., of all unbiased quadratic functions of the observations, these estimators have minimum variance).

Occasionally, the analysis of variance method produces a negative estimate of a variance component. Clearly, variance components are by definition nonnegative, so a negative estimate of a variance component is viewed with some concern. One course of action is to accept the estimate and use it as evidence that the true value of the variance component is zero, assuming that sampling variation led to the negative estimate. This has intuitive appeal, but suffers from theoretical difficulties, such as using zero in place of the negative estimate can disturb the statistical properties of other estimates. Another alternative is to reestimate the negative variance component with a method that always yields nonnegative estimates. Still another alternative is to consider the negative estimate as evidence that the assumed linear model is incorrect, and reexamine the problem. A good discussion of variance component estimation is given by Searle (1971a, b).

■**EXAMPLE 3-6.** A textile company weaves a fiber on a large number of looms. They would like the looms to be homogeneous so that they obtain a fiber product of uniform strength. The process engineer suspects that, in addition to the usual variation in strength within samples of fiber from the same loom, there may also be significant variations in strength between looms. To investigate this, he selects four looms at random, and makes four strength determinations on the fiber manufactured on each loom. This experiment is run in random order, and the data obtained are shown in Table 3-7. The analysis of variance is conducted, and is shown in Table 3-8. From this, we conclude that the looms in the plant differ significantly.

The variance components are estimated by $\hat{\sigma}^2 = 1.90$, and

$$\hat{\sigma}_\tau^2 = \frac{29.73 - 1.90}{4} = 6.96$$

Therefore, the variance of any observation on strength is estimated by $\hat{\sigma}^2 + \hat{\sigma}_\tau^2 = 1.90 + 6.96 = 8.86$. Most of this variability is attributable to differences *between* looms. ∎

Table 3-7 Strength Data for Example 3-6

Looms	Observations				$y_i.$
	1	2	3	4	
1	98	97	99	96	390
2	91	90	93	92	366
3	96	95	97	95	383
4	95	96	99	98	388

$$1527 = y_{..}$$

Table 3-8 Analysis of Variance for the Strength Data

Source of Variation	Sum of Squares	Degrees of Freedom	Mean Square	F_0
Looms	89.19	3	29.73	15.68[a]
Error	22.75	12	1.90	
Total	111.94	15		

[a]Significant at 5 percent.

We may easily find a confidence interval for the variance component σ^2. If the observations are normally and independently distributed, then $(N - a)MS_E/\sigma^2$ is distributed as χ^2_{N-a}. Thus

$$P\left(\chi^2_{1-(\alpha/2),N-a} \le \frac{(N-a)MS_E}{\sigma^2} \le \chi^2_{\alpha/2,N-a}\right) = 1 - \alpha$$

and a $100(1 - \alpha)$ percent confidence interval for σ^2 is

$$\frac{(N-a)MS_E}{\chi^2_{\alpha/2,N-a}} \le \sigma^2 \le \frac{(N-a)MS_E}{\chi^2_{1-(\alpha/2),N-a}}$$

Now consider the variance component σ_τ^2. The point estimator of σ_τ^2 is

$$\hat{\sigma}_\tau^2 = \frac{MS_{Treatments} - MS_E}{n}$$

The random variable $(a - 1)MS_{Treatments}/(\sigma^2 + n\sigma_\tau^2)$ is distributed as χ^2_{a-1} and $(N - a)MS_E/\sigma^2$ is distributed as χ^2_{N-a}. Thus the probability distribution of $\hat{\sigma}_\tau^2$ is a linear combination of two chi-square random variables, say

$$u_1\chi^2_{a-1} - u_2\chi^2_{N-a}$$

where

$$u_1 = \frac{\sigma^2 + n\sigma_\tau^2}{n(a-1)}, \quad \text{and} \quad u_2 = \frac{\sigma^2}{n(N-a)}$$

Unfortunately, a closed-form expression for the distribution of this linear combination of chi-square random variables cannot be obtained. Thus, an exact confidence interval for σ_τ^2 cannot be constructed. Approximate procedures are given in Graybill (1961) and Searle (1971a).

3-6 FITTING RESPONSE CURVES IN THE ONE-WAY CLASSIFICATION

The single factor investigated in the one-way classification analysis of variance can be either *quantitative* or *qualitative*. A quantitative factor is one whose levels can be associated with points on a numerical scale, such as temperature, pressure, or time. Qualitative factors, on the other hand, are factors in which the levels cannot be arranged in order of magnitude. Operators, batches of raw material, and shifts are typical qualitative factors because there is no reason to rank them in any particular numerical order.

Insofar as the initial design and analysis of the experiment are concerned, both types of factors are treated identically. The experimenter is interested in determining the differences, if any, between the levels of the factors. If the factor is qualitative, such as operators, it is meaningless to consider the response for a subsequent run at an intermediate level of the factor. However, with a quantitative factor such as time, the experimenter is usually interested in the entire range of values used, particularly the response from a subsequent run at an intermediate factor level. That is, if the levels 1.0 hours, 2.0 hours, and 3.0 hours are used in the experiment, we may wish to predict the response at 2.5 hours. In general, this is done for a quantitative factor by finding a mathematical relationship between the independent variable (the factor) and the response. The general procedure for this problem is called *regression analysis*, which is discussed in Chapter 13. However, if the levels of the factor are equally spaced, a simple procedure using the orthogonal contrast coefficients in Appendix Table VIII may be employed.

The procedure consists of computing a linear, quadratic, cubic, etc. effect and sum of squares for the factor. Each effect is a single-degree-of-freedom contrast, computed from the treatment totals at the a factor levels as in Section 3-4, and the corresponding sum of squares is found from Equation 3-17. It is possible to extract polynomial effects up through order $a-1$, if there are a factor levels used in the experiment.

The process is illustrated in Table 3-9 using the data of Example 3-1. In this problem the independent factor, percentage of cotton, is equally spaced at five levels.

Table 3-9 Calculation of Polynomial Effects for Example 3-1

Percentage of Cotton	Treatment Totals $y_i.$	Orthogonal Contrast Coefficients (c_i)			
		Linear	Quadratic	Cubic	Quartic
15	49	-2	2	-1	1
20	77	-1	-1	2	-4
25	88	0	-2	0	6
30	108	1	-1	-2	-4
35	54	2	2	1	1
Effects: $\left(\sum_{i=1}^{a} c_i y_i. \right)$		41	-155	-57	-109
Sums of squares: $\left[\dfrac{\left(\sum_{i=i}^{a} c_i y_i. \right)^2}{n \sum_{i=1}^{a} c_i^2} \right]$		$\dfrac{(41)^2}{5(10)} =$ 33.62	$\dfrac{(-155)^2}{5(14)} =$ 343.21	$\dfrac{(-57)^2}{5(10)} =$ 64.98	$\dfrac{(-109)^2}{5(70)} =$ 33.95

The sums of squares for the linear, quadratic, cubic, and quartic effects of the factor form a partition of the treatment sum of squares, and can be incorporated into the analysis of variance, as shown in Table 3-10. Each effect has one degree of freedom, and can be tested by comparing its sum of squares to mean square error.

Table 3-10 Analysis of Variance

Source of Variation	Sum of Squares	Degrees of Freedom	Mean Square	F_0
Percentage of cotton	475.76	4	118.94	14.76[a]
(Linear)	(33.62)	1	33.62	4.17
(Quadratic)	(343.21)	1	343.21	42.58[a]
(Cubic)	(64.98)	1	64.98	8.06[b]
(Quartic)	(33.95)	1	33.95	4.21
Error	161.20	20	8.06	
Total		24		

[a] Significant at 1 percent.
[b] Significant at 5 percent.

From inspection of Table 3-10, we note that the quadratic and cubic effects of cotton percentage are statistically significant when compared to $F_{.05,1,20}$. Therefore, we will fit a cubic polynomial to the data, say

$$y = \alpha_0 + \alpha_1 P_1(x) + \alpha_2 P_2(x) + \alpha_3 P_3(x) + \epsilon$$

where $P_u(x)$ is a uth order *orthogonal polynomial*, which implies that if there are a levels of x we have $\sum_{j=1}^{a} P_u(x_j)P_s(x_j) = 0$ for $u \neq s$. The first five orthogonal polynomials are

$$P_0(x) = 1$$

$$P_1(x) = \lambda_1 \left[\frac{(x - \bar{x})}{d} \right]$$

$$P_2(x) = \lambda_2 \left[\left(\frac{x - \bar{x}}{d} \right)^2 - \left(\frac{a^2 - 1}{12} \right) \right]$$

$$P_3(x) = \lambda_3 \left[\left(\frac{x - \bar{x}}{d} \right)^3 - \left(\frac{x - \bar{x}}{d} \right) \left(\frac{3a^2 - 7}{20} \right) \right]$$

$$P_4(x) = \lambda_4 \left[\left(\frac{x - \bar{x}}{d} \right)^4 - \left(\frac{x - \bar{x}}{d} \right)^2 \left(\frac{3a^2 - 13}{14} \right) + \frac{3(a^2 - 1)(a^2 - 9)}{560} \right]$$

handwritten note: note: if evaluate $P_k(x_i)$, get the orthogonal contrast coeff C_{ki}

so $\sum_i \sum_j y_{ij} P_k(x_i) = \sum_i y_i \cdot C_{ki}$

and $\sum_i \sum_j (P_k(x_i))^2 = n \sum_i C_{ki}^2$

where d is the distance between the levels of x, a is the total number of levels, and $\{\lambda_i\}$ are constants such that the polynomials have integer values. Appendix Table VIII lists coefficients of the orthogonal polynomials and values of the λ_i for $a \leqslant 10$. The parameters in the orthogonal polynomial model may be estimated by

handwritten note: $\dfrac{\sum_{i=1}^{a} C_{ki} y_i \cdot}{n \sum_{i=1}^{a} C_{ki}^2} =$

$$\hat{\alpha}_k = \frac{\sum \sum y P_k(x)}{\sum \sum [P_k(x)]^2}, \quad k = 0, 1, \ldots, a - 1$$

A mathematical justification of this equation will be given in Section 13-7.

For the data in Example 3-1, we may estimate the model parameters as

$$\hat{\alpha}_0 = \frac{\sum \sum y P_0(x)}{\sum \sum [P_0(x)]^2} = \frac{\sum \sum y_{ij}}{25} = 15.0400$$

$$\hat{\alpha}_1 = \frac{\sum \sum y P_1(x)}{\sum \sum [P_1(x)]^2} = \frac{41}{5(10)} = 0.8200$$

$$\hat{\alpha}_2 = \frac{\sum \sum y P_2(x)}{\sum \sum [P_2(x)]^2} = \frac{-155}{5(14)} = -2.2143$$

$$\hat{\alpha}_3 = \frac{\sum \sum y P_3(x)}{\sum \sum [P_3(x)]^2} = \frac{-57}{5(10)} = -1.1400$$

If we wish to either add or delete terms from the model, it is not necessary to recompute the $\{\hat{\alpha}_i\}$ already in the model because of the orthogonality property of the polynomials $\{P_i(x)\}$.

Since there are $a = 5$ levels of x and the spacing between the levels is $d = 5$, the orthogonal polynomial model becomes

$$\hat{y} = 15.0400 + 0.8200(1)\left(\frac{x - 25}{5}\right) - 2.2143(1)\left[\left(\frac{x - 25}{5}\right)^2 - \left(\frac{5^2 - 1}{12}\right)\right]$$

$$- 1.1400(5/6)\left[\left(\frac{x - 25}{5}\right)^3 - \left(\frac{x - 25}{5}\right)\left(\frac{3(5)^2 - 7}{20}\right)\right]$$

where $\lambda_1 = 1$, $\lambda_2 = 1$, and $\lambda_3 = 5/6$ have been obtained from Appendix Table VIII. This equation can be simplified to

$$\hat{y} = 64.5936 - 9.0100x + 0.4814x^2 - 0.0076x^3$$

which can be used to predict the fabric strength for cotton content between 15 and 35 percent. For example, if we wish to predict the fabric strength at, say $x = 30$ percent, we would obtain

$$\hat{y} = 64.5936 - 9.0100(30) + 0.4814(30^2) - 0.0076(30^3)$$

$$= 22.35$$

In general, we would like to fit the lowest degree polynomial that adequately describes the data. The computational simplicity of orthogonal polynomials is helpful in this type of problem because it allows the experimenter to add statistically significant terms or delete insignificant ones easily, without affecting previously computed coefficients. For a further discussion of regression methods, refer to Chapter 13.

3-7 POWER OF THE ANALYSIS OF VARIANCE

In any hypothesis-testing situation the power of the test is usually of interest to the experimenter. In this section we will discuss the use of operating characteristic curves to evaluate the power of the analysis of variance. An important use of the curves is to guide the experimenter in selecting the number of replicates so that his design will be sensitive to important potential differences in the treatments.

We first consider the power of the fixed effects model for the case of equal sample sizes per treatment. The power of the test is

$$1 - \beta = P\{\text{Reject } H_0 | H_0 \text{ is false}\}$$

$$= P\{F_0 > F_{\alpha, a-1, N-a} | H_0 \text{ is false}\} \tag{3-27}$$

To evaluate the probability statement in (3-27) we need to know the distribution of the test statistic F_0 if the null hypothesis is false. It can be shown that, if H_0 is false, the statistic $F_0 = MS_{\text{Treatments}}/MS_E$ is distributed as a *noncentral F* random variable with $a - 1$ and $N - a$ degrees of freedom and noncentrality parameter δ. If $\delta = 0$ then the noncentral F distribution becomes the usual (central) F distribution.

Operating characteristic curves given in Chart V of the Appendix are used to calculate the probability statement in Equation 3-27. These curves plot the probability of type II error (β) against a parameter Φ, where

$$\Phi^2 = \frac{n \sum_{i=1}^{a} \tau_i^2}{a\sigma^2} \tag{3-28}$$

The quantity Φ^2 is related to the noncentrality parameter δ. Curves are available for $\alpha = .05$ and $\alpha = .01$, and a range of degrees of freedom for numerator and denominator.

In using the operating characteristic curves, the experimenter must define the difference in means that he is interested in detecting in terms of $\sum_{i=1}^{a} \tau_i^2$. Also, the error variance σ^2 is usually unknown in practice. The experimenter then must choose ratios of $\sum_{i=1}^{a} \tau_i^2/\sigma^2$ that he wishes to detect. Alternatively, if an estimate of σ^2 is available, one may replace σ^2 with the estimate. For example, if we were interested in the sensitivity of an experiment that has already been performed, we might use MS_E as the estimate of σ^2.

■**EXAMPLE 3-7.** Suppose that five means are being compared in an analysis of variance with $\alpha = .01$. The experimenter would like to know how many replicates to run if it is important to reject H_0 with probability at least .90 if $\sum_{i=1}^{5} \tau_i^2/\sigma^2 = 5.0$. The parameter Φ^2 is, in this case,

$$\Phi^2 = \frac{n \sum_{i=1}^{a} \tau_i^2}{a\sigma^2} = \frac{n}{5}(5) = n$$

and the operating characteristic curve for $a - 1 = 5 - 1 = 4$ with $N - a = a(n - 1) = 5(n - 1)$ error degrees of freedom in Appendix Chart V must be used. As a first guess, try $n = 4$ replicates. This yields $\Phi^2 = 4$, $\Phi = 2$, and $5(3) = 15$ error degrees of freedom. Consequently, from Chart V, we find that $\beta \simeq .38$. Therefore the power of the test is approximately $1 - \beta = 1 - .38 = .62$, which is less than the required .90, and so we conclude that $n = 4$ replicates are not sufficient. Proceeding in a similar manner, we can construct the following display:

n	Φ^2	Φ	$a(n-1)$	β	Power $(1-\beta)$
4	4	2.00	15	.38	.62
5	5	2.24	20	.18	.82
6	6	2.45	25	.06	.94

Thus, at least $n = 6$ replicates must be run in order to obtain a test with the required power. ■

The power of the test for the random effects model is

$$1 - \beta = P\{\text{Reject } H_0 | H_0 \text{ is false}\}$$

$$= P\{F_0 > F_{\alpha, a-1, N-a} | \sigma_\tau^2 > 0\} \tag{3-29}$$

Once again the distribution of the test statistic $F_0 = MS_{Treatments}/MS_E$ under the alternative hypothesis is needed. It can be shown that if H_1 is true $(\sigma_\tau^2 > 0)$ the distribution of F_0 is central F with $a - 1$ and $N - a$ degrees of freedom.

Since the power of the random effects model is based on the usual central F distribution, we could use the tables of the F distribution in the Appendix to evaluate (3-29). However, it is simpler to determine the power of the test through the use of operating characteristic curves. A set of these curves for various values of numerator degrees of freedom, denominator degrees of freedom, and α of .05 or .01 are provided in Chart VI of the Appendix. They plot the probability of type II error against the parameter λ, where

$$\lambda = \sqrt{1 + \frac{n\sigma_\tau^2}{\sigma^2}} \tag{3-30}$$

Since σ^2 is usually unknown, we may either use a prior estimate or define the value of σ_τ^2 that we are interested in detecting in terms of the ratio σ_τ^2/σ^2.

■**EXAMPLE 3-8.** Suppose we have five treatments selected at random with six observations per treatment and $\alpha = .05$, and we wish to determine the power of the test if σ_τ^2 is equal to σ^2. Since $a = 5$, $n = 6$, and $\sigma_\tau^2 = \sigma^2$ we may compute

$$\lambda = \sqrt{1 + 6(1)} = 2.646$$

From the operating characteristic curve with $a - 1 = 4$, $N - a = 25$ degrees of freedom, and $\alpha = .05$ we find that

$$\beta \simeq .20,$$

and thus the power is approximately .80.   ■

3-8 DEPARTURES FROM ASSUMPTIONS IN ANALYSIS OF VARIANCE

The assumptions underlying the analysis of variance are that treatment effects are additive, and that experimental errors are normally and independently distributed, with the same error variance. Generally, we can never be sure that all of these assumptions are satisfied, and often we may have reason to suspect that some are false. Extensive studies have been made concerning the consequences of departures from assumptions in the analysis of variance. For example see Bartlett (1947), Cochran (1947), Eisenhart (1947), and Box (1954a, b).

Cochran and Cox (1957) note that the effect is minor when the assumptions underlying the analysis of variance are not exactly satisfied. That is, slight departures from the assumptions are of little concern. They claim departures from assumptions affect both the significance level and power of statistical tests. For example, when the experimenter thinks he is testing at the 5 percent level, the test may actually be at the

7 percent or 8 percent level. Usually, the true type I error is larger than the specified one, and, as a result, too many significant differences in treatments are reported. The power is affected in that a more powerful test could be obtained if the correct statistical model were known. Consequently, error probabilities and confidence intervals should be considered as approximate.

The assumption of homogeneity of variance frequently may cause disturbances. This problem may arise if certain treatments are erratic in their effects, or if the data follows a nonnormal, skewed distribution, as in skewed distributions the variance tends to be a function of the mean. The usual approach to dealing with this situation is to transform the observations and apply the analysis of variance to the transformed data. In this approach, one should note that the conclusions of the analysis of variance apply to the *transformed* populations.

Considerable research has been devoted to the selection of an appropriate transformation. If the experimenter knows the theoretical distribution of the observations he may utilize this information in choosing a transformation. For example, if the observations follow the Poisson distribution then the square root transformation $y_{ij}^{*} = \sqrt{y_{ij}}$ or $y_{ij}^{*} = \sqrt{1 + y_{ij}}$ would be used. If the data follow the log-normal distribution then the logarithmic transformation $y_{ij}^{*} = \log y_{ij}$ is appropriate. For binomial data expressed as fractions the arcsine transformation $y_{ij}^{*} = \arcsin y_{ij}$ is useful. When there is no obvious transformation, the experimenter usually seeks a transformation that equalizes the variance regardless of the value of the mean. In factorial experiments, which we will discuss in Chapter 6, another approach is to select a transformation that minimizes the interaction mean square, resulting in an experiment that is easier to interpret. For more discussion of transformations, refer to Bartlett (1947), Box and Cox (1964), Dolby (1963), and Draper and Hunter (1969).

If the treatment effects are not additive, the transformations discussed above may also be useful. There are several causes of nonadditivity. For example, the true treatment effects may be multiplicative; that is, $y_{ij} = \mu \tau_i \epsilon_{ij}$, or interaction effects may exist that have not been included in the model. Transformations that are made for nonadditivity will also affect the distribution of the experimental errors. However, in most cases the transformation will bring the error distribution closer to normal.

The assumption of independence is generally critical. Proper randomization of the experiment introduces independence in the assignment of treatments to experimental units, and the resulting experimental errors may be regarded as independent. Usually, proper randomization is achieved by randomly assigning the order in which the observations are to be taken.

3-9 A TEST FOR THE EQUALITY OF SEVERAL VARIANCES

The analysis of variance is relatively robust to moderate departure from the assumption that the treatments have a common variance when the sample sizes are equal. However, this is not true if the sample sizes are unequal or if one variance is very much larger than the others. Therefore, it is frequently desirable to test the hypotheses

$$H_0: \sigma_1^2 = \sigma_2^2 = \cdots = \sigma_a^2$$

$$H_1: \text{above not true for at least one } \sigma_i^2$$

A widely used procedure is *Bartlett's test*. The procedure involves computing a statistic whose sampling distribution is closely approximated by the chi-square distribution with $a - 1$ degrees of freedom when the a random samples are from independent normal populations. The test statistic is

$$\chi_0^2 = 2.3026 \frac{q}{c} \tag{3-31}$$

where

$$q = (N - a) \log_{10} S_p^2 - \sum_{i=1}^{a} (n_i - 1) \log_{10} S_i^2$$

$$c = 1 + \frac{1}{3(a - 1)} \left(\sum_{i=1}^{a} (n_i - 1)^{-1} - (N - a)^{-1} \right)$$

$$S_p^2 = \frac{\sum_{i=1}^{a} (n_i - 1) S_i^2}{N - a}$$

and S_i^2 is the sample variance of the ith population.

The quantity q is large when the sample variances S_i^2 differ greatly and is equal to zero when all S_i^2 are equal. Therefore, we should reject H_0 on values of χ_0^2 that are too large, that is, we reject H_0 only when

$$\chi_0^2 > \chi_{\alpha, a-1}^2$$

where $\chi_{\alpha, a-1}^2$ is the upper α percentage point of the chi-square distribution with $a - 1$ degrees of freedom. Various studies have indicated that Bartlett's test is very sensitive to the normality assumption, and should not be applied when the normality assumption is doubtful. Other tests for equality of variance are reviewed by Anderson and McLean (1974).

■**EXAMPLE 3-9.** We shall apply Bartlett's test to the data of Example 3-1. We first compute the sample variances in each treatment and find that $S_1^2 = 11.2$, $S_2^2 = 9.8$, $S_3^2 = 4.3$, $S_4^2 = 6.8$, and $S_5^2 = 8.2$. Then

$$S_p^2 = \frac{4(11.2) + 4(9.8) + 4(4.3) + 4(6.8) + 4(8.2)}{20} = 8.06$$

$$q = 20 \log_{10} (8.06) - 4 [\log_{10} 11.2 + \log_{10} 9.8 + \log_{10} 4.3 + \log_{10} 6.8 + \log_{10} 8.2] = 0.45$$

$$c = 1 + \frac{1}{3(4)} \left(\frac{5}{4} - \frac{1}{20} \right) = 1.10$$

and the test statistic is

$$\chi_0^2 = 2.3026 \frac{(0.45)}{(1.10)} = 0.93$$

Since $\chi_{.05,4}^2 = 9.49$, we cannot reject the null hypothesis and conclude that the variances are homogeneous.   ∎

3-10   THE REGRESSION APPROACH TO ANALYSIS OF VARIANCE

Thus far, we have given an intuitive or heuristic development of the analysis of variance. However, it is possible to give a more formal development. The method will be useful later in understanding the basis for the statistical analysis of more complex designs. Called the *general regression significance test*, the procedure essentially consists of finding the reduction in the total sum of squares for fitting the model with all parameters included and the reduction in sum of squares when the model is restricted to the null hypothesis. The difference between these two sums of squares is the treatment sum of squares with which a test of the null hypothesis can be conducted.

A fundamental part of the procedure is writing the normal equations for the model. These equations may always be obtained by forming the least squares function and differentiating it with respect to each unknown parameter, as we did in Section 3-3.2. However, an easier method is available. The rules listed below allow the normal equations for *any* experimental design model to be written directly.

RULE 1. There is one normal equation for each parameter in the model to be estimated.

RULE 2. The right-hand side of any normal equation is just the sum of all observations that contain the parameter associated with that particular normal equation.

To illustrate this rule consider the one-way classification model. The first normal equation is for the parameter μ; therefore, the right-hand side is $y_{..}$, since *all* observations contain μ.

RULE 3. The left-hand side of any normal equation is the sum of all model parameters, where each parameter is multiplied by the number of times it appears in the total on the right-hand side. The parameters are written with a circumflex (^) to indicate that they are *estimators*, and not the true parameter values.

For example, consider the first normal equation in the one-way classification. According to the above rules, it would be

$$N\hat{\mu} + n\hat{\tau}_1 + n\hat{\tau}_2 + \cdots + n\hat{\tau}_a = y_{..}$$

since μ appears in all N observations, τ_1 appears only in the n observations taken under the first treatment, τ_2 appears only in the n observations taken under the second treatment, and so on. From Equation 3-12 we verify that the equation shown above is correct. The second normal equation would correspond to τ_1, and is

$$n\hat{\mu} + n\hat{\tau}_1 = y_{1.}$$

since only the observations in the first treatment contain τ_1 (this gives $y_1.$ as the right-hand side), μ and τ_1 appear exactly n times in $y_1.$, and all other τ_i appear zero times. In general, the left-hand side of any normal equation is the expected value of the right-hand side.

Now, consider finding the reduction in sum of squares by fitting a particular model to the data. By fitting a model to the data we "explain" some of the variability; that is, we reduce the unexplained variability by some amount. The reduction in the unexplained variability is always the sum of the parameter estimates, each multiplied by the right-hand side of the normal equation which corresponds to that parameter. For example, in the one-way classification, the reduction due to fitting the model $y_{ij} = \mu + \tau_i + \epsilon_{ij}$ is

$$R(\mu, \tau) = \hat{\mu} y_{..} + \hat{\tau}_1 y_{1.} + \hat{\tau}_2 y_{2.} + \cdots + \hat{\tau}_a y_a.$$

$$= \hat{\mu} y_{..} + \sum_{i=1}^{a} \hat{\tau}_i y_i. \tag{3-32}$$

The notation $R(\mu, \tau)$ means the reduction in sum of squares from fitting the model containing μ and $\{\tau_i\}$. $R(\mu, \tau)$ is also sometimes called the "regression" sum of squares for the model $y_{ij} = \mu + \tau_i + \epsilon_{ij}$. The number of degrees of freedom associated with a reduction in the sum of squares, such as $R(\mu, \tau)$, is always equal to the number of linearly independent normal equations. The remaining variability unaccounted for by the model is found from

$$SS_E = \sum_{i=1}^{a} \sum_{j=1}^{n} y_{ij}^2 - R(\mu, \tau) \tag{3-33}$$

This quantity is used in the denominator of the test statistic for $H_0 : \tau_i = 0$.

We will now illustrate the general regression significance test for the one-way classification, and show that it yields the usual one-way analysis of variance. The model is $y_{ij} = \mu + \tau_i + \epsilon_{ij}$, and the normal equations are found from the above rules as

$$N\hat{\mu} + n\hat{\tau}_1 + n\hat{\tau}_2 + \cdots + n\hat{\tau}_a = y_{..}$$

$$n\hat{\mu} + n\hat{\tau}_1 \qquad\qquad = y_{1.}$$

$$n\hat{\mu} \qquad + n\hat{\tau}_2 \qquad = y_{2.}$$

$$\vdots \qquad\qquad\qquad \vdots$$

$$n\hat{\mu} \qquad\qquad\qquad + n\hat{\tau}_a = y_a.$$

Compare these normal equations with those obtained in (3-12).

Applying the constraint $\sum_{i=1}^{a} \hat{\tau}_i = 0$, the estimators for μ and τ_i are

$$\hat{\mu} = \bar{y}_{..}, \quad \hat{\tau}_i = \bar{y}_i. - \bar{y}_{..}, \quad i = 1, 2, \ldots, a$$

The reduction in sum of squares due to fitting this model is found from (3-32) as

$$R(\mu, \tau) = \hat{\mu} y_{..} + \sum_{i=1}^{a} \hat{\tau}_i y_{i.}$$

$$= (\bar{y}_{..}) y_{..} + \sum_{i=1}^{a} (\bar{y}_{i.} - \bar{y}_{..}) y_{i.}$$

$$= \frac{y_{..}^2}{N} + \sum_{i=1}^{a} \bar{y}_{i.} y_{i.} - \bar{y}_{..} \sum_{i=1}^{a} y_{i.}$$

$$= \sum_{i=1}^{a} \frac{y_{i.}^2}{n}$$

which has a degrees of freedom, since there are a linearly independent normal equations. The error sum of squares is, from (3-33),

$$SS_E = \sum_{i=1}^{a} \sum_{j=1}^{n} y_{ij}^2 - R(\mu, \tau)$$

$$= \sum_{i=1}^{a} \sum_{j=1}^{n} y_{ij}^2 - \sum_{i=1}^{a} \frac{y_{i.}^2}{n}$$

and has $N - a$ degrees of freedom.

To find the sum of squares due to the treatment effects (the τ_i), we consider the model restricted to the null hypothesis, that is, $\tau_i = 0$ for all i. The reduced model is $y_{ij} = \mu + \epsilon_{ij}$. There is only one normal equation for this model:

$$N\hat{\mu} = y_{..}$$

and the estimator of μ is $\hat{\mu} = \bar{y}_{..}$. Thus, the reduction in sum of squares due to fitting only μ is

$$R(\mu) = (\bar{y}_{..})(y_{..}) = \frac{y_{..}^2}{N}$$

Since there is only one normal equation for this reduced model, $R(\mu)$ has one degree of freedom. The sum of squares due to the $\{\tau_i\}$, given that μ is already in the model, is the difference between $R(\mu, \tau)$ and $R(\mu)$, which is

$$R(\tau | \mu) = R(\mu, \tau) - R(\mu)$$

$$= \sum_{i=1}^{a} \frac{y_{i.}^2}{n} - \frac{y_{..}^2}{N}$$

with $a - 1$ degrees of freedom, which we recognize from Equation 3-9 as $SS_{\text{Treatments}}$. Making the usual normality assumption, the appropriate statistic for testing $H_0 : \tau_i = 0$

is

$$F_0 = \frac{R(\tau \mid \mu)/(a - 1)}{\left[\sum_{i=1}^{a} \sum_{j=1}^{n} y_{ij}^2 - R(\mu, \tau)\right]/(N - a)}$$

which is distributed as $F_{a-1, N-a}$ under the null hypothesis. This is, of course, the test statistic for the one-way classification analysis of variance.

3-11 PROBLEMS

3-1. The tensile strength of portland cement is being studied. Four different mixing techniques can be used economically. The following data have been collected.

Mixing Technique	Tensile Strength (lbs/in.^{2})			
1	3129	3000	2865	2890
2	3200	3300	2975	3150
3	2800	2900	2985	3050
4	2600	2700	2600	2765

(a) Test the hypothesis that mixing techniques affect the strength of the cement. Use $\alpha = .05$.

(b) Use Duncan's multiple range test to make comparisons between pairs of means. Estimate the treatment effects.

3-2. A textile mill has a large number of looms. Each loom is supposed to provide the same output of cloth per minute. To investigate this assumption five looms are chosen at random and their output noted at different times. The following data are obtained.

Loom	Output in Pounds per Minute				
1	14.0	14.1	14.2	14.0	14.1
2	13.9	13.8	13.9	14.0	14.0
3	14.1	14.2	14.1	14.0	13.9
4	13.6	13.8	14.0	13.9	13.7
5	13.8	13.6	13.9	13.8	14.0

(a) Is this a fixed or random effects experiment? Are the looms equal in output?

(b) Estimate the variability between looms.

(c) Estimate the experimental error variance.

(d) What is the probability of accepting H_0 if σ_τ^2 is four times the experimental error variance?

3-3. An experiment was run to determine whether four specific firing temperatures affect the density of a certain type of brick. The experiment led to the following data:

Temperature (°F)	Density					Y_i
100	21.8	21.9	21.7	21.6	21.7	108.7
125	21.7	21.4	21.5	21.4		86.0
150	21.9	21.8	21.8	21.6	21.5	108.6
175	21.9	21.7	21.8	21.4		86.8

(a) Does the firing temperature affect the density of the bricks?
(b) Estimate the components in the model equation.

3-4. A manufacturer of television sets is interested in the effect on tube conductivity of four different types of coating for color picture tubes. The following conductivity data are obtained.

Coating Type	Conductivity			
1	143	141	150	146
2	152	149	137	143
3	134	136	132	127
4	129	127	132	129

(a) Is there any difference in conductivity due to coating type? Use $\alpha = .05$.
(b) Estimate the overall mean and the treatment effects.
(c) Compute a 95 percent interval estimate of the mean coating type 4. Compute a 99 percent interval estimate of the mean difference between coating types 1 and 4.
(d) Test all pairs of means using Duncan's multiple range test, with $\alpha = .05$.
(e) Assuming that coating type 4 is currently in use, what are your recommendations to the manufacturer? We wish to minimize conductivity.

3-5. The response time in milleseconds was determined for three different types of circuits used in an automatic value shutoff mechanism. The results were:

Circuit Type	Response Time				
1	9	12	10	8	15
2	20	21	23	17	30
3	6	5	8	16	7

(a) Test the hypothesis that the three circuit types have the same response time. Use $\alpha = .01$.
(b) Use Duncan's multiple range test to compare pairs of treatment means.
(c) Construct a set of orthogonal contrasts, assuming that at the outset of the

experiment you suspected the response time of circuit type 2 to be different from the other two.

(d)   What is the power of this test for detecting $\sum_{i=1}^{3} \tau_i^2/\sigma^2 = 3.0$?

3-6.   Four different digital computer circuits are being studied in order to compare the amount of noise present. The following data have been obtained.

Circuit Type	Noise Observed				
1	19	20	19	30	8
2	80	61	73	56	80
3	47	26	25	35	50
4	95	46	83	78	97

(a)   Is the amount of noise present the same for all four circuits?
(b)   Estimate the components of the appropriate model for this problem. Estimate all possible *differences* between pairs of treatments. Are these treatment differences uniquely estimated? If so, why?
(c)   ‛Vhich circuit would you select for use?

3-7.   A manufacturer suspects that the batches of raw material furnished by his supplier differ significantly in calcium content. There are a large number of batches currently in the warehouse. Five of these are randomly selected for study. A chemist makes five determinations on each batch and obtains the following data.

Batch 1	Batch 2	Batch 3	Batch 4	Batch 5
23.46	23.59	23.51	23.28	23.29
23.48	23.46	23.64	23.40	23.46
23.56	23.42	23.46	23.37	23.37
23.39	23.49	23.52	23.46	23.32
23.40	23.50	23.49	23.39	23.38

(a)   Is there a significant variation in calcium content from batch to batch?
(b)   Estimate the components of variance.
(c)   If it is important to detect a ratio of σ_τ^2/σ^2 of 2.5 with a high probability, is the number of replicates adequate?

3-8.   Several ovens in a metal working shop are used to heat metal specimens. All ovens are supposed to operate at the same temperature, although it is suspected that this may not be true. Three ovens are selected at random and their temperatures on successive heats is noted. These data are shown below:

Oven	Temperature				
1	491.50	498.30	498.10	493.50	493.60
2	488.50	484.65	479.90	477.35	
3	490.10	484.80	488.25	473.00	471.85	478.65

(a) Is there a significant variation in temperature between ovens?

(b) Estimate the components of variance for this model.

3-9. Four chemists are asked to determine the percentage of methyl alcohol in a certain chemical compound. Each chemist makes three determinations, and the results are shown below:

Chemist	Percentage of Methyl Alcohol		
1	84.99	84.04	84.38
2	85.15	85.13	84.88
3	84.72	84.48	85.16
4	84.20	84.10	84.55

(a) Do chemists differ significantly? Use $\alpha = .05$.

(b) What is the power of this test for detecting $\sum_{i=1}^{4} \tau_i^2/\sigma^2 = 3.5$?

(c) If chemist 2 is a new employee, construct a meaningful set of orthogonal contrasts that we might have thought useful at the start of the experiment.

(d) If it is desired to detect $\sum_{i=1}^{4} \tau_i^2/\sigma^2 = 1.5$ with probability at least .95, what sample size should be used?

3-10. Three brands of batteries are under study. It is suspected that the life (in weeks) of the three brands is different. Five batteries of each brand are tested with the following results:

	Weeks of Life	
Brand 1	Brand 2	Brand 3
100	76	108
96	80	100
92	75	96
96	84	98
92	82	100

(a) Are the lives of these brands of batteries different?

(b) Estimate the components of the appropriate statistical model.

(c) Construct a 95 percent interval estimate on the mean life of battery brand 2. Construct a 99 percent interval estimate on the mean difference between the life of battery brands 2 and 3.

(d) Which brand would you select for use? If the manufacturer will replace without charge any battery which fails in less than 85 weeks, what percentage would he expect to replace?

3-11. Four catalysts that may affect the concentration of one component in a three-component liquid mixture are being investigated. The following concentrations are obtained.

	Catalyst		
1	2	3	4
58.2	56.3	50.1	52.9
57.2	54.5	54.2	49.9
58.4	57.0	55.4	50.0
55.8	55.3		51.7
54.9			

(a) Do the four catalysts have the same effect on the concentration?
(b) Estimate the components of the appropriate statistical model.
(c) Construct a 99 percent confidence interval estimate of the mean response for catalyst 1.

3-12. Suppose that four normal populations have common variance $\sigma^2 = 25$ and means $\mu_1 = 50$, $\mu_2 = 60$, $\mu_3 = 50$, and $\mu_4 = 60$. How many observations should be taken on each population so that the probability of rejecting the hypothesis of equality of means is at least .90? Use $\alpha = .05$.

3-13. Use Bartlett's test to determine if the assumption of homogeneous variances is reasonable in Problem 3-10.

3-14. Consider testing the equality of the means of two normal populations where the variances are unknown, but assumed equal. The appropriate test procedure is the pooled t test (see Section 2-4.1). Show that the pooled t test is equivalent to the one-way classification analysis of variance.

3-15. Show that the variance of the linear combination $\sum_{i=1}^{a} c_i y_i$ is $\sigma^2 \sum_{i=1}^{a} n_i c_i^2$.

3-16. In a fixed-effects experiment, suppose that there are n observations for each of four treatments. Let Q_1^2, Q_2^2, and Q_3^2 be single degree-of-freedom components for the orthogonal contrasts. Prove that $SS_{\text{Treatments}} = Q_1^2 + Q_2^2 + Q_3^2$.

3-17. Consider the data shown in Problem 3-5.
(a) Write out the least squares normal equations for this problem, and solve them for $\hat{\mu}$ and $\hat{\tau}_i$ making the usual constraint ($\sum_{i=1}^{3} \hat{\tau}_i = 0$). Estimate $\tau_1 - \tau_2$.
(b) Solve the equations in (a) using the constraint $\hat{\tau}_3 = 0$. Are the estimators $\hat{\tau}_i$ and $\hat{\mu}$ the same as you found in (a)? Why? Now estimate $\tau_1 - \tau_2$ and compare your answer with (a). What statement can you make about estimating contrasts in the τ_i's?
(c) Estimate $\mu + \tau_1$, $2\tau_1 - \tau_2 - \tau_3$ and $\mu + \tau_1 + \tau_2$, using the two solutions to the normal equations. Compare the results obtained in each case.

3-18. Apply the general regression significance test to the data in Example 3-1. Show that the results yield the same sums of squares for error and treatments that are found by the usual analysis of variance.

Chapter Four
Randomized Blocks, Latin Squares, and Related Designs

4-1 THE RANDOMIZED COMPLETE BLOCK DESIGN

In many experimental problems, it is necessary to design the experiment so that variability arising from extraneous sources can be systematically controlled. As an example, suppose we wish to determine whether or not four different tips produce different readings on a hardness testing machine. The machine operates by pressing the tip into a metal specimen, and from the depth of the resulting depression, the hardness of the specimen can be determined. The experimenter has decided to obtain four observations for each tip. There is only one factor—tip type—and a completely randomized single-factor design would consist of randomly assigning each one of the 4 X 4 = 16 runs to an *experimental unit*, that is, a metal specimen, and observing the hardness reading that results. Thus, 16 different metal specimens would be required in this experiment, one for each run in the design.

There is a potentially serious problem with a completely randomized experiment in this design situation. If the metal specimens differ slightly in their hardness, such as might happen if they come from different heats, then the experimental units (the specimens) will contribute to the variability observed in the hardness data. As a result, the experimental error will reflect *both* random error *and* variability between specimens.

We would like to make the experimental error as small as possible, that is, we would like to remove the variability between specimens from the experimental error. A design that would accomplish this requires the experimenter to test each tip once on each of four specimens. This design, shown in Table 4-1, is called a *randomized complete block design*. The word "complete" indicates that each block (specimen) contains all the treatments (tips). By using this design, the blocks or specimens are a more homogeneous experimental unit on which to compare the tips. Effectively, this design strategy improves the accuracy of the comparisons between tips. Within a block, the order in which the four tips are tested is randomly determined. Notice the similarity of this design problem to the one of Section 1-4 which the paired *t* test was discussed. The randomized complete block design is a generalization of that concept.

Table 4-1   Randomized Complete Block Design for the Hardness Testing Experiment

	Specimen			
	1	2	3	4
Type of tip used	4	2	3	1
	3	1	2	2
	1	4	1	4
	2	3	4	3

The randomized complete block design is perhaps the most widely used experimental design. Situations for which the randomized complete block design is appropriate are numerous, and easily detected with practice. Units of test equipment or machinery are often different in their operating characteristics and would be a typical blocking factor. Batches of raw material, people, and time are also common sources of variability in an experiment that can be systematically controlled through blocking.

4-1.1 Statistical Analysis

Suppose we have, in general, *a* treatments (which are to be compared) and *b* blocks. The randomized complete block design is shown in Figure 4-1. There is one observation per treatment in each block and the order in which the treatments are run within each block is determined randomly. Because the only randomization of treatments is within the blocks, we often say that the blocks represent a *restriction* on randomization.

The statistical model for this design is

$$y_{ij} = \mu + \tau_i + \beta_j + \epsilon_{ij} \quad \begin{cases} i = 1, 2, \ldots, a \\ j = 1, 2, \ldots, b \end{cases} \tag{4-1}$$

Block 1	Block 2	Block b
y_{11}	y_{12}	y_{1b}
y_{21}	y_{22}	y_{2b}
y_{31}	y_{32}	y_{3b}
.	.	.
.	.	.
.	.	.
y_{a1}	y_{a2}	y_{ab}

Figure 4-1. The randomized complete block design.

where μ is an overall mean, τ_i is the effect of the ith treatment, β_j is the effect of the jth block, and ϵ_{ij} is the usual NID$(0, \sigma^2)$ random error term. Treatments and blocks will be considered initially as fixed factors. Furthermore, the treatment and block effects are defined as deviations from the overall mean so that

$$\sum_{i=1}^{a} \tau_i = 0$$

and

$$\sum_{j=1}^{b} \beta_j = 0$$

We are interested in testing the equality of the treatment effects. Thus the appropriate hypotheses are

$$H_0: \tau_1 = \tau_2 = \cdots = \tau_a = 0$$

$$H_1: \tau_i \neq 0 \text{ at least one } i$$

Let $y_{i.}$ be the total of all observations taken under treatment i, $y_{.j}$ be the total of all observations in block j, $y_{..}$ be the grand total of all observations, and $N = ab$ be the total number of observations. Expressed mathematically,

$$y_{i.} = \sum_{j=1}^{b} y_{ij}, \qquad i = 1, 2, \ldots, a \qquad (4\text{-}2)$$

$$y_{.j} = \sum_{i=1}^{a} y_{ij}, \qquad j = 1, 2, \ldots, b \qquad (4\text{-}3)$$

and

$$y_{..} = \sum_{i=1}^{a} \sum_{j=1}^{b} y_{ij} = \sum_{i=1}^{a} y_{i.} = \sum_{j=1}^{b} y_{.j} \qquad (4\text{-}4)$$

Similarly, $\bar{y}_{i.}$ is the average of the observations taken under treatment i, $\bar{y}_{.j}$ is the average of the observations in block j, and $\bar{y}_{..}$ is the grand average of all observations. That is,

$$\bar{y}_{i.} = y_{i.}/b, \qquad \bar{y}_{.j} = y_{.j}/a, \qquad \bar{y}_{..} = y_{..}/N \qquad (4\text{-}5)$$

We may express the total corrected sum of squares as

$$\sum_{i=1}^{a}\sum_{j=1}^{b}(y_{ij} - \bar{y}_{..})^2 = \sum_{i=1}^{a}\sum_{j=1}^{b}[(\bar{y}_{i.} - \bar{y}_{..}) + (\bar{y}_{.j} - \bar{y}_{..}) + (y_{ij} - \bar{y}_{i.} - \bar{y}_{.j} + \bar{y}_{..})]^2 \quad (4\text{-}6)$$

Expanding the right-hand side of (4-6) yields

$$\sum_{i=1}^{a}\sum_{j=1}^{b}(y_{ij} - \bar{y}_{..})^2 = b\sum_{i=1}^{a}(\bar{y}_{i.} - \bar{y}_{..})^2 + a\sum_{j=1}^{b}(\bar{y}_{.j} - \bar{y}_{..})^2$$

$$+ \sum_{i=1}^{a}\sum_{j=1}^{b}(y_{ij} - \bar{y}_{i.} - \bar{y}_{.j} + \bar{y}_{..})^2 + 2\sum_{i=1}^{a}\sum_{j=1}^{b}(\bar{y}_{i.} - \bar{y}_{..})(\bar{y}_{.j} - \bar{y}_{..})$$

$$+ 2\sum_{i=1}^{a}\sum_{j=1}^{b}(\bar{y}_{i.} - \bar{y}_{..})(y_{ij} - \bar{y}_{i.} - \bar{y}_{.j} + \bar{y}_{..})$$

$$+ 2\sum_{i=1}^{a}\sum_{j=1}^{b}(\bar{y}_{.j} - \bar{y}_{..})(y_{ij} - \bar{y}_{i.} - \bar{y}_{.j} + \bar{y}_{..})$$

Simple but tedious algebra proves that the three cross-products are zero. Therefore,

$$\sum_{i=1}^{a}\sum_{j=1}^{b}(y_{ij} - \bar{y}_{..})^2 = b\sum_{i=1}^{a}(\bar{y}_{i.} - \bar{y}_{..})^2 + a\sum_{j=1}^{b}(\bar{y}_{.j} - \bar{y}_{..})^2$$

$$+ \sum_{i=1}^{a}\sum_{j=1}^{b}(y_{ij} - \bar{y}_{.j} - \bar{y}_{i.} + \bar{y}_{..})^2 \quad (4\text{-}7)$$

represent a partition of the total sum of squares. Expressing the sums of squares in (4-7) symbolically, we have

$$SS_T = SS_{\text{Treatments}} + SS_{\text{Blocks}} + SS_E \qquad (4\text{-}8)$$

Since there are N observations, SS_T has $N - 1$ degrees of freedom. There are a treatments and b blocks, so $SS_{\text{Treatments}}$ and SS_{Blocks} have $a - 1$ and $b - 1$ degrees of freedom, respectively. The error sum of squares is just a sum of squares between cells minus the sum of squares for treatments and blocks. There are ab cells with $ab - 1$ degrees of freedom between them, so SS_E has $ab - 1 - (a - 1) - (b - 1) = (a - 1)(b - 1)$ degrees of freedom. Furthermore, the degrees of freedom on the right-hand side of (4-8) add to the total on the left; therefore, making the usual normality assumptions on the errors, one may use Theorem 3-1 and find that $SS_{\text{Treatments}}/\sigma^2$, $SS_{\text{Blocks}}/\sigma^2$, and SS_E/σ^2 are independently distributed chi-square random variables. Each sum of squares divided by its degrees of freedom is a mean square. The expected value of the mean squares, if treatments and blocks are fixed, can be shown to be

$$E(MS_{\text{Treatments}}) = \sigma^2 + \frac{b \sum_{i=1}^{a} \tau_i^2}{a - 1}$$

$$E(MS_{\text{Blocks}}) = \sigma^2 + \frac{a \sum_{j=1}^{b} \beta_j^2}{b - 1}$$

$$E(MS_E) = \sigma^2$$

Therefore, to test the equality of treatment effects, we would use the test statistic

$$F_0 = \frac{MS_{\text{Treatments}}}{MS_E}$$

which is distributed as $F_{a-1,(a-1)(b-1)}$ if the null hypothesis is true. The critical region is the upper tail of the F distribution, and we would reject H_0 if $F_0 > F_{\alpha,a-1,(a-1)(b-1)}$.

We can also test $H_0: \beta_j = 0$ by comparing the statistic $F_0 = MS_{\text{Blocks}}/MS_E$ to $F_{\alpha,b-1,(a-1)(b-1)}$. This is a test of whether there are differences between blocks, and it is often run to determine if blocking is necessary in similar future experiments. One should be very careful in testing this hypothesis. The blocks represent a restriction on randomization; that is, the ab treatment combinations in the design have not been assigned in random order to a set of experimental units. In fact, the blocks frequently *are* the experimental units (as in the hardness testing experiment), and randomization has been restricted to *within* each unit only. Under these circumstances, the test given above for $H_0: \beta_j = 0$ is inappropriate because the blocks are structured and utilized in a nonrandom fashion. This is an important consideration because an experimenter occasionally uses a factor that he is interested in to represent the blocks and then applies a set of treatments corresponding to a *second* factor within the blocks. However, in this case, the randomization restriction due to blocking will often prevent the test on blocks from being meaningful. If the experimenter is really interested in two factors, then other design techniques must be used. Anderson and McLean (1974) present an interesting treatment of this subject.

The test procedure is usually summarized in an analysis of variance table, such as shown in Table 4-2. Computing formulas for the sums of squares may be obtained for the elements in (4-7) by expressing them in terms of treatment and block totals. These computing formulas are

$$SS_T = \sum_{i=1}^{a} \sum_{j=1}^{b} y_{ij}^2 - \frac{y_{..}^2}{N} \tag{4-9}$$

Table 4-2   Analysis of Variance for Randomized Complete Block Design

Source of Variation	Sum of Squares	Degrees of Freedom	Mean Square	F_0
Treatments	$\sum_{i=1}^{a} \dfrac{y_{i.}^2}{b} - \dfrac{y_{..}^2}{N}$	$a - 1$	$\dfrac{SS_{Treatments}}{a - 1}$	$\dfrac{MS_{Treatments}}{MS_E}$
Blocks	$\sum_{j=1}^{b} \dfrac{y_{.j}^2}{a} - \dfrac{y_{..}^2}{N}$	$b - 1$	$\dfrac{SS_{Blocks}}{b - 1}$	
Error	SS_E (by subtraction)	$(a - 1)(b - 1)$	$\dfrac{SS_E}{(a - 1)(b - 1)}$	
Total	$\sum\sum y_{ij}^2 - \dfrac{y_{..}^2}{N}$	$N - 1$		

$$SS_{Treatments} = \sum_{i=1}^{a} \frac{y_{i.}^2}{b} - \frac{y_{..}^2}{N} \tag{4-10}$$

$$SS_{Blocks} = \sum_{j=1}^{b} \frac{y_{.j}^2}{a} - \frac{y_{..}^2}{N} \tag{4-11}$$

and the error sum of squares is obtained by subtraction as

$$SS_E = SS_T - SS_{Treatments} - SS_{Blocks} \tag{4-12}$$

While we have described the test procedure considering treatments and blocks as fixed factors, the same computational procedure is used if either treatments or blocks (or both) are random. There are, of course, corresponding changes in the interpretation of the results. For example, if blocks are random, then we expect the comparisons among treatments to be the same throughout the *population* of blocks of which those used in the experiment were randomly selected. There will also be corresponding changes in the expected mean squares. For example, if blocks are random, then $E(MS_{Blocks}) = \sigma^2 + a\sigma_\beta^2$, where σ_β^2 is the variance of the block effects. In any event, however, $E(MS_{Treatments})$ will always be free of any block effect and the test statistic for between-treatment variability is always $F_0 = MS_{Treatments}/MS_E$.

■**EXAMPLE 4-1.**   Consider the hardness testing experiment described in Section 4-1. There are four tips and four available metal specimens. Each tip is tested once on each specimen, resulting in a randomized complete block design. The data obtained is shown in Table 4-3. Remember that the *order* in which the tips were tested on a particular specimen was determined randomly.

Table 4-3 Randomized Complete Block Design for
the Hardness Testing Experiment

Type of Tip	Specimen (Block)			
	1	2	3	4
1	9.3	9.4	9.6	10.0
2	9.4	9.3	9.8	9.9
3	9.2	9.4	9.5	9.7
4	9.7	9.6	10.0	10.2

To simplify the calculations, we will code the original data by subtracting 9.5 from
each observation and multiplying the result by 10. This yields the data in Table 4-4.

Table 4-4 Coded Data for the Hardness Testing Experiment

Type of Tip	Specimen (Block)				$y_i.$
	1	2	3	4	
1	-2	-1	1	5	3
2	-1	-2	3	4	4
3	-3	-1	0	2	-2
4	2	1	5	7	15
$y._j$	-4	-3	9	18	$20 = y_{..}$

The sums of squares are obtained as follows:

$$SS_T = \sum_{i=1}^{4} \sum_{j=1}^{4} y_{ij}^2 - \frac{y_{..}^2}{N}$$

$$= 154.00 - \frac{(20)^2}{16} = 129.00$$

$$SS_{\text{Treatments}} = \sum_{i=1}^{4} \frac{y_{i.}^2}{b} - \frac{y_{..}^2}{N}$$

$$= \frac{(3)^2 + (4)^2 + (-2)^2 + (15)^2}{4} - \frac{(20)^2}{16} = 38.50$$

$$SS_{\text{Blocks}} = \sum_{j=1}^{4} \frac{y_{.j}^2}{a} - \frac{y_{..}^2}{N}$$

$$= \frac{(-4)^2 + (-3)^2 + (9)^2 + (18)^2}{4} - \frac{(20)^2}{16} = 82.50$$

$$SS_E = SS_T - SS_{\text{Treatments}} - SS_{\text{Blocks}}$$

$$= 129.00 - 38.50 - 82.50 = 8.00$$

The analysis of variance is shown in Table 4-5. Using $\alpha = .05$, the critical value of F is $F_{.05,3,9} = 3.86$. Since $14.42 > 3.86$, we conclude that the type of tip affects the hardness reading. Also, specimens (blocks) seem to differ significantly, since the mean square for blocks is large relative to error.

Table 4-5   Analysis of Variance for the Hardness Testing Experiment

Source of Variation	Sum of Squares	Degrees of Freedom	Mean Square	F_0
Treatments (type of tip)	38.50	3	12.83	14.42
Blocks (specimens)	82.50	3	27.50	
Error	8.00	9	0.89	
Total	129.00	15		

It is interesting to observe the results the experimenter would have obtained had he not been aware of randomized block designs. Suppose he used only four specimens and randomly assigned the tips to each, resulting in (by chance) the same design as in Table 4-3. The incorrect analysis of this data as a completely randomized single-factor design is shown in Table 4-6. Since $F_{.05,3,12} = 3.49$, the hypothesis of equal hardness measurements from the four tips cannot be rejected. Thus, the randomized block design reduces the amount of noise in the data sufficiently for differences between the four tips to be detected.   ∎

Table 4-6   Incorrect Analysis of the Hardness Testing Experiment as a Completely Randomized Design

Source of Variation	Sum of Squares	Degrees of Freedom	Mean Square	F_0
Type of tip	38.50	3	12.83	1.70
Error	90.50	12	7.54	
Total	129.00	15		

If the treatments in a randomized block design are fixed, and the analysis indicates significant differences in treatments, one may investigate the treatment means further. Orthogonal contrasts or Duncan's multiple range test could be used for this purpose. If Duncan's multiple range test is used, the standard error of a treatment mean is

$$S_{\bar{y}_{i.}} = \sqrt{\frac{MS_E}{b}}$$

Frequently the experimenter must decide how many blocks to use in his design. For example, in the hardness testing problem, four blocks (specimens) were employed. In general, the experimenter should choose enough blocks so that an appropriate power of the test is obtained. Notice that as the number of blocks increases, the number of degrees of freedom for error also increases, making the test more powerful. The operating characteristic curves in the Appendix may be used for this purpose. If the factor is fixed, then Appendix Chart V is used with

$$\Phi^2 = \frac{b \sum_{i=1}^{a} \tau_i^2}{a\sigma^2}$$

where there are $a - 1$ numerator degrees of freedom and $(a - 1)(b - 1)$ denominator degrees of freedom. If the factor is random, then we use Appendix Chart VI with

$$\lambda = \sqrt{1 + \frac{b\sigma_\tau^2}{\sigma^2}}$$

where there are $a - 1$ numerator degrees of freedom and $(a - 1)(b - 1)$ denominator degrees of freedom. In practice, the error variance σ^2 is usually replaced by an estimate.

■**EXAMPLE 4-2.** In a randomized block design five treatments are to be compared in six blocks. The treatments are regarded as fixed. The experimenter wants to determine the power of the test at the $\alpha = .05$ level of significance if the true treatment means are $\mu_1 = 5.0$, $\mu_2 = 8.0$, $\mu_3 = 4.0$, $\mu_4 = 3.0$, and $\mu_5 = 5.0$. An estimate of σ^2 from a prior experiment is $\hat{\sigma}^2 = 4.0$. Since the mean of the ith treatment is $\mu_i = \mu + \tau_i$, and since we require $\sum_{i=1}^{5} \tau_i = 0$, we have

$$\mu = \frac{\sum_{i=1}^{5} \mu_i}{5} = 5.0$$

which implies that $\tau_1 = 0$, $\tau_2 = 3$, $\tau_3 = -1$, $\tau_4 = -2$, and $\tau_5 = 0$. Therefore

$$\Phi^2 = \frac{b \sum_{i=1}^{a} \tau_i^2}{a\sigma^2} = \frac{(6)(14)}{(5)(4)} = 4.2$$

and $\Phi = 2.049$. There are $(a - 1)(b - 1) = (5 - 1)(6 - 1) = 20$ degrees of freedom for error, and the probability of type II error is found from Chart V to be approximately .06. Therefore, the probability of correctly rejecting H_0: $\tau_i = 0$ with this design if in fact $\sum \tau_i^2 = 14$ and $\sigma^2 = 4.0$ is approximately .94. ■

Suppose an experiment is conducted in a randomized block design, and blocking was not really necessary. The randomized block design yields $(a - 1)(b - 1)$ degrees of freedom for error. If the experiment had been conducted as a completely randomized design with b replicates, we would have had $a(b - 1)$ degrees of freedom for error.

Thus, incorrectly blocking has cost $a(b - 1) - (a - 1)(b - 1) = b - 1$ degrees of freedom for error, and the test has been made less sensitive needlessly. However, if block effects really *are* important, then the experimental error may be so inflated that significant treatment differences would remain undetected. As an illustration, see the incorrect analysis of Example 4-1. As a general rule, when doubting the importance of block effects, the experimenter should block and gamble that the block effect does exist. If he is wrong, the slight loss in error degrees of freedom will have little effect on the outcome, so long as a moderate number of degrees of freedom for error are available.

Finally, note that the linear model for the randomized block design is completely *additive;* that is, we assume that there is no interaction between blocks and treatments. Essentially, this says that the effect of a change in the level of the factor of interest is consistent from block to block. If interaction between blocks and treatments is present, then the error mean square reflects both experimental error and the interaction effects. In general, this will inflate the error mean square and adversely affect the test on treatments. The exception to this is the case where blocks are random. In this situation, both the mean square for treatments and the mean square for error contain interaction effects, and, consequently, treatments may be tested in the usual manner by comparing the mean square for treatments to the error mean square. This procedure will not yield information on the interaction. A method of experimentation described in Chapter 6 is required to detect the presence of interaction.

4-1.2   Estimating Missing Values

When using the randomized complete block design, sometimes an observation in one of the blocks is missing. This may happen because of carelessness or error, or for reasons beyond our control, such as unavoidable damage to an experimental unit. A missing observation introduces a new problem into the analysis, since treatments are no longer *orthogonal* to blocks; that is, every treatment does not occur in every block. There are two general approaches to the missing value problem. The first is an approximate analysis in which the missing observation is estimated, and then the usual analysis of variance is performed proceeding just as if the estimated observation were real data, with the error degrees of freedom reduced by one. This approximate analysis is the subject of this section. The second approach is an exact analysis, to be discussed in Section 4-1.3.

Suppose the observation y_{ij} for treatment i in block j is missing. Denote the missing observation by x. As an illustration, suppose in the hardness testing experiment of Example 4-1 specimen 3 was broken while tip 2 was being tested and that data point could not be obtained. The data might appear as in Table 4-7.

In general, we will let $y'_{..}$ represent the grand total with one missing observation, $y'_{i.}$ represent the total for the treatment with one missing observation, and $y'_{.j}$ be the total for the block with one missing observation. Suppose we wish to estimate the missing observation x so that x will have a minimum contribution to the error sum of squares. Since $SS_E = \Sigma_{i=1}^{a} \Sigma_{j=1}^{b} (y_{ij} - \bar{y}_{i.} - \bar{y}_{.j} + \bar{y}_{..})^2$, this is equivalent to choosing

Table 4-7 Randomized Complete Block Design for the Hardness Testing Experiment with One Missing Value

Type of Tip	Specimen (Block)			
	1	2	3	4
1	-2	-1	1	5
2	-1	-2	x	4
3	-3	-1	0	2
4	2	1	5	7

x in order to minimize

$$SS_E = \sum_{i=1}^{a}\sum_{j=1}^{b} y_{ij}^2 - \frac{1}{b}\sum_{i=1}^{a}\left(\sum_{j=1}^{b} y_{ij}\right)^2 - \frac{1}{a}\sum_{j=1}^{b}\left(\sum_{i=1}^{a} y_{ij}\right)^2 + \frac{1}{ab}\left(\sum_{i=1}^{a}\sum_{j=1}^{b} y_{ij}\right)^2$$

or

$$SS_E = x^2 - \frac{1}{b}(y'_{i.} + x)^2 - \frac{1}{a}(y'_{.j} + x)^2 + \frac{1}{ab}(y'_{..} + x)^2 + R \qquad (4\text{-}13)$$

where R includes all terms not involving x. From $\partial SS_E/\partial x = 0$ we obtain

$$x = \frac{ay'_{i.} + by'_{.j} - y'_{..}}{(a-1)(b-1)} \qquad (4\text{-}14)$$

as the estimate of the missing observation.

For the data in Table 4-7 we find that $y'_{2.} = 1$, $y'_{.3} = 6$, and $y'_{..} = 17$. Therefore, from Equation 4-14,

$$x \equiv y_{23} = \frac{4(1) + 4(6) - 17}{(3)(3)} = 1.22$$

The usual analysis of variance may now be performed, using $y_{23} = 1.22$, and reducing the error degrees of freedom by one. The analysis of variance is shown in Table 4-8. Compare the results of this approximate analysis with the results obtained for the full data set (Table 4-5).

Table 4-8 Approximate Analysis of Variance for Example 4-1 with one Missing Value

Source of Variation	Sum of Squares	Degrees of Freedom	Mean Square	F_0
Type of tip	39.98	3	13.33	17.12[a]
Specimens (blocks)	79.53	3	26.51	
Error	6.22	8	0.78	
Total	125.73	14		

[a]Significant at 5 percent.

If several observations are missing, they may be estimated by writing the error sum of squares as a function of the missing values, differentiating with respect to each missing value, equating the results to zero, and solving the resulting equations. Alternatively, we may use Equation 4-14 iteratively to estimate the missing values. To illustrate the iterative approach, suppose that two values are missing. Arbitrarily estimate the first missing value, and then use this value along with the real data and (4-14) to estimate the second. Now (4-14) can be used to reestimate the first missing value, and following this, the second can be reestimated. This process is continued until convergence is obtained. In any missing value problem, the error degrees of freedom are reduced by one for each missing observation.

4-1.3 Estimating Model Parameters and the General Regression Significance Test

If both treatments and blocks are fixed, we may estimate the parameters in the randomized complete block design model by least squares. Recall that the linear statistical model is

$$y_{ij} = \mu + \tau_i + \beta_j + \epsilon_{ij} \begin{cases} i = 1, 2, \ldots, a \\ j = 1, 2, \ldots, b \end{cases}$$

Applying the rules in Section 3-10 for writing the normal equations for an experimental design model, we obtain

$$
\begin{aligned}
\mu: \quad & ab\hat{\mu} + b\hat{\tau}_1 + b\hat{\tau}_2 + \cdots + b\hat{\tau}_a + a\hat{\beta}_1 + a\hat{\beta}_2 + \cdots + a\hat{\beta}_b = y_{..} \\
\tau_1: \quad & b\hat{\mu} + b\hat{\tau}_1 \qquad\qquad\qquad\quad + \hat{\beta}_1 + \hat{\beta}_2 + \cdots + \hat{\beta}_b = y_{1.} \\
\tau_2: \quad & b\hat{\mu} \qquad + b\hat{\tau}_2 \qquad\qquad\quad + \hat{\beta}_1 + \hat{\beta}_2 + \cdots + \hat{\beta}_b = y_{2.} \\
& \quad\vdots \qquad\qquad\qquad\vdots \qquad\qquad\qquad\qquad\vdots \\
\tau_a: \quad & b\hat{\mu} \qquad\qquad\qquad + b\hat{\tau}_a + \hat{\beta}_1 + \hat{\beta}_2 + \cdots + \hat{\beta}_b = y_{a.} \\
\beta_1: \quad & a\hat{\mu} + \hat{\tau}_1 + \hat{\tau}_2 + \cdots + \hat{\tau}_a + a\hat{\beta}_1 \qquad\qquad\qquad = y_{.1} \\
\beta_2: \quad & a\hat{\mu} + \hat{\tau}_1 + \hat{\tau}_2 + \cdots + \hat{\tau}_a \qquad + a\hat{\beta}_2 \qquad\qquad = y_{.2} \\
& \quad\vdots \qquad\qquad\qquad\vdots \qquad\qquad\qquad\qquad\vdots \\
\beta_b: \quad & a\hat{\mu} + \hat{\tau}_1 + \hat{\tau}_2 + \cdots + \hat{\tau}_a \qquad\qquad\qquad + a\hat{\beta}_b = y_{.b}
\end{aligned}
\tag{4-15}
$$

Notice that the second through the $(a + 1)$st equations in (4-15) sum to the first normal equation, as do the last b equations. Thus there are two linear dependencies in the normal equations, implying that two constraints must be imposed to solve

(4-15). The usual constraints are

$$\sum_{i=1}^{a} \hat{\tau}_i = 0, \qquad \sum_{j=1}^{b} \hat{\beta}_j = 0 \tag{4-16}$$

Using these constraints, the normal equations simplify considerably. In fact, they become

$$ab\hat{\mu} = y_{..}$$
$$b\hat{\mu} + b\hat{\tau}_i = y_{i.}, \qquad i = 1, 2, \ldots, a \tag{4-17}$$
$$a\hat{\mu} + a\hat{\beta}_j = y_{.j}, \qquad j = 1, 2, \ldots, b$$

whose solution is

$$\hat{\mu} = \bar{y}_{..}$$
$$\hat{\tau}_i = \bar{y}_{i.} - \bar{y}_{..}, \qquad i = 1, 2, \ldots, a \tag{4-18}$$
$$\hat{\beta}_j = \bar{y}_{.j} - \bar{y}_{..}, \qquad j = 1, 2, \ldots, b$$

The general regression significance test can be used to develop the analysis of variance for the randomized complete block design. Using the solution to the normal equations given by (4-18), the reduction in sum of squares for fitting the complete model is

$$R(\mu, \tau, \beta) = \hat{\mu} y_{..} + \sum_{i=1}^{a} \hat{\tau}_i y_{i.} + \sum_{j=1}^{b} \hat{\beta}_j y_{.j}$$

$$= \bar{y}_{..} y_{..} + \sum_{i=1}^{a} (\bar{y}_{i.} - \bar{y}_{..}) y_{i.} + \sum_{j=1}^{b} (\bar{y}_{.j} - \bar{y}_{..}) y_{.j}$$

$$= \frac{y_{..}^2}{ab} + \sum_{i=1}^{a} \bar{y}_{i.} y_{i.} - \frac{y_{..}^2}{ab} + \sum_{j=1}^{b} \bar{y}_{.j} y_{.j} - \frac{y_{..}^2}{ab}$$

$$= \sum_{i=1}^{a} \frac{y_{i.}^2}{b} + \sum_{j=1}^{b} \frac{y_{.j}^2}{a} - \frac{y_{..}^2}{ab}$$

with $a + b - 1$ degrees of freedom and the error sum of squares is

$$SS_E = \sum_{i=1}^{a} \sum_{j=1}^{b} y_{ij}^2 - R(\mu, \tau, \beta)$$

$$= \sum_{i=1}^{a} \sum_{j=1}^{b} y_{ij}^2 - \sum_{i=1}^{a} \frac{y_{i.}^2}{b} - \sum_{j=1}^{b} \frac{y_{.j}^2}{a} + \frac{y_{..}^2}{ab}$$

$$= \sum_{i=1}^{a} \sum_{j=1}^{b} (y_{ij} - \bar{y}_{i.} - \bar{y}_{.j} + \bar{y}_{..})^2$$

with $(a - 1)(b - 1)$ degrees of freedom. Compare this last equation with SS_E in Equation 4-7.

To test the hypothesis H_0: $\tau_i = 0$ the reduced model is

$$y_{ij} = \mu + \beta_j + \epsilon_{ij}$$

which is just a one-way classification analysis of variance. By analogy with Equation 3-32, the reduction in sum of squares for fitting this model is

$$R(\mu, \beta) = \sum_{j=1}^{b} \frac{y_{.j}^2}{a}$$

which has b degrees of freedom. Therefore, the sum of squares due to $\{\tau_i\}$ after fitting μ and $\{\beta_j\}$ is

$$R(\tau | \mu, \beta) = R(\mu, \tau, \beta) - R(\mu, \beta)$$

$$= \sum_{i=1}^{a} \frac{y_{i.}^2}{b} + \sum_{j=1}^{b} \frac{y_{.j}^2}{a} - \frac{y_{..}^2}{ab} - \sum_{j=1}^{b} \frac{y_{.j}^2}{a}$$

$$= \sum_{i=1}^{a} \frac{y_{i.}^2}{b} - \frac{y_{..}^2}{ab}$$

which we recognize as the treatment sum of squares with $a - 1$ degrees of freedom (Equation 4-10).

The block sum of squares is obtained by fitting the reduced model

$$y_{ij} = \mu + \tau_i + \epsilon_{ij}$$

which is also a one-way classification. Again, by analogy with (3-32), the reduction in sum of squares for fitting this model is

$$R(\mu, \tau) = \sum_{i=1}^{a} \frac{y_{i.}^2}{b}$$

with a degrees of freedom. The sum of squares for blocks (β_j) after fitting μ and τ_i is

$$R(\beta | \mu, \tau) = R(\mu, \tau, \beta) - R(\mu, \tau)$$

$$= \sum_{i=1}^{a} \frac{y_{i.}^2}{b} + \sum_{j=1}^{b} \frac{y_{.j}^2}{a} - \frac{y_{..}^2}{ab} - \sum_{i=1}^{a} \frac{y_{i.}^2}{b}$$

$$= \sum_{j=1}^{b} \frac{y_{.j}^2}{a} - \frac{y_{..}^2}{ab}$$

with $b - 1$ degrees of freedom, which we have found previously as Equation 4-11.

We have developed the sums of squares for treatments, blocks, and error in the randomized complete block design using the general regression significance test. While we would not ordinarily use the general regression significance test to actually analyze data in a randomized complete block, the procedure occasionally proves useful in more general randomized block designs, such as those discussed in Chapter 5.

Exact Analysis of the Missing Value Problem

In Section 4-1.2 an approximate procedure for dealing with missing observations in a randomized block design was presented. This approximate analysis consists of estimating the missing value so that the error mean square is minimized. It can be shown that the approximate analysis produces a biased mean square for treatments, in the sense that $E(MS_{Treatments})$ is larger than $E(MS_E)$ if the null hypothesis is true. Consequently, too many significant results are reported.

The missing value problem may be analyzed exactly by using the general regression significance test. The missing value causes the design to be *unbalanced*; and because all treatments do not occur in all blocks, we say that treatments and blocks are not *orthogonal*. This method of analysis is also used in more general types of randomized block designs, and is discussed further in Chapter 5. Problem 4-21 asks the reader to perform the exact analysis for a randomized complete block design with one missing value.

4-2. THE LATIN SQUARE DESIGN $\quad$ (a partially balanced design)

In Section 4-1 we introduced the randomized complete block design as a design to reduce the residual error in an experiment by removing variability due to the experimental units. There are several other types of designs that utilize the blocking principle. For example, suppose that an experimenter is studying the effect of five different formulations of an explosive mixture used in the manufacture of dynamite on the observed explosive force. Each formulation is mixed from a batch of raw material that is only large enough for five formulations to be tested. Furthermore, the formulations are prepared by several operators, and there may be substantial differences in the skills and experience of the operators. Thus it would seem that there are two extraneous factors to be "averaged out" in the design; batches of raw material and operators. The appropriate design for this problem consists of testing each formulation exactly once in each batch of raw material and for each formulation to be prepared exactly once by each of five operators. The resulting design, shown in Table 4-9, is called a *Latin square* design. Notice that the design is a square arrangement, and that the five formulations (or treatments) are denoted by the Latin letters $A, B, C, D,$ and E; hence the name Latin square. We see that both batches of raw material (rows) and operators (columns) are orthogonal to treatments.

The Latin square design is used to eliminate two extraneous sources of variability; that is, to systematically allow blocking in two directions. Thus the rows and columns actually represent two restrictions on randomization. In general, a Latin square for p factors, or a $p \times p$ Latin square, is a square containing p rows and p columns. Each of the resulting p^2 cells contains one of the p letters that corresponds to the treatments,

Table 4-9 Latin Square Design for the Dynamite Formulation Problem

Batches of Raw Material	Operators				
	1	2	3	4	5
1	A = 24	B = 20	C = 19	D = 24	E = 24
2	B = 17	C = 24	D = 30	E = 27	A = 36
3	C = 18	D = 38	E = 26	A = 27	B = 21
4	D = 26	E = 31	A = 26	B = 23	C = 22
5	E = 22	A = 30	B = 20	C = 29	D = 31

and each letter occurs once and only once in each row and column. Several examples of Latin squares are shown below:

4 X 4	5 X 5	6 X 6
A B D C	A D B E C	A D C E B F
B C A D	D A C B E	B A E C F D
C D B A	C B E D A	C E D F A B
D A C B	B E A C D	D C F B E A
	E C D A B	F B A D C E
		E F B A D C

The statistical model for a Latin square is

$$y_{ijk} = \mu + \alpha_i + \tau_j + \beta_k + \epsilon_{ijk} \quad \begin{cases} i = 1, 2, \ldots, p \\ j = 1, 2, \ldots, p \\ k = 1, 2, \ldots, p \end{cases} \quad (4\text{-}19)$$

where y_{ijk} is the observation in the ith row and kth column for the jth treatment, μ is the overall mean, α_i is the ith row effect, τ_j is the jth treatment effect, β_k is the kth column effect, and ϵ_{ijk} is the random error. The model is completely *additive*, that is, there is no interaction between rows, columns, and treatments. Since there is only one observation in each cell, only two of the three subscripts i, j, and k are needed to denote a particular observation. For example referring to the dynamite formulation problem in Table 4-9, if $i = 2$ and $k = 3$ we automatically find $j = 4$ (formulation D), and if $i = 1$ and $j = 3$ (formulation C) we find $k = 3$. This is a consequence of each treatment appearing exactly once in each row and column.

The analysis of variance consists of partitioning the total sum of squares of the $N = p^2$ observations into components for rows, columns, treatments, and error, for example,

$$SS_T = SS_{\text{Rows}} + SS_{\text{Columns}} + SS_{\text{Treatments}} + SS_E \quad (4\text{-}20)$$

with degrees of freedom

$$\begin{array}{cccccc} Total & Rows & Columns & Treatments & Error \\ p^2 - 1 = & p - 1 + & p - 1 & + & p - 1 & + (p - 2)(p - 1). \end{array}$$

Under the usual assumption that ϵ_{ijk} is $NID(0, \sigma^2)$, each sum of squares on the right-hand side of (4-20) is, on division by σ^2, an independently distributed chi-square random variable. The appropriate statistic for testing for no treatment effect is

$$F_0 = \frac{MS_{Treatments}}{MS_E}$$

Table 4-10 Analysis of Variance for the Latin Square Design

Source of Variance	Sum of Squares	Degrees of Freedom	Mean Square	F_0
Treatments	$SS_{Treatments}$	$p - 1$	$\dfrac{SS_{Treatments}}{p - 1}$	$F_0 = \dfrac{MS_{Treatments}}{MS_E}$
	$= \displaystyle\sum_{j=1}^{p} \dfrac{y_{.j.}^2}{p} - \dfrac{y_{...}^2}{N}$			
Rows	SS_{Rows}	$p - 1$	$\dfrac{SS_{Rows}}{p - 1}$	
	$= \displaystyle\sum_{i=1}^{p} \dfrac{y_{i..}^2}{p} - \dfrac{y_{...}^2}{N}$			
Columns	$SS_{Columns}$	$p - 1$	$\dfrac{SS_{Columns}}{p - 1}$	
	$= \displaystyle\sum_{k=1}^{p} \dfrac{y_{..k}^2}{p} - \dfrac{y_{...}^2}{N}$			
Error	SS_E (by subtraction)	$(p - 2)(p - 1)$	$\dfrac{SS_E}{(p - 2)(p - 1)}$	
Total	SS_T	$p^2 - 1$		
	$= \displaystyle\sum_i \sum_j \sum_k y_{ijk}^2$			
	$- \dfrac{y_{...}^2}{N}$			

which is distributed as $F_{p-1,(p-2)(p-1)}$ under the null hypothesis. One may also test for no row effect and no column effect by forming the ratio of MS_{Rows} or $MS_{Columns}$ to MS_E. However, since the rows and columns represent restrictions on randomization, these tests may not be appropriate.

The computational procedure for the analysis of variance is shown in Table 4-10. From the computational formulas for the sums of squares we see that the analysis is a simple extension of the randomized block design, with the sum of squares resulting from rows obtained from the row totals.

■EXAMPLE 4-3.   Consider the dynamite formulation problem described above, where both batches of raw material and operators represent randomization restrictions. The design for this experiment, shown in Table 4-9, is a 5 × 5 Latin square. After coding by subtracting 25 from each observation, we have the data in Table 4-11.

Table 4-11   Coded Data for the Dynamite Formulation Problem

Batches of Raw Material	Operators					$y_i..$
	1	2	3	4	5	
1	$A = -1$	$B = -5$	$C = -6$	$D = -1$	$E = -1$	-14
2	$B = -8$	$C = -1$	$D = 5$	$E = 2$	$A = 11$	9
3	$C = -7$	$D = 13$	$E = 1$	$A = 2$	$B = -4$	5
4	$D = 1$	$E = 6$	$A = 1$	$B = -2$	$C = -3$	3
5	$E = -3$	$A = 5$	$B = -5$	$C = 4$	$D = 6$	7
$y..k$	-18	$18.$	-4	5	9	$10 = y...$

The sums of squares for total, batches (rows), and operators (columns), are computed as follows:

$$SS_T = \sum_i \sum_j \sum_k y_{ijk}^2 - \frac{y_{...}^2}{N}$$

$$= 680 - \frac{(10)^2}{25} = 676.00$$

$$SS_{Batches} = \sum_{i=1}^{p} \frac{y_{i..}^2}{p} - \frac{y_{...}^2}{N}$$

$$= \frac{(-14)^2 + 9^2 + 5^2 + 3^2 + 7^2}{5} - \frac{(10)^2}{25} = 68.00$$

$$SS_{Operators} = \sum_{k=1}^{p} \frac{y_{..k}^2}{p} - \frac{y_{...}^2}{N}$$

$$= \frac{(-18)^2 + 18^2 + (-4)^2 + 5^2 + 9^2}{5} - \frac{(10)^2}{25} = 150.00$$

The totals for the treatments (Latin letters) are given below:

Latin Letter	Treatment Total
A	$y_{.1.} = 18$
B	$y_{.2.} = -24$
C	$y_{.3.} = -13$
D	$y_{.4.} = 24$
E	$y_{.5.} = 5.$

The sum of squares due to formulations is computed from these totals as

$$SS_{\text{Formulations}} = \sum_{j=1}^{p} \frac{y_{.j.}^2}{p} - \frac{y_{...}^2}{N}$$

$$= \frac{18^2 + (-24)^2 + (-13)^2 + 24^2 + 5^2}{5} - \frac{(10)^2}{25} = 330.00$$

The error sum of squares is found by subtraction,

$$SS_E = SS_T - SS_{\text{Batches}} - SS_{\text{Operators}} - SS_{\text{Formulations}}$$

$$= 676.00 - 68.00 - 150.00 - 330.00 = 128.00$$

The analysis of variance is shown in Table 4-12. Formulations differ at the one percent level. Also, the mean square for operators is large relative to the error mean square. ∎

Table 4-12 Analysis of Variance for the Dynamite Formulation Experiment

Source of Variation	Sum of Squares	Degrees of Freedom	Mean Square	F_0
Formulations	330.00	4	82.50	7.73[a]
Batches of raw Material	68.00	4	17.00	
Operators	150.00	4	37.50	
Error	128.00	12	10.67	
Total	676.00	24		

[a]Significant at 1 percent.

A Latin square in which the first row and column consists of the letters written in alphabetical order is called a *standard* Latin square. The design used in Example 4-3 is a standard Latin square. A standard Latin square can always be obtained by writing the first row in alphabetical order, and then writing each successive row as the row of letters just above shifted one place to the left. Table 4-13 summarizes several important facts about Latin squares and standard squares.

As with any experimental design, the observations in the Latin square should be taken in random order. The proper randomization procedure is to select the particular square employed at random. As we see in Table 4-13, there are a large number of Latin

Table 4-13[a] Standard Squares and Number of Latin Squares of Various Sizes

Size	3 × 3	4 × 4	5 × 5	6 × 6	7 × 7	p × p
Examples of standard squares	A B C B C A C A B	A B C D B C D A C D A B D A B C	A B C D E B A E C D C D A E B D E B A C E C D B A	A B C D E F B C F A D E C F B E A D D E A B F C E A D F C B F D E C B A	A B C D E F G B C D E F G A C D E F G A B D E F G A B C E F G A B C D F G A B C D E G A B C D E F	A B C . . . P B C D . . . A C D E . . . B . . . . . . P A B . . . (P − 1)
Number of standard squares	1	4	56	9408	16,942,080	—
Total number of Latin squares	12	576	161,280	818,851,200	61,479,419,904,000	$p!(p-1)! \times$ (number of standard squares)

[a]Some of the information in the table is found in *Statistical Tables for Biological, Agricultural and Medical Research*, 4th edition, by R. A. Fisher and F. Yates, Oliver and Boyd, Edinburgh, 1953. Little is known about the properties of Latin squares larger than 7 × 7.

squares of a particular size, so it is impossible to enumerate all the squares and select one randomly. The usual procedure is to select a Latin square from a table of such designs, as in Fisher and Yates (1953), and then arrange the order of the rows, columns, and letters at random. This is discussed more completely in Fisher and Yates (1953).

Occasionally one observation in a Latin square will be missing. For a $p \times p$ Latin square, the missing value may be estimated by

$$y_{ijk} = \frac{p(y'_{i..} + y'_{.j.} + y'_{..k}) - 2y'_{...}}{(p - 2)(p - 1)} \qquad (4\text{-}21)$$

where the primes indicate totals for the row, column, and treatment with the missing value, and $y'_{...}$ is the grand total with the missing value.

Latin squares can be useful in situations where the rows and columns represent factors the experimenter actually wishes to study, and are not randomization restrictions. Thus, three factors (rows, columns, and letters), each at p levels, can be investigated in only p^2 runs. This design assumes that there is no interaction between the factors. More will be said later on the subject of interaction.

Crossover Designs and Designs Balanced for Residual Effects

Occasionally one encounters a problem in which time periods is a factor in the experiment. In general, there are p treatments to be tested in p time periods using np experimental units. For example, a human performance analyst was studying the effect of two replacement fluids on dehydration on 20 subjects. In the first period, half of the subjects (chosen at random) were given fluid 1 and the other half fluid 2. At the end of the period the response was measured, and a period of time allowed to pass in which any physiological effect of the fluids was eliminated. Then the experimenter had the subjects who took fluid 1 now take fluid 2 and those who took fluid 2 now take fluid 1. This design is called a *crossover* design. It is analyzed as a set of 10 Latin squares with two rows (time periods) and two treatments (fluid types). The two columns in each of the 10 squares correspond to subjects. Details of the statistical analysis of these designs are given by Cochran and Cox (1957), John (1971), and Anderson and McLean (1974).

It is also possible to employ Latin square type designs for experiments in which the treatments have a *residual* effect, that is, for example if the data for fluid 2 in period 2 still reflected some effect of fluid 1 taken in period 1. Designs balanced for residual effects are discussed in detail by Cochran and Cox (1957) and John (1971).

4-3 THE GRAECO-LATIN SQUARE DESIGN

Consider a $p \times p$ Latin square, and superimpose on it a second $p \times p$ Latin square in which the treatments are denoted by Greek letters. If the two squares, when superimposed, have the property that each Greek letter appears once and only once with

Table 4-14   4 × 4 Graeco-Latin
Square Design

Row	Column			
	1	2	3	4
1	Aα	Bβ	Cγ	Dδ
2	Bδ	Aγ	Dβ	Cα
3	Cβ	Dα	Aδ	Bγ
4	Dγ	Cδ	Bα	Aβ

each Latin letter, the two Latin squares are said to be *orthogonal*, and the design obtained is called a Graeco-Latin square. An example of a 4 × 4 Graeco-Latin square is shown in Table 4-14.

The Graeco-Latin square design can be used to systematically control three sources of extraneous variability; that is, to block in *three* directions. The design allows investigation of four factors (rows, columns, Latin letters, and Greek letters), each at p levels in only p^2 runs. Graeco-Latin squares exist for all $p \geqslant 3$ except $p = 6$.

The statistical model for the Graeco-Latin square design is

$$y_{ijkl} = \mu + \theta_i + \tau_j + \omega_k + \Psi_l + \epsilon_{ijkl} \begin{cases} i = 1, 2, \ldots, p \\ j = 1, 2, \ldots, p \\ k = 1, 2, \ldots, p \\ l = 1, 2, \ldots, p \end{cases} \tag{4-22}$$

where y_{ijkl} is the observation in row i and column l for Latin letter j and Greek letter k, θ_i is the effect of the ith row, τ_j is the effect of Latin letter treatment j, ω_k is the effect of Greek letter treatment k, Ψ_l is the effect of column l, and ϵ_{ijkl} is an NID $(0, \sigma^2)$ random error component. Only two of the four subscripts are necessary to completely identify an observation.

The analysis of variance is very similar to that of a Latin square. Since the Greek letters appear exactly once in each row and column and exactly once with each Latin letter, the factor represented by the Greek letters is orthogonal to rows, columns, and Latin letter treatments. Therefore, a sum of squares due to the Greek letter factor may be computed from the Greek letter totals and the experimental error is further reduced by this amount. The computational details are illustrated in Table 4-15. The null hypotheses of equal row, column, Latin letter, and Greek letter treatments would be tested by dividing the corresponding mean square by mean square error. The rejection region is the upper tail point of the $F_{p-1,(p-3)(p-1)}$ distribution.

■EXAMPLE 4-4.   Suppose that in the dynamite formulation experiment of Example 4-3 an additional factor, test assemblies, could be of importance. Let there be five test assemblies denoted by the Greek letters α, β, γ, δ, and ε. The resulting 5 × 5 Graeco-Latin square design is shown in Table 4-16.

Table 4-15   Analysis of Variance for a Graeco-Latin Square Design

Source of Variation	Sum of Squares	Degrees of Freedom
Latin letter treatments	$SS_L = \sum_{j=1}^{p} \dfrac{y_{.j..}^2}{p} - \dfrac{y_{....}^2}{N}$	$p - 1$
Greek letter treatments	$SS_G = \sum_{k=1}^{p} \dfrac{y_{..k.}^2}{p} - \dfrac{y_{....}^2}{N}$	$p - 1$
Rows	$SS_{Rows} = \sum_{i=1}^{p} \dfrac{y_{i...}^2}{p} - \dfrac{y_{....}^2}{N}$	$p - 1$
Columns	$SS_{Columns} = \sum_{l=1}^{p} \dfrac{y_{...l}^2}{p} - \dfrac{y_{....}^2}{N}$	$p - 1$
Error	SS_E (by subtraction)	$(p - 3)(p - 1)$
Total	$SS_T = \sum_i \sum_j \sum_k \sum_l y_{ijkl}^2 - \dfrac{y_{....}^2}{N}$	$p^2 - 1$

Notice that since the totals for batches of raw material (rows), operators (columns), and formulations (Latin letters) are identical to Example 4-3, we have

$$SS_{Batches} = 68.00 \qquad SS_{Operators} = 150.00 \qquad \text{and} \qquad SS_{Formulations} = 330.00$$

The totals for the test assemblies (Greek letters) are:

Greek Letter	Test Assembly Total
α	$y_{..1.} = 10$
β	$y_{..2.} = -6$
γ	$y_{..3.} = -3$
δ	$y_{..4.} = -4$
ϵ	$y_{..5.} = 13.$

Thus the sum of squares due to test assemblies is

$$SS_{Assemblies} = \sum_{k=1}^{p} \frac{y_{..k.}^2}{p} - \frac{y_{....}^2}{N}$$

$$= \frac{10^2 + (-6)^2 + (-3)^2 + (-4)^2 + 13^2}{5} - \frac{(10)^2}{25} = 62.00$$

The complete analysis of variance is summarized in Table 4-17. Formulations are significantly different at one percent. Comparing Tables 4-17 and 4-12, we observe

Table 4-16 Graeco-Latin Square Design for the Dynamite Formulation Problem

Batches of Raw Material	Operators					$y_{i...}$
	1	2	3	4	5	
1	$A\alpha = -1$	$B\gamma = -5$	$C\epsilon = -6$	$D\beta = -1$	$E\delta = -1$	-14
2	$B\beta = -8$	$C\delta = -1$	$D\alpha = 5$	$E\gamma = 2$	$A\epsilon = 11$	9
3	$C\gamma = -7$	$D\epsilon = 13$	$E\beta = 1$	$A\delta = 2$	$B\alpha = -4$	5
4	$D\delta = 1$	$E\alpha = 6$	$A\gamma = 1$	$B\epsilon = -2$	$C\beta = -3$	3
5	$E\epsilon = -3$	$A\beta = 5$	$B\delta = -5$	$C\alpha = 4$	$D\gamma = 6$	7
$y_{...l}$	-18	18	-4	5	9	$10 = y_{....}$

Table 4-17 Analysis of Variance for the Dynamite Formulation Problem

Source of Variation	Sum of Squares	Degrees of Freedom	Mean Square	F_0
Formulations	330.00	4	82.50	10.00[a]
Batches of raw material	68.00	4	17.00	
Operators	150.00	4	37.50	
Test assemblies	62.00	4	15.50	
Error	66.00	8	8.25	
Total	676.00	24		

[a]Significant at 1 percent.

that removing the variability due to test assemblies has decreased the experimental error. However, in decreasing the experimental error we have also reduced the error degrees of freedom from 12 (in the Latin square design, Example 4-3) to 8. Thus our estimate of error has fewer degrees of freedom and the test may be less sensitive. ■

The concept of orthogonal pairs of Latin squares forming a Graeco-Latin square can be extended somewhat. A $p \times p$ *hypersquare* is a design in which three or more orthogonal $p \times p$ Latin squares are superimposed. In general, up to $p + 1$ factors could be studied if a complete set of $p - 1$ orthogonal Latin squares is available. Such a design would utilize all $(p + 1)(p - 1) = p^2 - 1$ degrees of freedom, so that an independent estimate of the error variance is necessary. Of course, there must be no interactions between the factors when using hypersquares.

4-4 PROBLEMS

4-1. A chemist wishes to test the effect of four chemical agents on the strength of a particular type of cloth. Because there might be variability from one bolt to another, the chemist decides to use a randomized block design with the bolts of cloth considered as blocks. He selects five bolts and applies all four chemicals in random order to each bolt. The resulting tensile strengths are shown below. Analyze the data and draw appropriate conclusions.

	Bolt				
Chemical	1	2	3	4	5
1	73	68	74	71	67
2	73	67	75	72	70
3	75	68	78	73	68
4	73	71	75	75	69

4-2. Three different washing solutions are being compared to study their effectiveness in retarding bacteria growth in five-gallon milk containers. The analysis is done in a laboratory, and only three trials can be run on any day. As days could represent a potential source of variability, the experimenter decides to use a randomized block design. Observations are taken for four days, and the data are shown below. Analyze the data and draw conclusions.

	Days			
Solution	1	2	3	4
1	13	22	18	39
2	16	24	17	44
3	5	4	1	22

4-3. Analyze the data in Problem 4-2 using the general regression significance test.

4-4. Assuming that chemical types and bolts are fixed, estimate the model parameters τ_i and β_j in Problem 4-1.

4-5. Draw an operating characteristic curve for the design in Problem 4-2. Does the test seem to be sensitive to small differences in the treatment effects?

4-6. Suppose that the observation for chemical type 2 and bolt 3 is missing in Problem 4-1. Analyze the problem by estimating the missing value. Perform the exact analysis and compare the results.

4-7. *Two Missing Values in a Randomized Block.* Suppose that in Problem 4-1 the observations for chemical type 2 and bolt 3 and chemical type 4 and bolt 4 are missing.
   (a) Analyze the design by iteratively estimating the missing values, as described in Section 4-1.2.
   (b) Differentiate SS_E with respect to the two missing values, equate the results to zero, and solve for estimates of the missing values. Analyze the design using these two estimates of the missing values.
   (c) Derive general formulas for estimating two missing values when the observations are in *different* blocks.

(d) Derive general formulas for estimating two different values when the observations are in the *same* block.

4-8. An industrial engineer is conducting an experiment on eye focus time. He is interested in the effect of the distance of the object from the eye on the focus time. Four different distances are of interest. He has five subjects available for the experiment. Because there may be differences between individuals, he decides to conduct the experiment in a randomized block design. The data obtained are shown below. Analyze the data and draw appropriate conclusions.

	Subject				
Distance (ft)	1	2	3	4	5
4	10	6	6	6	6
6	7	6	6	1	6
8	5	3	3	2	5
10	6	4	4	2	3

4-9. The effect of five different catalysts (A, B, C, D, E) on the reaction time of a chemical process is being studied. Each batch of new material is only large enough to permit five runs to be made. Furthermore, each run requires approximately $1\frac{1}{2}$ hours, so that only five runs can be made in one day. The experimenter decides to run the experiment as a Latin square, so that day and batch effects may be systematically controlled. He obtains the data shown below. Analyze this data and draw conclusions.

	Day				
Batch	1	2	3	4	5
1	$A = 8$	$B = 7$	$D = 1$	$C = 7$	$E = 3$
2	$C = 11$	$E = 2$	$A = 7$	$D = 3$	$B = 8$
3	$B = 4$	$A = 9$	$C = 10$	$E = 1$	$D = 5$
4	$D = 6$	$C = 8$	$E = 6$	$B = 6$	$A = 10$
5	$E = 4$	$D = 2$	$B = 3$	$A = 8$	$C = 8$

4-10. An industrial engineer is investigating the effect of four assembly methods (A, B, C, D) on the assembly time for a color television component. Four operators are selected for the study. Furthermore, the engineer knows that each assembly method produces such fatigue that the time required for the last assembly may be greater than the time required for the first, regardless of method. That is, a trend develops in the required assembly time. To account for this source of variability, the engineer uses the Latin square design shown on the top of page 97. Analyze the data and draw appropriate conclusions.

Order of	Operator			
Assembly	1	2	3	4
1	$C = 10$	$D = 14$	$A = 7$	$B = 8$
2	$B = 7$	$C = 18$	$D = 11$	$A = 8$
3	$A = 5$	$B = 10$	$C = 11$	$D = 9$
4	$D = 10$	$A = 10$	$B = 12$	$C = 14$

4-11. Suppose that, in Problem 4-9, the observation from batch 3 on day 4 is missing. Estimate the missing value from Equation (4-21) and perform the analysis using the value.

4-12. Consider a $p \times p$ Latin square with rows (α_i), columns (β_k), and treatments (τ_j) fixed. Obtain least squares estimates of the model parameters α_i, β_k, and τ_j.

4-13. Derive the missing value formula (4-21) for the Latin square design.

4-14. *Designs Involving Several Latin Squares.* [Cochran and Cox (1957), John (1971)]. The $p \times p$ Latin square contains only p observations for each treatment. To obtain more replications the experimenter may use several squares, say n. It is immaterial whether the squares used are the same or different. The appropriate model is

$$y_{ijkh} = \mu + \rho_h + \alpha_{i(h)} + \tau_j + \beta_{k(h)} + (\tau\rho)_{jh} + \epsilon_{ijkh} \begin{cases} i = 1, 2, \ldots, p \\ j = 1, 2, \ldots, p \\ k = 1, 2, \ldots, p \\ h = 1, 2, \ldots, n \end{cases}$$

where y_{ijkh} is the observation on treatment j in row i and column k of the hth square. Note that $\alpha_{i(h)}$ and $\beta_{k(h)}$ are the row and column effects in the hth square, ρ_h is the effect of the hth square, and $(\tau\rho)_{jh}$ is the interaction between treatments and squares.

(a) Set up the normal equations for this model, and solve for estimates of the model parameters. Assume that appropriate side conditions on the parameters are $\Sigma_h \hat{\rho}_h = 0$, $\Sigma_i \hat{\alpha}_{i(h)} = 0$ and $\Sigma_k \hat{\beta}_{k(h)} = 0$ for each h, $\Sigma_j \hat{\tau}_j = 0$, $\Sigma_j (\hat{\tau\rho})_{jh} = 0$ for each h, and $\Sigma_h (\hat{\tau\rho})_{jh} = 0$ for each j.

(b) Write down the analysis of variance table for this design.

4-15. Discuss how the operating characteristics curves in the Appendix may be used with the Latin square design.

4-16. Suppose that in Problem 4-9 the data taken on day 5 were incorrectly analyzed and had to be discarded. Develop an appropriate analysis for the remaining data.

4-17. The yield of a chemical process was measured using five batches of raw material, five acid concentrations, five standing times (A, B, C, D, E), and five catalyst concentrations (α, β, γ, δ, ϵ). The Graeco-Latin square shown below was used. Analyze the data and draw conclusions.

	Acid Concentration				
Batch	1	2	3	4	5
1	$A\alpha = 26$	$B\beta = 16$	$C\gamma = 19$	$D\delta = 16$	$E\epsilon = 13$
2	$B\gamma = 18$	$C\delta = 21$	$D\epsilon = 18$	$E\alpha = 11$	$A\beta = 21$
3	$C\epsilon = 20$	$D\alpha = 12$	$E\beta = 16$	$A\gamma = 25$	$B\delta = 13$
4	$D\beta = 15$	$E\gamma = 15$	$A\delta = 22$	$B\epsilon = 14$	$C\alpha = 17$
5	$E\delta = 10$	$A\epsilon = 24$	$B\alpha = 17$	$C\beta = 17$	$D\gamma = 14$

4-18. Suppose that in Problem 4-10 the engineer suspects that the workplaces used by the four operators may represent an additional source of variation. A fourth factor, workplace (α, β, γ, δ), may be introduced and another experiment conducted, yielding the Graeco-Latin square below. Analyze the data and draw conclusions.

Order of	Operator			
Assembly	1	2	3	4
1	$C\beta = 11$	$B\gamma = 10$	$D\delta = 14$	$A\alpha = 8$
2	$B\alpha = 8$	$C\delta = 12$	$A\gamma = 10$	$D\beta = 12$
3	$A\delta = 9$	$D\alpha = 11$	$B\beta = 7$	$C\gamma = 15$
4	$D\gamma = 9$	$A\beta = 8$	$C\alpha = 18$	$B\delta = 6$

4-19. Construct a 5×5 hypersquare for studying the effects of five factors. Exhibit the analysis of variance table for this design.

4-20. Consider the data in Problems 4-10 and 4-18. Suppressing the Greek letters in 4-18, analyze the data using the method developed in Problem 4-14.

4-21. Consider the randomized block design with one missing value in Table 4-7. Analyze this data by using the exact analysis of the missing value problem discussed in Section 4-1.3. Compare your results to the approximate analysis of this data in Table 4-8.

Chapter Five
Incomplete Block Designs

5-1 INTRODUCTION

In certain experiments using randomized block designs we may not be able to run all the treatment combinations in each block. Situations like this usually occur due to shortages of experimental apparatus or facilities, or the physical size of the block. For example, in the hardness testing experiment (Example 4-1), suppose that because of their size, each specimen could be used only for testing three tips. Therefore each tip cannot be tested on each specimen. For this type of problem it is possible to use randomized block designs in which every treatment is not present in every block. These designs are known as *randomized incomplete block designs*, and are the subject of this chapter.

5-2 BALANCED INCOMPLETE BLOCK DESIGNS

When all treatment comparisons are equally important, the treatment combinations used in each block should be selected in a balanced manner; that is, so that any pair of treatments occur together the same number of times as any other pair. Thus, a *balanced incomplete block design* is an incomplete block design in which any two treatments appear together an equal number of times. Suppose that there are a treatments, and each block can hold exactly k ($k < a$) treatments. A balanced incomplete block design may be constructed by taking $\binom{a}{k}$ blocks and assigning a different combination of treatments to each block. Frequently, however, balance can be obtained with fewer than $\binom{a}{k}$ blocks. Tables of balanced incomplete block designs are given in Fisher and Yates (1953), Davies (1956), and Cochran and Cox (1957).

As an example, a chemical engineer thinks that the time of reaction for a chemical process is a function of the type of catalyst employed. Four catalysts are currently being investigated. The experimental procedure consists of selecting a batch of raw material, loading the pilot plant, applying each catalyst in a separate run of the pilot plant, and observing the reaction time. Since variations in the batches of raw material may affect the performance of the catalysts, the engineer decided to use batches of raw material as blocks. However, each batch is only large enough to permit three catalysts to be run. Therefore, a randomized incomplete block design must be used. The balanced incomplete block design for this experiment, along with the observations recorded, is shown in Table 5-1. The order in which the catalysts were run in each block was randomized.

Table 5-1   Balanced Incomplete Block Design for Catalyst Experiment

Treatment	Block (Batch of Raw Material)				
(Catalyst)	1	2	3	4	$y_{i.}$
1	73	74	—	71	218
2	—	75	67	72	214
3	73	75	68	—	216
4	75	—	72	75	222
$y_{.j}$	221	224	207	218	$870 = y_{..}$

5-2.1   Statistical Analysis

As usual, we assume that there are a treatments and b blocks. In addition, we assume that each block contains k treatments, that each treatment occurs r times in the design (or is replicated r times), and that there are $N = ar = bk$ total observations. Furthermore, the number of times each pair of treatments appears in the same block is
over all the blocks

$$\lambda = \frac{r(k-1)}{a-1}$$

If $a = b$, the design is said to be *symmetric*.

The parameter λ must be an integer. To derive the relationship for λ, consider any treatment, say treatment 1. Since treatment 1 appears in r blocks and there are $k - 1$

other treatments in each of those blocks, there are $r(k - 1)$ observations in a block containing treatment 1. These $r(k - 1)$ observations have to also represent the remaining $a - 1$ treatments λ times. Therefore, $\lambda(a - 1) = r(k - 1)$.

The statistical model is

$$y_{ij} = \mu + \tau_i + \beta_j + \epsilon_{ij} \tag{5-1}$$

where y_{ij} is the ith observation in the jth block, μ is the overall mean, τ_i is the effect of the ith treatment, β_j is the effect of the jth block, and ϵ_{ij} is the NID$(0, \sigma^2)$ random error component. The total variation in the data is expressed by the total corrected sum of squares

$$SS_T = \sum_i \sum_j y_{ij}^2 - \frac{y_{..}^2}{N} \tag{5-2}$$

Total variability may be partitioned into

$$SS_T = SS_{\text{Treatments(adjusted)}} + SS_{\text{Blocks}} + SS_E$$

adjusted for blocks

where the sum of squares for treatments is adjusted to separate the treatment and the block effects. This adjustment is necessary because each treatment is represented in a different set of r blocks. Thus differences between unadjusted treatment totals y_1, y_2, ..., y_a are also affected by differences between blocks.

The block sum of squares is

$$SS_{\text{Blocks}} = \sum_{j=1}^{b} \frac{y_{.j}^2}{k} - \frac{y_{..}^2}{N} \tag{5-3}$$

where $y_{.j}$ is the total in the jth block. SS_{Blocks} has $b - 1$ degrees of freedom. The adjusted treatment sum of squares is

$$SS_{\text{Treatments(adjusted)}} = \frac{k \sum_{i=1}^{a} Q_i^2}{\lambda a} \tag{5-4}$$

where Q_i is the adjusted total for the ith treatment, computed as

$$Q_i = y_{i.} - \frac{1}{k} \sum_{j=1}^{b} n_{ij} y_{.j}, \quad i = 1, 2, \ldots, a \tag{5-5}$$

with $n_{ij} = 1$ if treatment i appears in block j and $n_{ij} = 0$ otherwise. Thus $(1/k) \cdot \sum_{j=1}^{b} n_{ij} y_{.j}$ is the average of the block totals containing treatment i. The adjusted treatment totals will always sum to zero. $SS_{\text{Treatments(adjusted)}}$ has $a - 1$ degrees of freedom. The error sum of squares is computed by subtraction as

$$SS_E = SS_T - SS_{\text{Treatments(adjusted)}} - SS_{\text{Blocks}} \tag{5-6}$$

and has $N - a - b + 1$ degrees of freedom.

The appropriate statistic for testing the equality of the treatment effects is

$$F_0 = \frac{MS_{Treatments\,(adjusted)}}{MS_E}$$

The analysis of variance is summarized in Table 5-2

■**EXAMPLE 5-1.**   Consider the data in Table 5-1 for the catalyst experiment. This is a balanced incomplete block design with $a = 4$, $b = 4$, $k = 3$, $r = 3$, $\lambda = 2$, and $N = 12$. The analysis of this data is as follows. The total sum of squares is

$$SS_T = \sum_i \sum_j y_{ij}^2 - \frac{y_{..}^2}{12}$$

$$= 63,156 - \frac{(870)^2}{12} = 81.00$$

The block sum of squares is found from Equation 5-3 as

$$SS_{Blocks} = \sum_{j=1}^{4} \frac{y_{.j}^2}{3} - \frac{y_{..}^2}{12}$$

$$= \frac{(221)^2 + (207)^2 + (224)^2 + (218)^2}{3} - \frac{(870)^2}{12} = 55.00$$

To compute the treatment sum of squares adjusted for blocks, we first determine the adjusted treatment totals using Equation 5-5 as

$$Q_1 = (218) - \tfrac{1}{3}(221 + 224 + 218) = -9/3$$
$$Q_2 = (214) - \tfrac{1}{3}(207 + 224 + 218) = -7/3$$
$$Q_3 = (216) - \tfrac{1}{3}(221 + 207 + 224) = -4/3$$
$$Q_4 = (222) - \tfrac{1}{3}(221 + 207 + 218) = 20/3$$

The adjusted sum of squares for treatments is computed from (5-4) as

$$SS_{Treatments\,(adjusted)} = \frac{k \sum_{i=1}^{4} Q_i^2}{\lambda a}$$

$$= \frac{3\,[(-9/3)^2 + (-7/3)^2 + (-4/3)^2 + (20/3)^2]}{(2)\,(4)} = 22.75$$

The error sum of squares is obtained by subtraction as

$$SS_E = SS_T - SS_{Treatments\,(adjusted)} - SS_{Blocks}$$

$$= 81.00 - 22.75 - 55.00 = 3.25$$

Table 5-2 Analysis of Variance for the Balanced Incomplete Block Design

Source of Variation	Sum of Squares	Degrees of Freedom	Mean Square	F_0
Treatments (adjusted)	$\dfrac{k \sum Q_i^2}{\lambda a}$	$a - 1$	$\dfrac{SS_{\text{Treatments(adj)}}}{a - 1}$	$F_0 = \dfrac{MS_{\text{Treatments(adj)}}}{MS_E}$
Blocks	$\sum \dfrac{y_{.j}^2}{k} - \dfrac{y_{..}^2}{N}$	$b - 1$	$\dfrac{SS_{\text{Blocks}}}{b - 1}$	
Error	SS_E (by subtraction)	$N - a - b + 1$	$\dfrac{SS_E}{N - a - b + 1}$	
Total	$\sum\sum y_{ij}^2 - \dfrac{y_{..}^2}{N}$	$N - 1$		

Table 5-3 Analysis of Variance for Example 5-1

Source of Variation	Sum of Squares	Degrees of Freedom	Mean Square	F_0
Treatments (adjusted for blocks)	22.75	3	7.58	11.66
Blocks	55.00	3	—	
Error	3.25	5	0.65	
Total	81.00	11		

The analysis of variance is shown in Table 5-3. Since $F_0 > F_{.05, 3, 5} = 5.41$ we conclude that the catalyst employed has a significant effect on the time of reaction. ■

If the factor under study is fixed, then tests on individual treatment means may be of interest. If orthogonal contrasts are employed, the contrasts must be made on the *adjusted* treatment totals (the $\{Q_i\}$) rather than the $\{y_{i.}\}$. The contrast sum of squares is

$$SS_C = \frac{k \left(\sum_{i=1}^{a} c_i Q_i \right)^2}{\lambda a \sum_{i=1}^{a} c_i^2}$$

where $\{c_i\}$ are the contrast coefficients. Duncan's multiple range test may be used to compare all pairs of adjusted treatment means, which we will show in Section 5-2.2 are estimated by $\hat{\tau}_i = kQ_i/(\lambda a)$. The appropriate standard error is

$$S = \sqrt{\frac{kMS_E}{\lambda a}} \tag{5-7}$$

As an example, suppose we apply Duncan's multiple range test to the data in Example 5-1. The adjusted treatment means, in ascending order, are

$$\hat{\tau}_1 = kQ_1/(\lambda a) = (3)(-9/3)/(2)(4) = -9/8$$

$$\hat{\tau}_2 = kQ_2/(\lambda a) = (3)(-7/3)/(2)(4) = -7/8$$

$$\hat{\tau}_3 = kQ_3/(\lambda a) = (3)(-4/3)/(2)(4) = -4/8$$

$$\hat{\tau}_4 = kQ_4/(\lambda a) = (3)(20/3)/(2)(4) = 20/8$$

From Appendix Table VII with $\alpha = .05$ and five error degrees of freedom, we obtain $r_{.05}(2, 5) = 3.64$, $r_{.05}(3, 5) = 3.74$, and $r_{.05}(4, 5) = 3.79$. Substituting $MS_E = 0.65$ from Table 5-3 into (5-7) yields

$$S = \sqrt{\frac{kMS_E}{\lambda a}} = \sqrt{\frac{3(0.65)}{2(4)}} = 0.49$$

Thus

$$R_2 = r_{.05} (2, 5) S = (3.64)(0.49) = 1.80$$

$$R_3 = r_{.05} (3, 5) S = (3.74)(0.49) = 1.83$$

$$R_4 = r_{.05} (4, 5) S = (3.79)(0.49) = 1.86$$

and the comparisons yield

$$4 \text{ vs. } 1 = 20/8 - (-9/8) = 29/8 > 1.86 \ (R_4)$$

$$4 \text{ vs. } 2 = 20/8 - (-7/8) = 27/8 > 1.83 \ (R_3)$$

$$4 \text{ vs. } 3 = 20/8 - (-4/8) = 24/8 > 1.80 \ (R_2)$$

$$3 \text{ vs. } 1 = -4/8 - (-9/8) = 5/8 < 1.83 \ (R_3)$$

$$3 \text{ vs. } 2 = -4/8 - (-7/8) = 3/8 < 1.80 \ (R_2)$$

$$2 \text{ vs. } 1 = -7/8 - (-9/8) = 2/8 < 1.80 \ (R_2)$$

The test results are summarized in Figure 5-1. Catalyst 4 differs from the other three, while all other pairs of means do not differ.

$\hat{\tau}_1$	$\hat{\tau}_2$	$\hat{\tau}_3$	$\hat{\tau}_4$
$-9/8$	$-7/8$	$-4/8$	$20/8$

Figure 5-1.   Results of Duncan's multiple range test.

In the analysis we have elected to partition the total sum of squares into an adjusted sum of squares for treatments, an unadjusted sum of squares for blocks, and an error sum of squares. Sometimes we would like to assess the block effects. To do this, we require an alternate partitioning of SS_T, that is,

$$SS_T = SS_{\text{Treatments}} + SS_{\text{Blocks(adjusted)}} + SS_E$$

Here $SS_{\text{Treatments}}$ is unadjusted. If the design is symmetric, that is, if $a = b$, then a simple formula may be obtained for $SS_{\text{Blocks (adjusted)}}$. The adjusted block totals are

$$Q'_j = y_{.j} - \frac{1}{r} \sum_{i=1}^{a} n_{ij} y_{i.}, \quad j = 1, 2, \ldots, b \tag{5-8}$$

and

$$SS_{\text{Blocks (adjusted)}} = \frac{r \sum_{j=1}^{b} (Q'_j)^2}{\lambda b} \tag{5-9}$$

The balanced incomplete block design in Example 5-1 is symmetric, since $a = b = 4$. Therefore,

$$Q_1' = (221) - \frac{1}{3}(218 + 216 + 222) = 7/3$$

$$Q_2' = (224) - \frac{1}{3}(218 + 214 + 216) = 24/3$$

$$Q_3' = (207) - \frac{1}{3}(214 + 216 + 222) = -31/3$$

$$Q_4' = (218) - \frac{1}{3}(218 + 214 + 222) = 0$$

and

$$SS_{\text{Blocks (adjusted)}} = \frac{3[(7/3)^2 + (-31/3)^2 + (24/3)^2 + (0)^2]}{(2)(4)} = 66.08$$

Also,

$$SS_{\text{Treatments}} = \frac{(218)^2 + (214)^2 + (216)^2 + (222)^2}{3} - \frac{(870)^2}{12} = 11.67$$

A summary of the analysis of variance for the symmetric balanced incomplete block design is given in Table 5-4. Notice that the sums of squares associated with the mean squares in Table 5-4 do not add to the total sum of squares, that is,

$$SS_T \neq SS_{\text{Treatments (adjusted)}} + SS_{\text{Blocks (adjusted)}} + SS_E$$

This is a consequence of the nonorthogonality of treatments and blocks.

Table 5-4 Analysis of Variance Table for Example 5-1, Including Both Treatments and Blocks

Source of Variation	Sum of Squares	Degrees of Freedom	Mean Square	F_0
Treatments (adjusted)	22.75	3	7.58	11.66
Treatments (unadjusted)	11.67	3		
Blocks (unadjusted)	55.00	3		
Blocks (adjusted)	66.08	3	22.03	33.90
Error	3.25	5	0.65	
Total	81.00	11		

5-2.2 Least Squares Estimation of the Parameters

Consider estimating the treatment effects for the balanced incomplete block model. The least squares normal equations are

$$\mu: \quad N\hat{\mu} + r\sum_{i=1}^{a}\hat{\tau}_i + k\sum_{j=1}^{b}\hat{\beta}_j = y_{..}$$

$$\tau_i: \quad r\hat{\mu} + r\hat{\tau}_i + \sum_{j=1}^{b} n_{ij}\hat{\beta}_j = y_{i.}, \qquad i = 1, 2, \ldots, a \qquad (5\text{-}10)$$

$$\beta_j: \quad k\hat{\mu} + \sum_{i=1}^{a} n_{ij}\hat{\tau}_i + k\hat{\beta}_j = y_{.j}, \qquad j = 1, 2, \ldots, b$$

Imposing $\Sigma\,\hat{\tau}_i = \Sigma\,\hat{\beta}_j = 0$, we find that $\hat{\mu} = \bar{y}_{...}$. Furthermore, using the equations for $\{\beta_j\}$ to eliminate the block effect from the equations for $\{\tau_i\}$ we obtain

note $\sum n_{ij} = r$ $\sum_c n_{ij} = k$

$$rk\hat{\tau}_i - r\hat{\tau}_i - \sum_{j=1}^{b}\sum_{\substack{p=1 \\ p \neq i}}^{a} n_{ij}\,n_{pj}\,\hat{\tau}_p = ky_{i.} - \sum_{j=1}^{b} n_{ij}\,y_{.j} \qquad (5\text{-}11)$$

Note that the right-hand side of (5-11) is kQ_i, where Q_i is the ith adjusted treatment total (see Equation 5-4). Now, since $\Sigma_{j=1}^{b}\,n_{ij}\,n_{pj} = \lambda$ if $p \neq i$ and $n_{pj}^2 = n_{pj}$ (because $n_{pj} = 0$ or 1) we may rewrite (5-11) as

$$r(k-1)\hat{\tau}_i - \lambda\sum_{\substack{p=1 \\ p \neq i}}^{a}\hat{\tau}_p = kQ_i, \qquad i = 1, 2, \ldots, a \qquad (5\text{-}12)$$

Finally, note that the constraint $\Sigma_{i=1}^{a}\,\hat{\tau}_i = 0$ implies that $\Sigma_{\substack{p=1 \\ p \neq i}}^{a}\,\hat{\tau}_p = -\hat{\tau}_i$ and recall that $r(k-1) = \lambda(a-1)$ to obtain

$$\lambda a\hat{\tau}_i = kQ_i, \qquad i = 1, 2, \ldots, a \qquad (5\text{-}13)$$

Therefore, the least squares estimators of the treatment effects in the balanced incomplete block model are

$$\hat{\tau}_i = \frac{kQ_i}{\lambda a}, \qquad i = 1, 2, \ldots, a \qquad (5\text{-}14)$$

As an illustration, consider the balanced incomplete block design in Example 5-1. Since $Q_1 = -9/3$, $Q_2 = -7/3$, $Q_3 = -4/3$, and $Q_4 = 20/3$, we obtain

$$\hat{\tau}_1 = \frac{3(-9/3)}{(2)(4)} = -9/8, \qquad \hat{\tau}_2 = \frac{3(-7/3)}{(2)(4)} = -7/8$$

$$\hat{\tau}_3 = \frac{3(-4/3)}{(2)(4)} = -4/8, \qquad \hat{\tau}_4 = \frac{3(20/3)}{(2)(4)} = 20/8$$

as we found in Section 5-2.1.

5-3   RECOVERY OF INTERBLOCK INFORMATION IN THE BALANCED INCOMPLETE BLOCK DESIGN

The analysis of the balanced incomplete block design given in Section 5-2 is usually called the *intrablock* analysis, because block differences are eliminated and all contrasts in the treatment effects can be expressed as comparisons between observations in the same block.   This analysis is appropriate regardless of whether the blocks are fixed or random.   Yates (1940) noted that if the block effects are uncorrelated random variables with zero means and variance σ_β^2, then one may obtain additional information about the treatment effects τ_i.   Yates called the method of obtaining this additional information the *interblock* analysis.

Consider the block totals $y_{.j}$ as a collection of b observations.   The model for these observations is [following John (1971)] :

$$y_{.j} = k\mu + \sum_{i=1}^{a} n_{ij}\tau_i + \left(k\beta_j + \sum_{i=1}^{a} \epsilon_{ij}\right) \tag{5-15}$$

where the term in parentheses may be regarded as error.   The interblock estimators of μ and τ_i are found by minimizing the least squares function

$$L = \sum_{j=1}^{b}\left(y_{.j} - k\mu - \sum_{i=1}^{a} n_{ij}\tau_i\right)^2$$

This yields the following least squares normal equations

$$\mu: \qquad N\tilde{\mu} + r\sum_{i=1}^{a}\tilde{\tau}_i = y_{..}$$

$$\tau_i:\ kr\tilde{\mu} + r\tilde{\tau}_i + \lambda\sum_{\substack{p=1\\p\neq i}}^{a}\tilde{\tau}_p = \sum_{j=1}^{b} n_{ij}y_{.j}, \qquad i = 1, 2, \ldots, a \tag{5-16}$$

where $\tilde{\mu}$ and $\tilde{\tau}_i$ denote the *interblock* estimators.   Imposing the constraint $\Sigma_{i=1}^{a}\ \tilde{\tau}_i = 0$ we obtain the solution to (5-16) as

$$\tilde{\mu} = \bar{y}_{..} \tag{5-17}$$

$$\tilde{\tau}_i = \frac{\displaystyle\sum_{j=1}^{b} n_{ij}y_{.j} - kr\bar{y}_{..}}{r - \lambda}, \qquad i = 1, 2, \ldots, a \tag{5-18}$$

It is possible to show that the interblock estimators $\{\tilde{\tau}_i\}$ and the intrablock estimators $\{\hat{\tau}_i\}$ are uncorrelated.

The interblock estimators $\{\tilde{\tau}_i\}$ can differ from the intrablock estimators $\{\hat{\tau}_i\}$. For example, the interblock estimators for the balanced incomplete block design in Example 5-1 are computed as follows:

$$\tilde{\tau}_1 = \frac{663 - (3)\,(3)\,(72.50)}{3 - 2} = 10.50$$

$$\tilde{\tau}_2 = \frac{649 - (3)\,(3)\,(72.50)}{3 - 2} = -3.50$$

$$\tilde{\tau}_3 = \frac{652 - (3)\,(3)\,(72.50)}{3 - 2} = -0.50$$

$$\tilde{\tau}_4 = \frac{646 - (3)\,(3)\,(72.50)}{3 - 2} = -6.50$$

Note that the values of $\sum_{j=1}^{b} n_{ij}\, y_{.j}$ were used previously on page 102 in computing the adjusted treatment totals in the intrablock analysis.

Now suppose we wish to combine the interblock and intrablock estimators to obtain a single unbiased minimum variance estimate of each τ_i. It is possible to show that both $\hat{\tau}_i$ and $\tilde{\tau}_i$ are unbiased, and also that

$$V(\hat{\tau}_i) = \frac{k\,(a - 1)}{\lambda a^2}\, \sigma^2 \quad \text{(intrablock)}$$

and

$$V(\tilde{\tau}_i) = \frac{k\,(a - 1)}{a\,(r - \lambda)}\, (\sigma^2 + k\sigma_\beta^2) \quad \text{(interblock)}$$

We will use a linear combination of the two estimators, say

$$\tau_i^* = \alpha_1\, \hat{\tau}_i + \alpha_2\, \tilde{\tau}_i \tag{5-19}$$

to estimate τ_i. For this estimation method, the minimum variance unbiased combined estimator τ_i^* should have weights $\alpha_1 = u_1/(u_1 + u_2)$ and $\alpha_2 = u_2/(u_1 + u_2)$, where $u_1 = 1/V(\hat{\tau}_i)$ and $u_2 = 1/V(\tilde{\tau}_i)$. Thus the optimal weights are inversely proportional to the variances of $\hat{\tau}_i$ and $\tilde{\tau}_i$. This implies that the best combined estimator is

$$\tau_i^* = \frac{\hat{\tau}_i\, \dfrac{k\,(a-1)}{a\,(r-\lambda)}\,(\sigma^2 + k\sigma_\beta^2) + \tilde{\tau}_i\, \dfrac{k\,(a-1)}{\lambda a^2}\, \sigma^2}{\dfrac{k\,(a-1)}{\lambda a^2}\, \sigma^2 + \dfrac{k\,(a-1)}{a\,(r-\lambda)}\,(\sigma^2 + k\sigma_\beta^2)}, \quad i = 1, 2, \ldots, a$$

which can be simplified to

$$\tau_i^* = \frac{kQ_i(\sigma^2 + k\sigma_\beta^2) + \left(\displaystyle\sum_{j=1}^{b} n_{ij}\, y_{.j} - kr\bar{y}_{..}\right)\sigma^2}{(r - \lambda)\sigma^2 + \lambda a(\sigma^2 + k\sigma_\beta^2)}, \quad i = 1, 2, \ldots, a \tag{5-20}$$

Unfortunately, (5-20) cannot be used to estimate the τ_i because the variances σ^2 and σ_β^2 are unknown. The usual approach is to estimate σ^2 and σ_β^2 from the data and replace these parameters in (5-20) by the estimates. The estimate usually taken for σ^2 is the error mean square from the intrablock analysis of variance, or the *intrablock error*. Thus

$$\hat{\sigma}^2 = MS_E$$

The estimate of σ_β^2 is found from the mean square for blocks adjusted for treatments. In general, for a balanced incomplete block design, this mean square is

$$MS_{\text{Blocks(adjusted)}} = \left(\frac{k \sum\limits_{i=1}^{a} Q_i^2}{\lambda a} + \sum\limits_{j=1}^{b} \frac{y_{.j}^2}{k} - \sum\limits_{i=1}^{a} \frac{y_{i.}^2}{r} \right) \Big/ (b-1) \qquad (5\text{-}21)$$

and its expected value [which is derived in Graybill (1961)] is

$$E[MS_{\text{Blocks(adjusted)}}] = \sigma^2 + \frac{a(r-1)}{b-1} \sigma_\beta^2$$

Thus, if $MS_{\text{Blocks(adjusted)}} > MS_E$ the estimate of $\hat{\sigma}_\beta^2$ is

$$\hat{\sigma}_\beta^2 = \frac{[MS_{\text{Blocks(adjusted)}} - MS_E](b-1)}{a(r-1)} \qquad (5\text{-}22)$$

and if $MS_{\text{Blocks(adjusted)}} \leqslant MS_E$ we set $\hat{\sigma}_\beta^2 = 0$. This results in the combined estimator

$$\tau_i^* = \begin{cases} \dfrac{kQ_i(\hat{\sigma}^2 + k\hat{\sigma}_\beta^2) + \left(\sum\limits_{j=1}^{b} n_{ij}\, y_{.j} - kr\bar{y}_{..} \right)\hat{\sigma}^2}{(r-\lambda)\,\hat{\sigma}^2 + \lambda a(\hat{\sigma}^2 + k\hat{\sigma}_\beta^2)}, & \hat{\sigma}_\beta^2 > 0 \qquad (5\text{-}23a) \\[3em] \dfrac{y_{i.} - (1/a)\, y_{..}}{r}, & \hat{\sigma}_\beta^2 = 0 \qquad (5\text{-}23b) \end{cases}$$

We will compute the combined estimates for the data in Example 5-1. From Table 5-4 we obtain $\hat{\sigma}^2 = MS_E = 0.65$ and $MS_{\text{Blocks(adjusted)}} = 22.03$. [Note that in computing $MS_{\text{Blocks(adjusted)}}$ we make use of the fact that this is a symmetric design. In general, we must use (5-21)]. Since $MS_{\text{Blocks(adjusted)}} > MS_E$, we use (5-22) to estimate σ_β^2 as

$$\hat{\sigma}_\beta^2 = \frac{(22.03 - 0.65)(3)}{4(3-1)} = 8.02$$

Therefore, we may substitute $\hat{\sigma}^2 = 0.65$ and $\hat{\sigma}_\beta^2 = 8.02$ into (5-23a) to obtain the combined estimates listed on page 111. For convenience, the intrablock and interblock estimates are also given. In this example, the combined estimates are close to the intrablock estimates because the variance of the interblock estimates is relatively large.

Parameter	Intrablock Estimate	Interblock Estimate	Combined Estimate
τ_1	-1.12	10.50	-1.09
τ_2	-0.88	-3.50	-0.88
τ_3	-0.50	-0.50	-0.50
τ_4	2.50	-6.50	2.47

5-4 PARTIALLY BALANCED INCOMPLETE BLOCK DESIGNS

While we have concentrated on the balanced case, there are a number of other types of incomplete block designs that occasionally prove useful. Balanced incomplete block designs do not exist for all combinations of parameters that we might wish to employ, since the constraint that λ be an integer can force either the number of blocks, or the block size to be excessively large. For example, if there are eight treatments and the block size is 3, then for λ to be integer the smallest number of replications is $r = 21$. This leads to a design of 56 blocks, which is clearly too large for most practical problems. To reduce the number of blocks required in cases such as this the experimenter can employ *partially balanced* incomplete block designs, in which some pairs of treatments appear together λ_1 times, some pairs appear together λ_2 times, . . . , and the remaining pairs appear together λ_m times. Pairs of treatments that appear together λ_i times are called *ith associates*. The design is then said to have m *associate classes*.

An example of a partially balanced incomplete block design is shown in Table 5-5. Some treatments appear together $\lambda_1 = 2$ times (such as treatments 1 and 2), while the others appear together only $\lambda_2 = 1$ times (such as treatments 4 and 5). Thus the design has two associate classes. We will describe the intrablock analysis for these designs.

can m = 0 ?

Table 5-5 A Partially Balanced Incomplete Block Design with Two Associate Classes

Block	Treatment Combinations		
1	1	2	3
2	3	4	5
3	2	5	6
4	1	2	4
5	3	4	6
6	1	5	6

$a = 6$
$k = 3$
$b = 6$
$r = 3$

A partially balanced incomplete block design with two associate classes is described by the following parameters:

1. There are a treatments arranged in b blocks. Each block contains k runs and each treatment appears in r blocks.
2. Two treatments which are ith associates appear together in λ_i blocks $i = 1, 2$.
3. Each treatment has exactly n_i ith associates, $i = 1, 2$. The number n_i is independent of the treatment chosen. *for any of the ith associates*
4. If two treatments are ith associates, then the number of treatments that are jth associates of one treatment and kth associates of the other treatment is p^i_{jk}, $(i, j, k = 1, 2)$. It is convenient to write the p^i_{jk} as (2×2) matrices with p^i_{jk} the jkth element of the ith matrix.

For the design in Table 5-5 we may readily verify that $a = 6$, $b = 6$, $k = 3$, $r = 3$, $\lambda_1 = 2$, $\lambda_2 = 1$, $n_1 = 1$, $n_2 = 4$,

$$\{p^1_{jk}\} = \begin{bmatrix} 0 & 0 \\ 0 & 4 \end{bmatrix} \quad \text{and} \quad \{p^2_{jk}\} = \begin{bmatrix} 0 & 1 \\ 1 & 2 \end{bmatrix}$$

We will show how to determine the p^1_{jk}. Consider any two treatments that are first associates, say 1 and 2. For treatment 1, the only first associate is 2 and the second associates are 3, 4, 5, and 6. For treatment 2, the only first associate is 1 and the second associates are 3, 4, 5, and 6. Combining this information produces Table 5-6. Counting the number of treatments in the cells of this table we clearly have $\{p^1_{jk}\}$ given above. The elements p^2_{jk} are determined similarly.

Table 5-6 Relationship of Treatments to 1 and 2

	Treatment 2	
Treatment 1	1st Associate	2nd Associate
1st associate		
2nd associate		3, 4, 5, 6

The linear statistical model for the partially balanced incomplete block design with two associate classes is

$$y_{ij} = \mu + \tau_i + \beta_j + \epsilon_{ij} \tag{5-24}$$

where μ is the overall mean, τ_i is the ith treatment effect, β_j is the jth block effect, and ϵ_{ij} is the NID$(0, \sigma^2)$ random error component. We will compute a total sum of squares, a block sum of squares (unadjusted), and a treatment sum of squares (adjusted). As before, we call

$$Q_i = y_i. - \frac{1}{k} \sum_{j=1}^{b} n_{ij} y_{.j}$$

the adjusted total for the ith treatment. We will also define:

$$S_1(Q_i) = \sum_s Q_s, \quad s \text{ and } i \text{ are first associates}$$

$$\Delta = k^{-2} \{ (rk - r + \lambda_1)(rk - r + \lambda_2) + (\lambda_1 - \lambda_2)[r(k-1)(p_{12}^1 - p_{12}^2) + \lambda_2 p_{12}^1 - \lambda_1 p_{12}^2] \}$$

$$c_1 = (k\Delta)^{-1} [\lambda_1(rk - r + \lambda_2) + (\lambda_1 - \lambda_2)(\lambda_2 p_{12}^1 - \lambda_1 p_{12}^2)]$$

$$c_2 = (k\Delta)^{-1} [\lambda_2(rk - r + \lambda_1) + (\lambda_1 - \lambda_2)(\lambda_2 p_{12}^1 - \lambda_1 p_{12}^2)]$$

The estimate of the ith treatment effect is

$$\hat{\tau}_i = \frac{1}{r(k-1)} [(k - c_2)Q_i + (c_1 - c_2)S_1(Q_i)]$$

and the adjusted treatment sum of squares is

$$SS_{\text{Treatments (adjusted)}} = \sum_{i=1}^{a} \hat{\tau}_i Q_i \qquad (5\text{-}25)$$

The analysis of variance is summarized in Table 5-7. To test $H_0: \tau_i = 0$ we use $F_0 = MS_{\text{Treatment (adjusted)}} / MS_E$.

Table 5-7 Analysis of Variance for the Partially Balanced Incomplete Block Design with Two Associate Classes

Source of Variation	Sum of Squares	Degrees of Freedom
Treatments (adjusted)	$\sum_{i=1}^{a} \hat{\tau}_i Q_i$	$a - 1$
Blocks	$\frac{1}{k}\sum_{j=1}^{b} y_{\cdot j}^2 - \frac{y_{\cdot\cdot}^2}{bk}$	$b - 1$
Error	Subtraction	$bk - b - a + 1$
Total	$\sum_i \sum_j y_{ij}^2 - \frac{y_{\cdot\cdot}^2}{bk}$	$bk - 1$

We may show that the variance of any contrast of the form $\hat{\tau}_u - \hat{\tau}_v$ is

$$V(\hat{\tau}_u - \hat{\tau}_v) = \frac{2(k - c_i)\sigma^2}{r(k-1)}$$

where treatments u and v are ith associates ($i = 1, 2$). This indicates that comparisons between treatments are not all estimated with the same precision. This is a consequence of the partial balance of the design.

We have given only the intrablock analysis. For details of the interblock analysis refer to Bose and Shimamoto (1952) or John (1971). The latter reference contains a good discussion of the general theory of incomplete block designs. An extensive table of partially balanced incomplete block designs with two associate classes has been given by Bose, Clatworthy, and Shrikhande (1954).

5-5 YOUDEN SQUARES   *is balanced*

These are "incomplete" Latin square designs, in which the number of columns does not equal the number of rows and treatments. For example, consider the design shown in Table 5-8. Notice that if we append the column (E, A, B, C, D) to this design, the result is a 5×5 Latin square. Most of these designs were developed by W. J. Youden, and consequently are called *Youden squares*.

Table 5-8   Youden Square for Five
Treatments (A, B, C, D, E)

	Column			
Row	1	2	3	4
1	A	B	C	D
2	B	C	D	E
3	C	D	E	A
4	D	E	A	B
5	E	A	B	C

Although a Youden square is always a Latin square from which at least one column (or row or diagonal) is missing, it is not necessarily true that every Latin square with more than one column (or row or diagonal) deleted is a Youden square. Arbitrary removal of more than one column (say) from a Latin square may destroy its balance. In general, a Youden square is a symmetric balanced incomplete block design in which rows correspond to blocks and each treatment occurs exactly once in each column or "position" of the block. Thus, it is possible to construct Youden squares from all symmetric balanced incomplete block designs, as shown by Smith and Hartley (1948). A table of Youden squares is given in Davies (1956), and other types of incomplete Latin squares are discussed by Cochran and Cox (1957, Chapter 13).

The linear model for a Youden square is

a column
$$y_{ijh} = \mu + \alpha_i + \tau_j + \beta_h + \epsilon_{ijh}$$

where μ is the overall mean, α_i is the ith block effect, τ_j is the jth treatment effect, β_h is the hth position effect, and ϵ_{ijh} is the usual NID$(0, \sigma^2)$ error term. Since positions occur exactly once in each block, and once with each treatment, positions are orthogonal to blocks and treatments. The analysis of the Youden square is similar to the analysis of a balanced incomplete block design, except that a sum of squares between the position totals may also be calculated.

■**EXAMPLE 5-2.** An industrial engineer is studying the effect of five illumination levels on the occurrence of defects in an assembly operation. Because time may be a factor in the experiment, he has decided to run the experiment in five blocks, where each block is a day of the week. However, the department in which the experiment is conducted has four work stations, and these stations represent a potential source of variability. The engineer decides to run a Youden square with five rows (days or blocks), four columns (work stations), and five treatments (the illumination levels). The coded data are shown in Table 5-9.

Table 5-9 Coded Data for Example 5-2

Day (Block)	Work Station *←positions/column*				$y_{i..}$	Treatment Totals
	1	2	3	4		
1	$A = 3$	$B = 1$	$C = -2$	$D = 0$	2	$y_{.1.} = 12\ (A)$
2	$B = 0$	$C = 0$	$D = -1$	$E = 7$	6	$y_{.2.} = 2\ (B)$
3	$C = -1$	$D = 0$	$E = 5$	$A = 3$	7	$y_{.3.} = -4\ (C)$
4	$D = -1$	$E = 6$	$A = 4$	$B = 0$	9	$y_{.4.} = -2\ (D)$
5	$E = 5$	$A = 2$	$B = 1$	$C = -1$	7	$y_{.5.} = 23\ (E)$
$y_{..k}$	6	9	7	9	$31 = y_{...}$	

Considering this design as a balanced incomplete block, we find $a = b = 5$, $r = k = 4$, and $\lambda = 3$. Also,

$$SS_T = \sum_i \sum_j \sum_k y_{ijk}^2 - \frac{y_{...}^2}{N} = 183.00 - \frac{(31)^2}{20} = 134.95$$

The adjusted sum of squares for treatments is found as follows:

$$Q_A = 12 - \tfrac{1}{4}(2 + 7 + 9 + 7) = 23/4$$

$$Q_B = 2 - \tfrac{1}{4}(2 + 6 + 9 + 7) = -16/4$$

$$Q_C = -4 - \tfrac{1}{4}(2 + 6 + 7 + 7) = -38/4$$

$$Q_D = -2 - \tfrac{1}{4}(2 + 6 + 7 + 9) = -32/4$$

$$Q_E = 23 - \tfrac{1}{4}(6 + 7 + 9 + 7) = 63/4$$

$$SS_{\text{Treatments (adjusted)}} = \frac{k \sum_{i=1}^{a} Q_i^2}{\lambda a}$$

$$= \frac{4[(23/4)^2 + (-16/4)^2 + (-38/4)^2 + (-32/4)^2 + (63/4)^2]}{(3)(5)} = 120.37$$

adjusted for Blocks, not for positions because each work station contains each treatment

Also,

$$SS_{\text{Days}} = \sum_{i=1}^{b} \frac{y_{i..}^2}{k} - \frac{y_{...}^2}{N} = \frac{(2)^2 + (6)^2 + (7)^2 + (9)^2 + (7)^2}{4} - \frac{(31)^2}{20} = 6.70$$

$$SS_{\text{Stations}} = \sum_{b=1}^{k} \frac{y_{..h}^2}{b} - \frac{y_{...}^2}{N} = \frac{(6)^2 + (9)^2 + (7)^2 + (9)^2}{5} - \frac{(31)^2}{20} = 1.35$$

and

$$SS_E = SS_T - SS_{\text{Treatments (adjusted)}} - SS_{\text{Days}} - SS_{\text{Stations}}$$

$$= 134.95 - 120.37 - 6.70 - 1.35 = 6.53$$

Block or day effects may be assessed by computing the adjusted sum of squares for blocks. This yields

$$Q_1' = 2 - \tfrac{1}{4}(12 + 2 - 4 - 2) = 0/4$$

$$Q_2' = 6 - \tfrac{1}{4}(2 - 3 - 2 + 23) = 5/4$$

$$Q_3' = 7 - \tfrac{1}{4}(12 - 4 - 2 + 23) = -1/4$$

$$Q_4' = 9 - \tfrac{1}{4}(12 + 2 - 2 + 23) = 1/4$$

$$Q_5' = 7 - \tfrac{1}{4}(12 + 2 - 4 + 23) = -5/4$$

$$SS_{\text{Days (adjusted)}} = \frac{r \sum_{j=1}^{b} Q_j'^2}{\lambda b} = \frac{4[(0/4)^2 + (5/4)^2 + (-1/4)^2 + (1/4)^2 + (-5/4)^2]}{(3)(5)} = 0.87$$

The complete analysis of variance is shown in Table 5-10. Illumination levels are significant at 1 percent.   ∎

Table 5-10   Analysis of Variance for Example 5-2

Source of Variation	Sum of Squares	Degrees of Freedom	Mean Square	F_0
Illumination level, adjusted	120.37	4	30.09	36.87[a]
Days, unadjusted	6.70	4	—	
Days, adjusted	(0.87)	(4)	0.22	
Work station	1.35	3	0.45	
Error	6.53	8	0.82	
Total	134.95	19		

[a]Significant at 1 percent.

5-6 LATTICE DESIGNS

Consider a balanced incomplete block design with k^2 treatments arranged in $b = k(k + 1)$ blocks with k runs per block and $r = k + 1$ replicates. Such a design is called a *balanced lattice*. An example is shown in Table 5-11 for $k^2 = 9$ treatments in 12 blocks of three runs each. Notice that the blocks can be grouped into sets such that each set contains a complete replicate. The analysis of variance for the balanced lattice design proceeds like that of a balanced incomplete block design, except that a sum of squares for replicates is computed and removed from the sum of squares for blocks. Replicates will have k degrees of freedom and blocks will have $k^2 - 1$ degrees of freedom.

Table 5-11   A 3 X 3 Balanced Lattice Design

Block	Replicate I			Block	Replicate III		
1	1	2	3	7	1	5	9
2	4	5	6	8	7	2	6
3	7	8	9	9	4	8	3
Block	Replicate II			Block	Replicate IV		
4	1	4	7	10	1	8	6
5	2	5	8	11	4	2	9
6	3	6	9	12	7	5	3

Lattice designs are frequently used in situations where there are a large number of treatment combinations. In order to reduce the size of the design, the experimenter may resort to *partially* balanced lattices. We briefly describe some of these designs here. Two replicates of a design for k^2 treatments in $2k$ blocks of k runs each is called a *simple lattice*. For example, consider the first two replicates of the design in Table 5-11. The partial balance is easily seen, as for example, treatment 2 appears in the same block with treatments 1, 3, 5, and 8, but does not appear at all with treatments, 4, 6, 7, and 9. A lattice design with k^2 treatments in $3k$ blocks grouped into three replicates is called a *triple lattice*. An example would be the first three replicates in Table 5-11. A lattice design for k^2 treatments in $4k$ blocks arranged in four replicates is called a *quadruple lattice*.

There are a number of other types of lattice designs that occasionally prove useful. For example, the *cubic lattice* design can be used for k^3 treatments in k^2 blocks of k runs. A lattice design for $k(k + 1)$ treatments in $k + 1$ blocks of size k is called a *rectangular lattice*. Details of the analysis of lattice designs and tables of plans are given in Cochran and Cox (1957).

5-7 PROBLEMS

5-1. An engineer is studying the mileage performance characteristics of five types of gasoline additives. In the road test he wishes to use cars as blocks; however, because of a time constraint he must use an incomplete block design. He runs the balanced design with five blocks shown below. Analyze the data and draw conclusions.

	Car				
Additive	1	2	3	4	5
1		17	14	13	12
2	14	14		13	10
3	12		13	12	9
4	13	11	11	12	
5	11	12	10		8

5-2. Construct orthogonal contrasts for the data in Problem 5-1. Compute the sum of squares for each contrast.

5-3. Seven different hardwood concentrations are being studied to determine their effect on the strength of the paper produced. However, the pilot plant can only produce three runs each day. As days may differ, the analyst uses the balanced incomplete block design shown below. Analyze the data and draw conclusions.

Hardwood Concentration (%)	Days						
	1	2	3	4	5	6	7
2	114				120		117
4	126	120				119	
6		137	117				134
8	141		129	149			
10		145		150	143		
12			120		118	123	
14				136		130	127

5-4. Analyze the data in Example 5-1 using the general regression significance test.

5-5. Prove that $k \sum_{i=1}^{a} Q_i^2 /(\lambda a)$ is the adjusted sum of squares for treatments in a balanced incomplete block design.

5-6. An experimenter wishes to compare four treatments in blocks of two runs. Find a balanced incomplete block design for this experiment with 6 blocks.

5-7. An experimenter wishes to compare eight treatments in blocks of four runs. Find a balanced incomplete block design with 14 blocks and $\lambda = 3$.

5-8. Perform the interblock analysis for the design in Problem 5-1.

5-9. Perform the interblock analysis for the design in Problem 5-3.

5-10. Verify that a balanced incomplete block design with parameters $a = 8$, $r = 8$, $k = 4$, and $b = 16$ does not exist.

5-11. Show that the variance of the intrablock estimators $\{\hat{\tau}_i\}$ is $k(a - 1)\sigma^2/(\lambda a^2)$.

5-12. Consider the partially balanced incomplete block design below. Perform the intrablock analysis discussed in Section 5-4.

			Block			
Treatment	1	2	3	4	5	6
1	14			10		16
2	10		12	15		
3	20	24			19	
4		16		11	10	
5		13	17			12
6			9		10	8

5-13. Perform the intrablock analysis for the partially balanced incomplete block design shown below.

					Block				
Treatment	1	2	3	4	5	6	7	8	9
1	164	153	160						
2	118							104	112
3	150			135			141		
4		137				125			121
5			103		98			109	
6		120		108	100				
7			100			92	94		
8				123		130		115	
9					119		111		109

5-14. Using the design in Table 5-5, verify the following relationships between the parameters of a partially balanced incomplete block design with two associate classes:

$$n_1 + n_2 = a - 1 \qquad n_1\lambda_1 + n_2\lambda_2 = r(k - 1)$$
$$p_{11}^1 + p_{12}^2 = n_1 \qquad p_{21}^1 + p_{22}^1 = n_2$$
$$p_{11}^1 + p_{12}^1 = n_1 - 1 \qquad p_{21}^2 + p_{22}^2 = n_2 - 1$$
$$n_1 p_{12}^1 = n_2 p_{11}^2 \qquad n_1 p_{22}^1 = n_2 p_{12}^2$$

5-15. Analyze the Youden square design shown below.

	Position			
Blocks	1	2	3	4
1	$A = 2$	$B = 9$	$C = 0$	$D = 14$
2	$B = 6$	$A = 5$	$E = 5$	$C = 3$
3	$C = 1$	$D = 9$	$A = 0$	$E = 7$
4	$D = 8$	$E = 8$	$B = 10$	$A = 4$
5	$E = 7$	$C = 6$	$D = 11$	$B = 10$

5-16. Consider the array

$$
\begin{array}{cccc}
1 & 2 & 3 & 4 \\
5 & 6 & 7 & 8 \\
9 & 10 & 11 & 12 \\
13 & 14 & 15 & 16
\end{array}
$$

Show that the rows and columns of this arrangement form a partially balanced incomplete block design for 16 treatments in eight blocks. Verify that this is also a simple lattice design.

5-17. *Extended Incomplete Block Designs.* Occasionally the block size obeys the relationship $a < k < 2a$. An extended incomplete block design consists of a single replicate of each treatment in each block along with an incomplete block design with $k^* = k - a$. In the balanced case, the incomplete block design will have parameters $k^* = k - a$, $r^* = r - b$, and λ^*. Write out the statistical analysis. [Hint: in the extended incomplete block design we have $\lambda = 2r - b + \lambda^*$.]

Chapter Six
Introduction to Factorial Experiments

6-1 ELEMENTARY DEFINITIONS
AND PRINCIPLES

Many experiments require a study of the effects of two or more factors. It can be shown that, in general, *factorial* experiments are the most efficient designs for this type of analysis. By a factorial experiment we mean that in each complete trial or replication of the experiment all possible combinations of the levels of the factors are investigated. For example, if there are a levels of factor A and b levels of factor B, then each replicate contains all ab treatment combinations. When factors are arranged in a factorial experiment they are often said to be *crossed*.

defn of factorial design

The effect of a factor is defined to be the change in response produced by a change in the level of the factor. This is frequently called a *main effect* because it refers to the primary factors of interest in the experiment. For example, consider the data in Table 6-1. The main effect of factor A could be thought of as the difference between the average response at the first level of A and the average response at the second level of A. Numerically, this is

$$A = \frac{40 + 52}{2} - \frac{20 + 30}{2} = 21$$

That is, increasing factor A from level 1 to level 2 causes an average response increase of 21 units. Similarly, the main effect of B is

$$B = \frac{30 + 52}{2} - \frac{20 + 40}{2} = 11$$

If the factors appear at more than two levels, the above definition must be modified, since there are many ways to express the differences between the average responses. This point is discussed more completely later.

Table 6-1 A Factorial Experiment

		Factor B	
		B_1	B_2
Factor A	A_1	20	30
	A_2	40	52

In some experiments, we may find that the difference in response between the levels of one factor is not the same at all levels of the other factors. When this occurs, there is an *interaction* between the factors. For example, consider the data in Table 6-2. At the first level of factor B the A effect is

$$A = 50 - 20 = 30$$

and at the second level of factor B the A effect is

$$A = 12 - 40 = -28$$

Since the effect of A depends on the level chosen for factor B, we see that there is interaction between A and B.

Table 6-2 A Factorial Experiment with Interaction

		Factor B	
		B_1	B_2
Factor A	A_1	20	40
	A_2	50	12

These ideas may be illustrated graphically. Figure 6-1 plots the response data in Table 6-1 against factor A for both levels of factor B. Note that the B_1 and B_2 lines are approximately parallel, indicating a lack of interaction between factors A and B.

Figure 6-1. A factorial experiment without interaction.

Similarly, Figure 6-2 plots the response data in Table 6-2. Here we see that the B_1 and B_2 lines are not parallel. This indicates an interaction between factors A and B. Graphs such as these are frequently very useful in interpreting significant interactions and in reporting results to nonstatistically trained management. However, they should not be utilized as the sole technique of data analysis, as their interpretation is subjective and their appearance is often misleading.

Note that when an interaction is large, the corresponding main effects have little practical meaning. For the data of Table 6-2, we would estimate the main effect of A to be

$$A = \frac{50 + 12}{2} - \frac{20 + 40}{2} = 1$$

which is very small, and we are tempted to conclude that there is no effect due to A. However, when we examined the effects of A at *different levels of factor B*, we saw that this is not the case. Factor A has an effect, but it *depends on the level of factor B*. That is, knowledge of the AB interaction is more useful than knowledge of the main effect. A significant interaction can tend to *mask* the significance of main effects. This is clearly indicated by the data in Table 6-2. In the presence of significant interaction, the experimenter can examine the levels of one factor, say A, with levels of the other factors fixed to assist in drawing conclusions about the main effect of A.

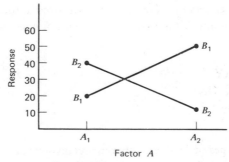

Figure 6-2. A factorial experiment with interaction.

6-2 THE ADVANTAGE OF FACTORIALS

The advantage of factorial experiments can be easily illustrated. Suppose we have two factors A and B, each at two levels. Denote the levels of the factors by A_1, A_2, B_1, and B_2. Information on both factors could be obtained by varying the factors one at a time, as shown in Table 6-3. The effect of changing factor A is given by $A_2 B_1 - A_1 B_1$. Since experimental error is present, it is desirable to take two observations (say) at each treatment combination, and estimate the effects of the factors using average responses. Thus, a total of six observations are required.

Table 6-3 One-Factor-at-a-Time Method

Factor B

		B_1	B_2
Factor A	A_1	$A_1 B_1$	$A_1 B_2$
	A_2	$A_2 B_1$	

If a factorial experiment had been performed, an additional treatment combination, $A_2 B_2$, would have been taken. Now using just *four* observations, two estimates of the A effect can be made; $A_2 B_1 - A_1 B_1$, and $A_2 B_2 - A_1 B_2$. Similarly, two estimates of the B effect can be made. These two estimates of each main effect could be averaged to produce average main effects which are *just as precise* as those from the single factor experiment, but only four total observations are required.

Now suppose interaction is present. If the one factor at a time design found that $A_1 B_2$ and $A_2 B_1$ gave better response that $A_1 B_1$, a logical conclusion would be that $A_2 B_2$ would be even better. However, if interaction is present, this conclusion may be *seriously in error*. For example, refer to the data in Table 6-2.

In summary, note that factorial designs have several advantages. They are more efficient than one factor at a time experiments. Furthermore, a factorial design is necessary when interactions may be present, to avoid misleading conclusions. Finally, factorial designs allow effects of a factor to be estimated at several levels of the other factors, yielding conclusions that are valid over a range of experimental conditions.

6-3 THE TWO-WAY CLASSIFICATION ANALYSIS OF VARIANCE

One of the simplest types of factorial experiments involves only two factors or sets of treatments. There are a levels of factor A and b levels of factor B, and these are arranged in a factorial design; that is, each replicate of the experiment contains all ab treatment combinations. Assume there are n replicates of the experiment, and let y_{ijk} represent the observation taken under the ith level of factor A and the jth level of factor B in the kth replicate. The data will appear as in Table 6-4. The order in which the abn observations are taken is selected at random, so that this design is a completely randomized design.

Table 6-4   Data Arrangement for a Two-Factor Factorial Design

Factor B

		1	2	$\cdots$	b
	1	$y_{111}, y_{112},$ $\cdots, y_{11n}$	$y_{121}, y_{122},$ $\cdots, y_{12n}$		$y_{1b1}, y_{1b2},$ $\cdots, y_{1bn}$
	2	$y_{211}, y_{212},$ $\cdots, y_{21n}$	$y_{221}, y_{222},$ $\cdots, y_{22n}$		$y_{2b1}, y_{2b2},$ $\cdots, y_{2bn}$
Factor A	$\vdots$				
	a	$y_{a11}, y_{a12},$ $\cdots, y_{a1n}$	$y_{a21}, y_{a22},$ $\cdots, y_{a2n}$		$y_{ab1}, y_{ab2},$ $\cdots, y_{abn}$

The observations may be described by the linear statistical model

$$y_{ijk} = \mu + \tau_i + \beta_j + (\tau\beta)_{ij} + \epsilon_{ijk} \quad \begin{cases} i = 1, 2, \ldots, a \\ j = 1, 2, \ldots, b \\ k = 1, 2, \ldots, n \end{cases} \quad (6\text{-}1)$$

where μ is the overall mean effect, τ_i is the true effect of the ith level of factor A, β_j is the true effect of the jth level of factor B, $(\tau\beta)_{ij}$ is the effect of the interaction between τ_i and β_j, and ϵ_{ijk} is a random error component. Both factors are initially assumed fixed, and the treatment effects are defined as deviations from the overall mean, so $\sum_{i=1}^{a} \tau_i = 0$ and $\sum_{j=1}^{b} \beta_j = 0$. Similarly, the interaction effects are fixed and defined so that $\sum_{i=1}^{a} (\tau\beta)_{ij} = \sum_{j=1}^{b} (\tau\beta)_{ij} = 0$. Since there are n replicates of the experiment, there are abn total observations.

We are interested in estimating the parameters in the model, Equation 6-1, and testing various hypotheses about them. The appropriate hypothesis testing procedure will be the analysis of variance. More specifically, as we are considering *two* controllable sources of variation (A and B), the procedure is called the two-way classification analysis of variance.

6-3.1 Statistical Analysis of the Fixed Effects Model

Let $y_{i..}$ denote the total of all observations under the ith level of factor A, $y_{.j.}$ denote the total of all observations under the jth level of factor B, $y_{ij.}$ denote the total of all observations under the ith level of factor A *and* the jth level of factor B, and $y_{...}$ denote the grand total of all the observations. Define $\bar{y}_{i..}, \bar{y}_{.j.}, \bar{y}_{ij.}$ and $\bar{y}_{...}$ as the corresponding treatment, cell and grand averages. Expressed mathematically

$$y_{i..} = \sum_{j=1}^{b} \sum_{k=1}^{n} y_{ijk} \qquad \bar{y}_{i..} = \frac{y_{i..}}{bn} \qquad i = 1, 2, \ldots, a$$

$$y_{.j.} = \sum_{i=1}^{a} \sum_{k=1}^{n} y_{ijk} \qquad \bar{y}_{.j.} = \frac{y_{.j.}}{an} \qquad j = 1, 2, \ldots, b$$

$$\qquad\qquad (6\text{-}3)$$

$$y_{ij.} = \sum_{k=1}^{n} y_{ijk} \qquad \bar{y}_{ij.} = \frac{y_{ij.}}{n} \qquad \begin{aligned} i &= 1, 2, \ldots, a \\ j &= 1, 2, \ldots, b \end{aligned}$$

$$y_{...} = \sum_{i=1}^{a} \sum_{j=1}^{b} \sum_{k=1}^{n} y_{ijk} \qquad \bar{y}_{...} = \frac{y_{...}}{abn}$$

The total corrected sum of squares may be expressed as

$$\sum_{i=1}^{a} \sum_{j=1}^{b} \sum_{k=1}^{n} (y_{ijk} - \bar{y}_{...})^2 = \sum_{i=1}^{a} \sum_{j=1}^{b} \sum_{k=1}^{n} [(\bar{y}_{i..} - \bar{y}_{...}) + (\bar{y}_{.j.} - \bar{y}_{...})$$

$$+ (\bar{y}_{ij.} - \bar{y}_{i..} - \bar{y}_{.j.} + \bar{y}_{...}) + (y_{ijk} - \bar{y}_{ij.})]^2$$

$$= bn \sum_{i=1}^{a} (\bar{y}_{i..} - \bar{y}_{...})^2 + an \sum_{j=1}^{b} (\bar{y}_{.j.} - \bar{y}_{...})^2$$

$$+ n \sum_{i=1}^{a} \sum_{j=1}^{b} (\bar{y}_{ij.} - \bar{y}_{i..} - \bar{y}_{.j.} + \bar{y}_{...})^2 + \sum_{i=1}^{a} \sum_{j=1}^{b} \sum_{k=1}^{n} (y_{ijk} - \bar{y}_{ij.})^2 \quad (6\text{-}4)$$

since the six cross-products on the right-hand side are zero. Notice that the total sum of squares has been partitioned into a sum of squares due to "rows" or factor A (SS_A), a sum of squares due to "columns" or factor B (SS_B), a sum of squares due to the interaction between A and B (SS_{AB}), and a sum of squares due to error (SS_E). From the last component on the right-hand side of (6-4) we see that there must be at least two replicates ($n \geq 2$) to obtain an error sum of squares that is not confounded with interaction.

We may write (6-4) symbolically as

$$SS_T = SS_A + SS_B + SS_{AB} + SS_E \qquad\qquad (6\text{-}5)$$

observations per cell or per treatment combination

The number of degrees of freedom associated with each sum of squares is

Effect	Degrees of Freedom
A	$a - 1$
B	$b - 1$
AB Interaction	$(a - 1)(b - 1)$
Error	$ab(n - 1)$
Total	$abn - 1$

We may justify this allocation of the $abn - 1$ total degrees of freedom to the sums of squares as follows: The main effects A and B have a and b levels, respectively; therefore they have $a - 1$ and $b - 1$ degrees of freedom as shown. The interaction degrees of freedom is simply the number of degrees of freedom for cells (which is $ab - 1$), minus the number of degrees of freedom for the two main effects A and B; that is, $ab - 1 - (a - 1) - (b - 1) = (a - 1)(b - 1)$. Within each of the ab cells there are $n - 1$ degrees of freedom between the n replicates, thus there are $ab(n - 1)$ degrees of freedom for error. Note that the number of degrees of freedom on the right-hand side of (6-5) adds to the total number of degrees of freedom. If we assume ϵ_{ijk} is NID$(0, \sigma^2)$ and apply Cochran's theorem (Theorem 3-1) each sum of squares on the right-hand side of (6-5) when divided by σ^2 is independently distributed as χ^2 with the indicated number of degrees of freedom.

Assuming that factors A and B are fixed, the expected values of the mean squares are

$$E(MS_A) = E\left(\frac{SS_A}{a - 1}\right) = \sigma^2 + \frac{bn \sum\limits_{i=1}^{a} \tau_i^2}{a - 1}$$

$$E(MS_B) = E\left(\frac{SS_B}{b - 1}\right) = \sigma^2 + \frac{an \sum\limits_{j=1}^{b} \beta_j^2}{b - 1}$$

$$E(MS_{AB}) = E\left(\frac{SS_{AB}}{(a - 1)(b - 1)}\right) = \sigma^2 + \frac{n \sum\limits_{i=1}^{a} \sum\limits_{j=1}^{b} (\tau\beta)_{ij}^2}{(a - 1)(b - 1)}$$

and

$$E(MS_E) = E\left(\frac{SS_E}{ab(n - 1)}\right) = \sigma^2$$

Therefore, to test the hypotheses $H_0: \tau_i = 0$ (no row factor effects), $H_0: \beta_j = 0$ (no column factor effects), and $H_0: (\tau\beta)_{ij} = 0$ (no interaction effects) we would divide the corresponding mean square by mean square error. This ratio will follow an F distribution with appropriate numerator degrees of freedom and $ab(n - 1)$ denominator de-

Table 6-5 The Analysis of Variance Table for the Two-Way Classification, Fixed Effects Model

Source of Variation	Sum of Squares	Degrees of Freedom	Mean Square	F_0
A treatments	SS_A	$a - 1$	$MS_A = \dfrac{SS_A}{a - 1}$	$F_0 = \dfrac{MS_A}{MS_E}$
B treatments	SS_B	$b - 1$	$MS_B = \dfrac{SS_B}{b - 1}$	$F_0 = \dfrac{MS_B}{MS_E}$
Interaction	SS_{AB}	$(a - 1)(b - 1)$	$MS_{AB} = \dfrac{SS_{AB}}{(a - 1)(b - 1)}$	$F_0 = \dfrac{MS_{AB}}{MS_E}$
Error	SS_E	$ab(n - 1)$	$MS_E = \dfrac{SS_E}{ab(n - 1)}$	
Total	SS_T	$abn - 1$		

grees of freedom, and the critical region will be located in the upper tail. The test procedure is arranged in an analysis of variance table, such as shown in Table 6-5.

Computational formulas for the sums of squares in Equation 6-5 may be obtained easily. The total sum of squares is computed as usual by

$$SS_T = \sum_{i=1}^{a} \sum_{j=1}^{b} \sum_{k=1}^{n} y_{ijk}^2 - \frac{y_{...}^2}{abn} \tag{6-6}$$

The sums of squares for main effects are

$$SS_A = \sum_{i=1}^{a} \frac{y_{i..}^2}{bn} - \frac{y_{...}^2}{abn} \tag{6-7}$$

and

$$SS_B = \sum_{j=1}^{b} \frac{y_{.j.}^2}{an} - \frac{y_{...}^2}{abn} \tag{6-8}$$

It is convenient to obtain the SS_{AB} in two stages. First we compute the sum of squares between the ab cell totals, called the sum of squares due to "subtotals"

$$SS_{\text{subtotals}} = \sum_{i=1}^{a} \sum_{j=1}^{b} \frac{y_{ij.}^2}{n} - \frac{y_{...}^2}{abn}$$

cells or treatment combinations

This sum of squares also contains SS_A and SS_B. Therefore, the second step is to compute SS_{AB} as

$$SS_{AB} = SS_{\text{subtotals}} - SS_A - SS_B \tag{6-9}$$

We may compute SS_E by subtraction as

$$SS_E = SS_T - SS_{AB} - SS_A - SS_B \tag{6-10}$$

or

$$SS_E = SS_T - SS_{subtotals}$$

Tests of hypotheses about τ_i, β_j, and $(\tau\beta)_{ij}$ use numerator mean squares that are statistically independent. However, the F ratios themselves are *not* independent because they have the same denominator mean square, that is, MS_F. Therefore, consideration of the significance levels for the hypotheses about τ_i, β_j, and $(\tau\beta)_{ij}$ is a problem of *simultaneous* statistical inference. We will ignore this problem, as it is beyond the scope of this book. The interested reader is referred to Miller (1966).

■**EXAMPLE 6-1.** The maximum output voltage of a particular type of storage battery is thought to be influenced by the material used in the plates and the temperature in the location at which the battery is installed. Four replicates of a factorial experiment are run in the laboratory for three materials and three temperatures, and results are shown in Table 6-6. The circled numbers in the cells are the cell totals $\{y_{ij.}\}$. The order in which the 36 observations are taken is randomly determined.
The sums of squares are computed as follows.

$$SS_T = \sum_{i=1}^{a} \sum_{j=1}^{b} \sum_{k=1}^{n} y_{ijk}^2 - \frac{y_{...}^2}{abn}$$

$$= (130)^2 + (155)^2 + (74)^2 + \cdots + (60)^2 - \frac{(3799)^2}{36} = 77,646.96$$

$$SS_{material} = \sum_{i=1}^{a} \frac{y_{i..}^2}{bn} - \frac{y_{...}^2}{abn}$$

$$= \frac{(998)^2 + (1300)^2 + (1501)^2}{(3)(4)} - \frac{(3799)^2}{36} = 10,683.72$$

$$SS_{temperature} = \sum_{j=1}^{b} \frac{y_{.j.}^2}{an} - \frac{y_{...}^2}{abn}$$

$$= \frac{(1738)^2 + (1291)^2 + (770)^2}{(3)(4)} - \frac{(3799)^2}{36} = 39,118.72$$

$$SS_{interaction} = \sum_{i=1}^{a} \sum_{j=1}^{b} \frac{y_{ij.}^2}{n} - \frac{y_{...}^2}{abn} - SS_{material} - SS_{temperature}$$

$$= \frac{(539)^2 + (229)^2 + \cdots + (342)^2}{4} - \frac{(3799)^2}{36} - 10,683.72$$

$$- 39,118.72 = 9,613.77$$

Table 6-6   Maximum Output Voltage Data for Example 6-1

Material Type	Temperature (°F)									$y_{i..}$
	50			65			80			
1	130	155	(539)	34	40	(229)	20	70	(230)	998
	74	180		80	75		82	58		
2	150	188	(623)	136	122	(479)	25	70	(198)	1300
	159	126		106	115		58	45		
3	138	110	(576)	174	120	(583)	96	104	(342)	1501
	168	160		150	139		82	60		
$y_{.j.}$	1738			1291			770			$3799 = y_{...}$

and

$$SS_E = SS_T - SS_{material} - SS_{temperature} - SS_{interaction}$$

$$= 77,646.96 - 10,683.72 - 39,118.72 - 9,613.77 = 18,230.75$$

The analysis of variance is shown in Table 6-7.   Since $F_{.05,4,27} = 2.73$ we conclude that there is a significant interaction between material types and temperature.   Furthermore, $F_{.05,2,27} = 3.35$, so the main effects of material type and temperature   are also significant.

To assist in interpreting the results of this experiment, it is helpful to construct a graph of the average responses at each treatment combination.   This graph is shown in Figure 6-3.   The significant interaction is indicated by the lack of parallelism of the lines.   In general, higher output voltage is attained at low temperature, regardless of material type.   Changing from low to intermediate temperature, output voltage with material type 3 actually *increases*, while it decreases for types 1 and 2.   From intermediate to high temperature, output voltages decrease for material types 2 and 3, and are essentially unchanged for type 1.   ∎

Table 6-7   Analysis of Variance for Battery Voltage Data

Source of Variation	Sum of Squares	Degrees of Freedom	Mean Square	F_0
Material types	10,683.72	2	5,341.86	7.91
Temperature	39,118.72	2	19,558.36	28.97
Interaction	9,613.77	4	2,403.44	3.56
Error	18,230.75	27	675.21	
Total	77,646.96	35		

Figure 6-3. Material type—temperature plot for Example 6-1.

It is possible to make comparisons between the individual means of either factor using Duncan's multiple range test, since both factors have been assumed fixed. If interaction is significant, however, then comparisons between the means of a factor (e.g., A) may be obscured by the AB interaction. One approach to this situation is to fix factor B at a specific level and apply Duncan's multiple range test to the means of factor A at that level. To illustrate, suppose that in Example 6 1 we are interested in detecting differences between the means of the three material types. Since interaction is significant, we will make this comparison at just one level of temperature, say level 2 (65 degrees). We will assume that the best estimate of the error variance is the MS_E from the analysis of variance table, utilizing the assumption that the experimental error variance is the same over all treatment combinations. The three material type means, arranged in ascending order are

$$\bar{y}_{12.} = 57.25 \quad \text{(material type 1)}$$

$$\bar{y}_{22.} = 119.75 \quad \text{(material type 2)}$$

$$\bar{y}_{32.} = 145.75 \quad \text{(material type 3)}$$

The standard error of these treatment means is

$$S_{\bar{y}_{i2.}} = \sqrt{\frac{MS_E}{n}} = \sqrt{\frac{675.21}{4}} = 12.99$$

since each mean contains $n = 4$ observations. From Appendix Table VII we obtain the values $r_{.05}(2, 27) \simeq 2.91$ and $r_{.05}(3, 27) \simeq 3.06$. The least significant ranges are

$$R_2 = r_{.05}(2, 27) S_{\bar{y}_{i2.}} = (2.91)(12.99) = 37.80$$

$$R_3 = r_{.05}(3, 27) S_{\bar{y}_{i2.}} = (3.06)(12.99) = 39.75$$

and the comparisons yield

$$3 \text{ vs. } 1 = 145.75 - 57.25 = 88.50 > 39.75 \,(R_3)$$

$$3 \text{ vs. } 2 = 145.75 - 119.75 = 26.00 < 37.80 \,(R_2)$$

$$2 \text{ vs. } 1 = 119.75 - 57.25 = 62.50 > 37.80 \,(R_2)$$

This analysis indicates that at the temperature level 65 degrees, the mean output voltage is the same for material types 2 and 3, and that the mean output voltage for material type 1 is significantly lower in comparison to both types 2 and 3.

If interaction is significant, the experimenter could compare *all ab* cell means to determine which ones differ significantly. In this analysis, differences between cell means includes interaction effects, as well as both main effects. In Example 6-1, this would give 36 comparisons between all possible pairs of the nine cell means.

Finally, note that only results for a *balanced* design have been given. The treatment of unbalanced factorials is considerably more difficult. In some instances missing value formulas, such as those in Chapter 4, may be useful. Usually, however, an analysis based on the general regression significance test must be used. Unbalanced designs are discussed extensively by Searle (1971a).

6-3.2 Estimating the Model Parameters

The parameters in the two-way classification analysis of variance model

$$y_{ijk} = \mu + \tau_i + \beta_j + (\tau\beta)_{ij} + \epsilon_{ijk} \tag{6-11}$$

may be estimated by least squares. Since the model has $1 + a + b + ab$ parameters to be estimated, there are $1 + a + b + ab$ normal equations. Using the method of Section 3-10, it is not difficult to show that the normal equations are

$$\mu: \quad abn\hat{\mu} + bn\sum_{i=1}^{a}\hat{\tau}_i + an\sum_{j=1}^{b}\hat{\beta}_j + n\sum_{i=1}^{a}\sum_{j=1}^{b}\widehat{(\tau\beta)}_{ij} = y_{...} \tag{6-12a}$$

$$\tau_i: \quad bn\hat{\mu} + bn\hat{\tau}_i + n\sum_{j=1}^{b}\hat{\beta}_j + n\sum_{j=1}^{b}\widehat{(\tau\beta)}_{ij} = y_{i..}, \qquad i = 1, 2, \ldots, a \tag{6-12b}$$

$$\beta_j: \quad an\hat{\mu} + n\sum_{i=1}^{a}\hat{\tau}_i + an\hat{\beta}_j + n\sum_{i=1}^{a}\widehat{(\tau\beta)}_{ij} = y_{.j.}, \qquad j = 1, 2, \ldots, b \tag{6-12c}$$

$$(\tau\beta)_{ij}: \quad n\hat{\mu} + n\hat{\tau}_i + n\hat{\beta}_j + n\widehat{(\tau\beta)}_{ij} = y_{ij.}, \qquad \begin{cases} i = 1, 2, \ldots, a \\ j = 1, 2, \ldots, b \end{cases} \tag{6-12d}$$

For convenience, we have shown the parameter corresponding to each normal equation on the left in (6-12).

As we attempt to solve the normal equations, we notice that the a equations in (6-12b) sum to (6-12a), and that the b equations of (6-12c) sum to (6-12a). Also summing (6-12d) over j for a particular i will give Equation 6-12b, and summing (6-12d) over i for a particular j will give Equation 6-12c. Therefore, there are $a + b + 1$ linear dependencies in this system of equations and no unique solution will exist. In order to obtain a solution, we impose the constraints

$$\sum_{i=1}^{a} \hat{\tau}_i = 0 \tag{6-13a}$$

$$\sum_{j=1}^{b} \hat{\beta}_j = 0 \tag{6-13b}$$

$$\sum_{i=1}^{a} \widehat{(\tau\beta)}_{ij} = 0, \quad j = 1, 2, \ldots, b \tag{6-13c}$$

and

$$\sum_{j=1}^{b} \widehat{(\tau\beta)}_{ij} = 0, \quad i = 1, 2, \ldots, a \tag{6-13d}$$

Equations 6-13a and 6-13b constitute two constraints, while (6-13c) and (6-13d) form $a + b - 1$ independent constraints. Therefore, we have $a + b + 1$ total constraints, the number needed.

Applying these constraints, the normal equations (6-12) simplify considerably, and we obtain the solution

$$\hat{\mu} = \bar{y}_{...}$$

$$\hat{\tau}_i = \bar{y}_{i..} - \bar{y}_{...}, \qquad i = 1, 2, \ldots, a$$

$$\hat{\beta}_j = \bar{y}_{.j.} - \bar{y}_{...}, \qquad j = 1, 2, \ldots, b \tag{6-14}$$

$$\widehat{(\tau\beta)}_{ij} = \bar{y}_{ij.} - \bar{y}_{i..} - \bar{y}_{.j.} + \bar{y}_{...}, \quad \begin{cases} i = 1, 2, \ldots, a \\ j = 1, 2, \ldots, b \end{cases}$$

Notice the considerable intuitive appeal of this solution to the normal equations. Row treatment effects are estimated by the row average minus the grand average, column treatments are estimated by the column average minus the grand average, and the ijth interaction is estimated by the ijth cell average minus the grand average, the ith row effect, and the jth column effect.

Because constraints (6-13) have been used to solve the normal equations, the model parameters are not uniquely estimated. However, certain important *functions* of the model parameters *are* estimable, that is, uniquely estimated regardless of the constraint chosen. An example is $\tau_i - \tau_u + (\tau\beta)_{i.} - (\tau\beta)_{u.}$, which might be thought of as the "true" difference between the ith and uth levels of factor A. Notice that the true difference between the levels of any main effect includes an "average" interaction effect. It is this result that disturbs the tests on main effects in the presence of interaction, as noted earlier. In general, any function of the model parameters that is a linear combination of the left-hand side of the normal equations is estimable. This property was also noted in Chapter 3 when we were discussing the one-way classification.

6-3.3 The Power of the Test

The operating characteristic curves in Chart V of the Appendix can be used to find the power of the tests in factorial experiments. The appropriate values of Φ^2, numerator degrees of freedom, and denominator degrees of freedom are shown in Table 6-8. To illustrate the use of the operating characteristic curves, suppose that in Example 6-1 we wished to determine if four replicates are sufficient to give a high probability of rejection if material types differ by as much as $\Sigma_{i=1}^{3} \tau_i^2 = 800$. From Table 6-8 we find that

$$\Phi^2 = \frac{3(4)(800)}{3(675.21)} = 4.74$$

using $MS_E = 675.21$ as an estimate of σ^2. From Chart V in the Appendix with $\alpha = .05$, 2 numerator degrees of freedom, and 27 denominator degrees of freedom we find that $\beta = .09$ approximately. Thus four replicates gives a power of approximately .91; that is, we have a 91 percent chance of correctly rejecting $H_0 : \tau_i = 0$ if the treatments differ by the amount specified.

Table 6-8 Operating Characteristic Curve Parameters for Chart V in the Appendix for the Fixed Effects Model, Two-Way Classification

Factor	Φ^2	Numerator Degrees of Freedom	Denominator Degrees of Freedom
A	$\dfrac{bn \sum_{i=1}^{a} \tau_i^2}{a\sigma^2}$	$a - 1$	$ab(n - 1)$
B	$\dfrac{an \sum_{j=1}^{b} \beta_j^2}{b\sigma^2}$	$b - 1$	$ab(n - 1)$
AB	$\dfrac{n \sum_{i=1}^{a} \sum_{j=1}^{b} (\tau\beta)_{ij}^2}{\sigma^2 [(a - 1)(b - 1) + 1]}$	$(a - 1)(b - 1)$	$ab(n - 1)$

6-4 RANDOM AND MIXED MODELS

In previous sections we have assumed that both factors A and B are fixed. That is, the experimenter has specifically chosen the a levels of factor A and the b levels of factor B used in the design, and, consequently, the inferences drawn from the analysis of variance are applicable only to the levels of A and B actually used. In this section we consider two alternate models: the random effects model, in which the levels of A and B are randomly selected, and the mixed model, in which one factor is fixed and the other random.

6-4.1 The Random Effects Model

Consider the situation in which the levels of factors A and B are chosen at random from larger populations. This is the *random effects* or *components of variance* model. Because the a levels of A and the b levels of B were randomly chosen, our inferences will be valid about *all* levels in the populations under study. We may represent the observations by the linear model

$$y_{ijk} = \mu + \tau_i + \beta_j + (\tau\beta)_{ij} + \epsilon_{ijk} \begin{cases} i = 1, 2, \ldots, a \\ j = 1, 2, \ldots, b \\ k = 1, 2, \ldots, n \end{cases} \quad (6\text{-}15)$$

where the model parameters τ_i, β_j, $(\tau\beta)_{ij}$, and ϵ_{ijk} are random variables. Specifically, we assume that τ_i is NID$(0, \sigma_\tau^2)$, β_j is NID$(0, \sigma_\beta^2)$, $(\tau\beta)_{ij}$ is NID$(0, \sigma_{\tau\beta}^2)$, and ϵ_{ijk} is NID$(0, \sigma^2)$. The variance of any observation is

$$V(y_{ijk}) = \sigma_\tau^2 + \sigma_\beta^2 + \sigma_{\tau\beta}^2 + \sigma^2$$

and σ_τ^2, σ_β^2 $\sigma_{\tau\beta}^2$, and σ^2 are called *variance components*. The hypotheses that we are interested in testing are $H_0: \sigma_\tau^2 = 0$, $H_0: \sigma_\beta^2 = 0$, and $H_0: \sigma_{\tau\beta}^2 = 0$. Notice the similarity to the one-way classification random effects model.

The basic analysis of variance remain unchanged; that is, SS_A, SS_B, SS_{AB}, SS_T, and SS_E are all calculated as in the fixed effects case. However, to form the test statistics we must examine the expected mean squares. It may be shown that

$$E(MS_A) = \sigma^2 + n\sigma_{\tau\beta}^2 + bn\sigma_\tau^2$$

$$E(MS_B) = \sigma^2 + n\sigma_{\tau\beta}^2 + an\sigma_\beta^2 \quad (6\text{-}16)$$

$$E(MS_{AB}) = \sigma^2 + n\sigma_{\tau\beta}^2$$

and

$$E(MS_E) = \sigma^2$$

From the expected mean squares, we see that the appropriate statistic for testing H_0: $\sigma_{\tau\beta}^2 = 0$ is

$$F_0 = \frac{MS_{AB}}{MS_E} \qquad (6\text{-}17)$$

since under H_0 both numerator and denominator of F_0 have expectation σ^2, and only if H_0 is false is $E(MS_{AB})$ greater than $E(MS_E)$. The ratio F_0 is distributed as $F_{(a-1)(b-1), ab(n-1)}$. Similarly, for testing H_0: $\sigma_\tau^2 = 0$ we would use

$$F_0 = \frac{MS_A}{MS_{AB}} \qquad (6\text{-}18)$$

which is distributed as $F_{a-1,(a-1)(b-1)}$, and for testing H_0: $\sigma_\beta^2 = 0$ the statistic is

$$F_0 = \frac{MS_B}{MS_{AB}} \qquad (6\text{-}19)$$

which is distributed as $F_{b-1,(a-1)(b-1)}$. These are all upper-tail, one-tail tests. Notice that these test statistics are not the same as those used if both factors A and B are fixed. The expected mean squares are always used as a guide to test statistic construction.

■**EXAMPLE 6-2.** Suppose that in Example 6-1 a large number of material types and many different temperatures could have been chosen, and that the ones given in Table 6-6 were selected at random. The model is now a random effects model, and the analysis of variance is shown in Table 6-9. We see that the first four columns of the analysis of variance table are identical to Example 6-1. Now, however, the F ratios are computed according to Equations 6-17, 6-18, and 6-19. Interaction is still significant at the 5 percent level, and since $F_{.05,2,4} = 6.94$, we conclude that the effect of temperature is significant, while material types do not significantly affect the output voltage. ■

Table 6-9 Analysis of Variance for Example 6-2, Random Effects Model

Source of Variation	Sum of Squares	Degrees of Freedom	Mean Square	F_0
Material types	10,683.72	2	5,341.86	2.22
Temperature	39,118.72	2	19,558.36	8.13[a]
Interaction	9,613.77	4	2,403.44	3.56[a]
Error	18,230.75	27	675.21	
Total	77,646.96	35		

[a]Significant at 5 percent.

The variance components may be estimated by the *analysis of variance method*; that is, by equating the observed mean squares in the lines of the analysis of variance table to their expected values and solving for the variance components. This yields

$$\hat{\sigma}^2 = MS_E$$

$$\hat{\sigma}^2_{\tau\beta} = \frac{MS_{AB} - MS_E}{n}$$

$$\hat{\sigma}^2_\beta = \frac{MS_B - MS_{AB}}{an} \tag{6-20}$$

$$\hat{\sigma}^2_\tau = \frac{MS_A - MS_{AB}}{bn}$$

To illustrate the method, we will estimate the variance components for the experimental situation described in Example 6-2. From Equations 6-20 we obtain

$$\hat{\sigma}^2_\tau = \frac{5,341.86 - 2,403.44}{(3)(4)} = 244.87$$

$$\hat{\sigma}^2_\beta = \frac{19,558.36 - 2,403.44}{(3)(4)} = 1429.58$$

$$\hat{\sigma}^2_{\tau\beta} = \frac{2,403.44 - 675.21}{4} = 432.06$$

and

$$\hat{\sigma}^2 = 675.21$$

It is possible to obtain a confidence interval estimate of σ^2 using the chi-square distribution, since $ab(n-1)MS_E/\sigma^2$ is distributed as $\chi^2_{ab(n-1)}$. However, exact confidence intervals on the other variance components cannot be obtained for the reasons discussed in Section 3-5. Approximate procedures for confidence intervals for variance components are given in Searle (1971a, b).

6-4.2 Mixed Models

We now consider the situation where one of the factors, A, is fixed and the other, B, is random. This is called the *mixed model* analysis of variance. The linear statistical model is

$$y_{ijk} = \mu + \tau_i + \beta_j + (\tau\beta)_{ij} + \epsilon_{ijk} \begin{cases} i = 1, 2, \ldots, a \\ j = 1, 2, \ldots, b \\ k = 1, 2, \ldots, n \end{cases} \tag{6-21}$$

Here τ_i is a fixed effect, β_j is a random effect, the interaction $(\tau\beta)_{ij}$ is assumed to be a random effect, and ϵ_{ijk} is a random error. We also assume that the $\{\tau_i\}$ are fixed effects such that $\sum_{i=1}^{a} \tau_i = 0$ and β_j is a NID$(0, \sigma_\beta^2)$ random variable. The interaction effect, $(\tau\beta)_{ij}$, is a normal random variable with mean 0 and variance $[(a-1)/a]\sigma_{\tau\beta}^2$; however, summing the interaction component over the fixed factor equals zero. That is,

$$\sum_{i=1}^{a} (\tau\beta)_{ij} = (\tau\beta)_{.j} = 0, \qquad j = 1, 2, \ldots, b$$

This implies that certain interaction elements at different levels of the fixed factor are not independent. In fact, we may show (see Problem 6-22) that

$$\text{Cov } [(\tau\beta)_{ij}, (\tau\beta)_{i'j}] = -\frac{1}{a} \sigma_{\tau\beta}^2, \qquad i \neq i'$$

The covariance between $(\tau\beta)_{ij}$ and $(\tau\beta)_{ij'}$ for $j \neq j'$ is zero, and the random error ϵ_{ijk} is NID$(0, \sigma^2)$.

In this model the variance of $(\tau\beta)_{ij}$ is defined as $[(a-1)/a]\sigma_{\tau\beta}^2$ rather than $\sigma_{\tau\beta}^2$ in order to simplify the expected mean squares. The assumption $(\tau\beta)_{.j} = 0$ also has an effect on the expected mean squares, which we may show are

$$E(MS_A) = \sigma^2 + n\sigma_{\tau\beta}^2 + \frac{bn \sum_{i=1}^{a} \tau_i^2}{a-1}$$

$$E(MS_B) = \sigma^2 + an\sigma_\beta^2 \qquad\qquad (6\text{-}22)$$

$$E(MS_{AB}) = \sigma^2 + n\sigma_{\tau\beta}^2$$

and

$$E(MS_E) = \sigma^2$$

Therefore, the appropriate test statistic for testing H_0: $\tau_i = 0$ is

$$F_0 = \frac{MS_A}{MS_{AB}}$$

which is distributed as $F_{a-1,(a-1)(b-1)}$. For testing H_0: $\sigma_\beta^2 = 0$, the test statistic is

$$F_0 = \frac{MS_B}{MS_E}$$

which is distributed as $F_{b-1, ab(n-1)}$. Finally, for testing H_0: $\sigma_{\tau\beta}^2 = 0$ we would use

$$F_0 = \frac{MS_{AB}}{MS_E}$$

which is distributed as $F_{(a-1)(b-1), ab(n-1)}$.

In the mixed model, it is possible to estimate the fixed factor effects as

$$\hat{\mu} = \bar{y}_{...}$$
$$\hat{\tau}_i = \bar{y}_{i..} - \bar{y}_{...}, \qquad i = 1, 2, \ldots a \tag{6-23}$$

The variance components σ_β^2, $\sigma_{\tau\beta}^2$, and σ^2 may be estimated using the analysis of variance method. Eliminating the first equation from (6-22) leaves three equations in three unknowns, whose solution is

$$\hat{\sigma}_\beta^2 = \frac{MS_B - MS_E}{an}$$
$$\hat{\sigma}_{\tau\beta}^2 = \frac{MS_{AB} - MS_E}{n} \tag{6-24}$$

and

$$\hat{\sigma}^2 = MS_E$$

This general approach can be used to estimate the variance components in *any* mixed model. After eliminating the mean squares containing fixed factors, there will always be a set of equations remaining that can be solved for the variance components.

Table 6-10 summarizes the analysis of variance for the two-factor mixed model. In mixed models it is possible to test hypotheses about individual treatment means for the fixed factor. In using procedures such as Duncan's multiple range test, care must be exercised to use the proper standard error of the treatment mean. Anderson and McLean (1974) show that the standard error of the fixed effect treatment mean is

$$\left[\frac{\text{Mean Square for Testing the Fixed Effect}}{\text{Number of Observations in each Treatment Mean}} \right]^{1/2}$$

Table 6-10 Analysis of Variance for the Two-Factor Mixed Model

Source of Variation	Sum of Squares	Degrees of Freedom	Expected Mean Square
Rows (A)	SS_A	$a - 1$	$\sigma^2 + n\sigma_{\tau\beta}^2 + bn \sum \tau_i^2/(a - 1)$
Columns (B)	SS_B	$b - 1$	$\sigma^2 + an\sigma_\beta^2$
Interaction	SS_{AB}	$(a - 1)(b - 1)$	$\sigma^2 + n\sigma_{\tau\beta}^2$
Error	SS_E	$ab(n - 1)$	σ^2
Total	SS_T	$abn - 1$	

Alternate Mixed Models
Several different versions of the mixed model have been proposed. These models differ from the "standard" mixed model discussed above in the assumptions made

about the random components β_j and $(\tau\beta)_{ij}$. One of these alternate models will now be briefly discussed.

Suppose that the $\{\tau_i\}$ are fixed effects such that $\Sigma_{i=1}^{a} \tau_i = 0$ and β_j is a NID$(0, \sigma_\beta^2)$ random variable. The interaction effects $(\tau\beta)_{ij}$ are NID$(0, \sigma_{\tau\beta}^2)$ and are also independent of the random effect β_j. The random error component ϵ_{ijk} is NID$(0, \sigma^2)$. Note that the primary difference between this mixed model and the previous model is the assumption about the independence of the interaction effects.

The expected mean squares for this model are shown in Table 6-11. Comparing this table to Table 6-10, we note that the only obvious difference is the presence of the variance component $\sigma_{\tau\beta}^2$ in the expected mean square for the random effect (actually, there are other differences because of the different definitions of the variance of the interaction effect in the two models, but they are minor). As a result, we would test $H_0: \sigma_\beta^2 = 0$ with the statistic

$$F_0 = \frac{MS_B}{MS_{AB}}$$

as contrasted with $F_0 = MS_B/MS_E$ in the standard model. This test should be more conservative using this model instead of the standard model, since MS_{AB} will generally be larger than MS_E.

The analysis of variance method may be used to estimate the variance components. Referring to Table 6-11, we find that the only change from Equation 6-24 is that

$$\hat{\sigma}_\beta^2 = \frac{MS_B - MS_{AB}}{an}$$

Both of these models are special cases of the mixed model proposed by Scheffé (1956a)(1959). This model assumes that the observations may be represented by

$$y_{ijk} = m_{ij} + \epsilon_{ijk} \begin{cases} i = 1, 2, \ldots, a \\ j = 1, 2, \ldots, b \\ k = 1, 2, \ldots, n \end{cases}$$

Table 6-11 Analysis of Variance for the Two-Factor Alternate Mixed Model

Source of Variation	Sum of Squares	Degrees of Freedom	Expected Mean Square
Rows (A)	SS_A	$a - 1$	$\sigma^2 + n\sigma_{\tau\beta}^2 + bn\,\Sigma\,\tau_i^2/(a-1)$
Columns (B)	SS_B	$b - 1$	$\sigma^2 + n\sigma_{\tau\beta}^2 + an\sigma_\beta^2$
Interaction	SS_{AB}	$(a-1)(b-1)$	$\sigma^2 + n\sigma_{\tau\beta}^2$
Error	SS_E	$ab(n-1)$	σ^2
Total	SS_T	$abn - 1$	

where m_{ij} and ϵ_{ijk} are independent random variables. The structure of m_{ij} is

$$m_{ij} = \mu + \tau_i + b_j + c_{ij}$$

$$E(m_{ij}) = \mu + \tau_i$$

$$\sum_{i=1}^{a} \tau_i = 0$$

and

$$c_{.j} = 0, \qquad j = 1, 2, \ldots, b$$

The variances and covariances of b_j and c_{ij} are expressed through the covariances of the m_{ij}. Furthermore, the parameters β_j and $(\tau\beta)_{ij}$ in other formulations of the mixed model can be related to b_j and c_{ij}. The statistical analysis of Scheffé's model is identical to our standard model, except that in general the statistic MS_A/MS_{AB} is not always distributed as F when $H_0 : \tau_i = 0$ is true.

In light of this multiplicity of mixed models, a logical question is which model should one use? Most statisticians tend to prefer the standard model and it is the most widely encountered in the literature. If the correlative structure of the random components is not large, then either mixed model is appropriate, and there are only minor differences between these models. When we subsequently refer to mixed models, we shall assume the standard model structure. However, if there are large correlations in the data, then Scheffé's model may have to be employed. The choice between models should always be dictated by the data. The article by Hocking (1973) is a clear summary of various mixed models.

6-4.3 Power of the Random and Mixed Models

The operating characteristic curves in the Appendix may be used for the two-way classification random and mixed models analysis of variance. Chart VI is used for the random model. The parameter λ, numerator degrees of freedom, and denominator degrees of freedom are shown in the top half of Table 6-12. For the mixed model, both Charts V and VI must be used. The appropriate values for Φ^2 and λ are shown in the bottom half of Table 6-12.

6-5 THE GENERAL FACTORIAL EXPERIMENT

The results for the two-way classification analysis of variance may be extended to the general case where there are a levels of factor A, b levels of factor B, c levels of factor C, and so on, arranged in a factorial experiment. In general, there will be $abc \ldots n$ total observations, if there are n replicates of the complete experiment. Once again,

Table 6-12 Operating Characteristic Curve Parameters for Charts V and VI in the Appendix for the Random and Mixed Models, Two-Way Classification

The Random Effects Model

Factor	λ	Numerator Degrees of Freedom	Denominator Degrees of Freedom
A	$\sqrt{1 + \dfrac{bn\, \sigma_\tau^2}{\sigma^2 + n\, \sigma_{\tau\beta}^2}}$	$a - 1$	$(a - 1)(b - 1)$
B	$\sqrt{1 + \dfrac{an\, \sigma_\beta^2}{\sigma^2 + n\, \sigma_{\tau\beta}^2}}$	$b - 1$	$(a - 1)(b - 1$
AB	$\sqrt{1 + \dfrac{n\, \sigma_{\tau\beta}^2}{\sigma^2}}$	$(a - 1)(b - 1)$	$ab(n - 1)$

The Mixed Model

Factor	Parameter	Numerator Degrees of Freedom	Denominator Degrees of Freedom	Appendix Chart
A (Fixed)	$\Phi^2 = \dfrac{bn\sum\limits_{i=1}^{a} \tau_i^2}{a\,[\sigma^2 + n\, \sigma_{\tau\beta}^2]}$	$a - 1$	$(a - 1)(b - 1)$	V
B (Random)	$\lambda = \sqrt{1 + \dfrac{an\, \sigma_\beta^2}{\sigma^2}}$	$b - 1$	$ab(n - 1)$	VI
AB	$\lambda = \sqrt{1 + \dfrac{n\, \sigma_{\tau\beta}^2}{\sigma^2}}$	$(a - 1)(b - 1)$	$ab(n - 1)$	VI

note that we must have at least two replicates ($n \geqslant 2$) in order to determine a sum of squares due to error if all possible interactions are included in the model.

If all factors in the experiment are fixed we may easily formulate and test hypotheses about the main effects and interactions. For a fixed effects model, test statistics for each main effect and interaction may be constructed by dividing the corresponding mean square for the effect or interaction by mean square error. All of these F tests will be upper-tail, one-tail tests. The number of degrees of freedom for any main effect are the number of levels of the factor minus one, and the number of degrees of freedom for an interaction are the product of the number of degrees of freedom associated with the individual components of the interaction.

For example, consider the three-factor analysis of variance model

$$y_{ijkl} = \mu + \tau_i + \beta_j + \gamma_k + (\tau\beta)_{ij} + (\tau\gamma)_{ik} + (\beta\gamma)_{jk}$$

$$+ (\tau\beta\gamma)_{ijk} + \epsilon_{ijkl}, \begin{cases} i = 1, 2, \ldots, a \\ j = 1, 2, \ldots, b \\ k = 1, 2, \ldots, c \\ l = 1, 2, \ldots, n \end{cases} \tag{6-25}$$

Assuming that A, B, and C are fixed, the analysis of variance table is shown in Table 6-13. The F tests on main effects and interactions follow directly from the expected mean squares.

We will give the computing formulas for the sums of squares in Table 6-13. The total sum of squares is found in the usual way as

$$SS_T = \sum_{i=1}^{a} \sum_{j=1}^{b} \sum_{k=1}^{c} \sum_{l=1}^{n} y_{ijkl}^2 - \frac{y_{....}^2}{abcn} \tag{6-26}$$

The sums of squares for the main effects are found from the totals for factors $A(y_{i...})$, $B(y_{.j..})$, and $C(y_{..k.})$ as follows:

$$SS_A = \sum_{i=1}^{a} \frac{y_{i...}^2}{bcn} - \frac{y_{....}^2}{abcn} \tag{6-27}$$

$$SS_B = \sum_{j=1}^{b} \frac{y_{.j..}^2}{acn} - \frac{y_{....}^2}{abcn} \tag{6-28}$$

$$SS_C = \sum_{k=1}^{c} \frac{y_{..k.}^2}{abn} - \frac{y_{....}^2}{abcn} \tag{6-29}$$

To compute the two-factor interaction sums of squares, the totals for the $A \times B$, $A \times C$, and $B \times C$ cells are needed. It is frequently helpful to collapse the original data table into three two-way tables in order to compute these quantities. The sums of squares are found from

$$SS_{AB} = \sum_{i=1}^{a} \sum_{j=1}^{b} \frac{y_{ij..}^2}{cn} - \frac{y_{....}^2}{abcn} - SS_A - SS_B$$

$$= SS_{\text{subtotals }(AB)} - SS_A - SS_B \tag{6-30}$$

$$SS_{AC} = \sum_{i=1}^{a} \sum_{k=1}^{c} \frac{y_{i.k.}^2}{bn} - \frac{y_{....}^2}{abcn} - SS_A - SS_C$$

$$= SS_{\text{subtotals }(AC)} - SS_A - SS_C \tag{6-31}$$

Table 6-13 The Analysis of Variance Table for the Three-Factor Fixed Effects Model

Source of Variation	Sum of Squares	Degrees of Freedom	Mean Square	Expected Mean Squares	F_0
A	SS_A	$a-1$	MS_A	$\sigma^2 + \dfrac{bcn\sum \tau_i^2}{a-1}$	$F_0 = \dfrac{MS_A}{MS_E}$
B	SS_B	$b-1$	MS_B	$\sigma^2 + \dfrac{acn\sum \beta_j^2}{b-1}$	$F_0 = \dfrac{MS_B}{MS_E}$
C	SS_C	$c-1$	MS_C	$\sigma^2 + \dfrac{abn\sum \gamma_k^2}{c-1}$	$F_0 = \dfrac{MS_C}{MS_E}$
AB	SS_{AB}	$(a-1)(b-1)$	MS_{AB}	$\sigma^2 + \dfrac{cn\sum\sum (\tau\beta)_{ij}^2}{(a-1)(b-1)}$	$F_0 = \dfrac{MS_{AB}}{MS_E}$
AC	SS_{AC}	$(a-1)(c-1)$	MS_{AC}	$\sigma^2 + \dfrac{bn\sum\sum (\tau\gamma)_{ik}^2}{(a-1)(c-1)}$	$F_0 = \dfrac{MS_{AC}}{MS_E}$
BC	SS_{BC}	$(b-1)(c-1)$	MS_{BC}	$\sigma^2 + \dfrac{an\sum\sum (\beta\gamma)_{jk}^2}{(b-1)(c-1)}$	$F_0 = \dfrac{MS_{BC}}{MS_E}$
ABC	SS_{ABC}	$(a-1)(b-1)(c-1)$	MS_{ABC}	$\sigma^2 + \dfrac{n\sum\sum\sum (\tau\beta\gamma)_{ijk}^2}{(a-1)(b-1)(c-1)}$	$F_0 = \dfrac{MS_{ABC}}{MS_E}$
Error	SS_E	$abc(n-1)$	MS_E	σ^2	
Total	SS_T	$abcn-1$			

and

$$SS_{BC} = \sum_{j=1}^{b} \sum_{k=1}^{c} \frac{y_{.jk.}^2}{an} - \frac{y_{....}^2}{abcn} - SS_B - SS_C$$

$$= SS_{\text{subtotals } (BC)} - SS_B - SS_C \tag{6-32}$$

Note that the two-factor subtotals sums of squares are found from the totals in each two-way table. The three-factor interaction sum of squares is computed from the three-way cell totals $\{y_{ijk.}\}$ as

$$SS_{ABC} = \sum_{i=1}^{a} \sum_{j=1}^{b} \sum_{k=1}^{c} \frac{y_{ijk.}^2}{n} - \frac{y_{....}^2}{abcn} - SS_A - SS_B - SS_C - SS_{AB} - SS_{AC} - SS_{BC} \tag{6-33a}$$

$$= SS_{\text{subtotals } (ABC)} - SS_A - SS_B - SS_C - SS_{AB} - SS_{AC} - SS_{BC} \tag{6-33b}$$

The error sum of squares may be found by subtracting the sum of squares for each main effect and interaction from the total sum of squares, or by

$$SS_E = SS_T - SS_{\text{subtotals } (ABC)} \tag{6-34}$$

■**EXAMPLE 6-3.** A soft drink bottler is studying the effect of percent carbonation (A), operating pressure in the filler (B), and line speed (C) on the volume of carbonated beverage packaged in each bottle. Three levels of carbonation, and two levels of pressure and speed are selected, and a factorial experiment with two replicates is conducted. The order in which the 24 observations are taken is determined randomly. The coded data that result from this experiment are shown in Table 6-14. The three-way cell totals $\{y_{ijk.}\}$ are shown in circles in this table.

The total corrected sum of squares is found from (6-26) as

$$SS_T = \sum_{i=1}^{a} \sum_{j=1}^{b} \sum_{k=1}^{c} \sum_{l=1}^{n} y_{ijkl}^2 - \frac{y_{....}^2}{abcn}$$

$$= 571 - \frac{(75)^2}{24} = 336.625$$

and the sums of squares for the main effects are calculated from (6-27), (6-28), and (6-29) as

$$SS_{\text{carbonation}} = \sum_{i=1}^{a} \frac{y_{i...}^2}{bcn} - \frac{y_{....}^2}{abcn}$$

$$= \frac{(-4)^2 + (20)^2 + (59)^2}{8} - \frac{(75)^2}{24} = 252.750$$

$$SS_{\text{pressure}} = \sum_{j=1}^{b} \frac{y_{.j..}^2}{acn} - \frac{y_{....}^2}{abcn}$$

$$= \frac{(21)^2 + (54)^2}{12} - \frac{(75)^2}{24} = 45.375$$

Table 6-14 Coded Data for Example 6-3

Percent Carbonation (A)	Operating Pressure (B)				
	25 psi		30 psi		
	Line Speed (C)		Line Speed (C)		
	200	250	200	250	$y_{i...}$
10	-3 ⃝-4 -1	-1 ⃝-1 0	-1 ⃝-1 0	1 ② 1	-4
12	0 ① 1	2 ③ 1	2 ⑤ 3	6 ⑪ 5	20
14	5 ⑨ 4	7 ⑬ 6	7 ⑯ 9	10 ㉑ 11	59
$B \times C$ Totals, $y_{.jk}$	6	15	20	34	$75 = y_{....}$
$y_{.j..}$	21		54		

$A \times B$ Totals
$y_{ij..}$

A	B	
	25	30
10	-5	1
12	4	16
14	22	37

$A \times C$ Totals
$y_{i.k.}$

A	C	
	200	250
10	-5	1
12	6	14
14	25	34

and

$$SS_{speed} = \sum_{k=1}^{c} \frac{y^2_{..k.}}{abn} - \frac{y^2_{....}}{abcn}$$

$$= \frac{(26)^2 + (49)^2}{12} - \frac{(75)^2}{24} = 22.042$$

To calculate the sums of squares for the two-factor interactions, we must find the two-way cell totals. For example, to find the carbonation-pressure or AB interaction, we need the totals for the $A \times B$ cells $\{y_{ij..}\}$ shown in Table 6-14. Using (6-30), we find the sums of squares as

$$SS_{AB} = \sum_{i=1}^{a} \sum_{j=1}^{b} \frac{y_{ij.}^2}{cn} - \frac{y_{....}^2}{abcn} - SS_A - SS_B$$

$$= \frac{(-5)^2 + (1)^2 + (4)^2 + (16)^2 + (22)^2 + (37)^2}{4} - \frac{(75)^2}{24} - 252.750 - 45.375$$

$$= 5.250$$

The carbonation-speed or AC interaction uses the $A \times C$ cell totals $\{y_{i.k.}\}$, shown in Table 6-14, and Equation 6-31:

$$SS_{AC} = \sum_{i=1}^{a} \sum_{k=1}^{c} \frac{y_{i.k.}^2}{bn} - \frac{y_{....}^2}{abcn} - SS_A - SS_C$$

$$= \frac{(-5)^2 + (1)^2 + (6)^2 + (14)^2 + (25)^2 + (34)^2}{4} - \frac{(75)^2}{24} - 252.750 - 22.042$$

$$= 0.583$$

The pressure-speed or BC interaction is found from the $B \times C$ cell totals $\{y_{.jk.}\}$ shown in Table 6-14 and Equation 6-32:

$$SS_{BC} = \sum_{j=1}^{b} \sum_{k=1}^{c} \frac{y_{.jk.}^2}{an} - \frac{y_{....}^2}{abcn} - SS_B - SS_C$$

$$= \frac{(6)^2 + (15)^2 + (20)^2 + (34)^2}{6} - \frac{(75)^2}{24} - 45.375 - 22.042$$

$$= 1.042$$

The three-factor interaction sum of squares is found from the $A \times B \times C$ cell totals $\{y_{ijk.}\}$, which are circled in Table 6-14. From Equation 6-33a we find

$$SS_{ABC} = \sum_{i=1}^{a} \sum_{j=1}^{b} \sum_{k=1}^{c} \frac{y_{ijk.}^2}{n} - \frac{y_{....}^2}{abcn} - SS_A - SS_B - SS_C - SS_{AB} - SS_{AC} - SS_{BC}$$

$$= \frac{(-4)^2 + (-1)^2 + (-1)^2 + (2)^2 + (1)^2 + (3)^2 + (5)^2 + (11)^2 + (9)^2 + (13)^2 + (16)^2 + (21)^2}{2}$$

$$- \frac{(75)^2}{24} - 252.750 - 45.375 - 22.042 - 5.250 - 0.583 - 1.042$$

$$= 1.083$$

Finally, noting that

$$SS_{\text{subtotals }(ABC)} = \sum_{i=1}^{a} \sum_{j=1}^{b} \sum_{k=1}^{c} \frac{y_{ijk.}^2}{n} - \frac{y_{....}^2}{abcn} = 328.125$$

Table 6-15   Analysis of Variance for Example 6-3

Source of Variation	Sum of Squares	Degrees of Freedom	Mean Square	F_0
Percent carbonation (A)	252.750	2	126.375	178.412[a]
Operating pressure (B)	45.375	1	45.375	64.059[a]
Line speed (C)	22.042	1	22.042	31.118[a]
AB	5.250	2	2.625	3.706[b]
AC	0.583	2	0.292	0.412
BC	1.042	1	1.042	1.471
ABC	1.083	2	0.542	0.765
Error	8.500	12	0.708	
Total	336.625	23		

[a] Significant at 1 percent.
[b] Significant at 10 percent.

we have

$$SS_E = SS_T - SS_{\text{subtotals }(ABC)}$$

$$= 336.625 - 328.125$$

$$= 8.500$$

The analysis of variance is summarized in Table 6-15. We see that percent carbonation, operating pressure, and line speed significantly affect the fill volume. The carbonation-pressure interaction is significant at 10 percent, indicating a mild interaction between these factors. ∎

We have indicated that if all the factors in a factorial experiment are fixed, then test statistic construction is straightforward. The statistic for testing any main effect or interaction is always formed by dividing the mean square for the main effect or interaction by mean square error. However, if the factorial experiment involves a random or mixed model, then test statistic construction is not immediately obvious. We must examine the expected mean squares to determine the correct tests. For example, recall the two-way classification random and mixed models in Section 6-4, where the appropriate test procedure followed directly from the expected mean squares. It seems, then, that we need a procedure for developing the expected value of a mean square. We will defer a complete discussion of this topic until Chapter 7, where a set of rules will be given for obtaining the sums of squares, degrees of freedom, and expected mean squares for a general class of designs.

6-6 POLYNOMIAL EFFECTS OF QUANTITATIVE FACTORS

In Section 3-6 a simple method was presented for partitioning an equally-spaced quantitative factor into single-degree-of-freedom components corresponding to the linear, quadratic, cubic, and higher effects of that factor. Furthermore, it was shown that orthogonal polynomials can be used to fit a response curve to the data, so that the response at intermediate factor levels can be predicted. The same principles discussed in Section 3-6 can be applied to factorial experiments, and are illustrated in the following two examples. For other interesting examples of this technique, refer to Hicks (1973) and Anderson and McLean (1974).

■**EXAMPLE 6-4. One Quantitative Factor and One Qualitative Factor.** Consider the experiment described in Example 6-1. Here temperature is quantitative and material type is qualitative. Furthermore, the three levels of temperature are equally-spaced. Consequently, we can compute a linear and quadratic temperature effect to study how temperature affects output voltage. The calculation of the linear and quadratic effects of temperature, denoted T_L and T_Q, and their sums of squares, is shown in Table 6-16.

Table 6-16 Calculation of Polynomial Effects of Temperature, Example 6-4

Temperature Levels (°F)	Treatment Totals $y_{.j.}$	Orthogonal Contrast Coefficients (c_j)	
		Linear	Quadratic
50	1738	−1	+1
65	1291	0	−2
80	770	+1	+1
Effects: $\left(\sum_{j=1}^{3} c_j y_{.j.}\right)$		$T_L = -968$	$T_Q = -74$
Sums of squares: $\left[\dfrac{\left(\sum_{j=1}^{3} c_j y_{.j.}\right)^2}{an \sum_{j=1}^{3} c_j^2}\right]$		$SS_{T_L} = \dfrac{(-968)^2}{(4)(3)(2)}$ $= 39{,}042.67$	$SS_{T_Q} = \dfrac{(-74)^2}{(4)(3)(6)}$ $= 76.05$

The interaction sum of squares may be partitioned into an interaction component between material type and the linear effect of temperature $(M \times T_L)$ and an interaction component between material type and the quadratic effect of temperature $(M \times T_Q)$. To compute these interaction components, the linear and quadratic effects of temperature are determined for each material type, and then the sums of squares between these effects are calculated. The process is shown in Table 6-17.

Note that $SS_{MT} = SS_{M \times T_L} + SS_{M \times T_Q}$. The analysis of variance is displayed in Table 6-18. The linear temperature effect is evident, with no indication of significant quadratic response to temperature. Furthermore, the significant interaction is due to the $M \times T_Q$ component.

Table 6-17 Polynomial Effects of the MT Interaction, Example 6-4

Temperature Levels ($^\circ$F)	Totals for Material Types			Orthogonal Contrast Coefficients (c_j)	
	$1(y_{1j.})$	$2(y_{2j.})$	$3(y_{3j.})$	Linear	Quadratic
50	539	623	576	-1	$+1$
65	229	479	583	0	-2
80	230	198	342	$+1$	$+1$

Effects: $M_1 \times T_L = -309$, $M_2 \times T_L = -425$, $M_3 \times T_L = -234$

$M_1 \times T_Q = 311$, $M_2 \times T_Q = -137$, $M_3 \times T_Q = -248$

Sums of squares:

$$SS_{M \times T_L} = \frac{(-309)^2 + (-425)^2 + (-234)^2}{4(2)} - \frac{(-968)^2}{12(2)} = 2{,}315.07$$

$$SS_{M \times T_Q} = \frac{(311)^2 + (-137)^2 + (-248)^2}{4(6)} - \frac{(-74)^2}{12(6)} = 7{,}298.70$$

Table 6-18 Analysis of Variance for Example 6-4

Source of Variation	Sum of Squares	Degrees of Freedom	Mean Square	F_0
Material types (M)	10,683.72	2	5,341.86	7.91[a]
Temperature (T)	39,118.72	2	19,559.36	28.97[a]
(linear)	(39,042.67)	1	39,042.67	57.82[a]
(quadratic)	(76.05)	1	76.05	0.12
MT interaction	9,613.77	4	2,403.44	3.56[b]
($M \times T_L$)	(2,315.07)	2	1157.54	1.71
($M \times T_Q$)	(7,298.70)	2	3649.35	5.41[b]
Error	18,230.75	27	675.21	
Total	77,646.96	35		

[a] Significant at 1 percent.
[b] Significant at 5 percent.

Since the effect of temperature is significant, the experimenter may wish to fit a response curve to the data, so that output voltage may be predicted as a function of temperature. However, the significant MT interaction implies that the output voltage response to temperature depends on which material type is used. Therefore, the experimenter would have to fit *three* response curves—one for each material type. ∎

If several factors in a factorial experiment are quantitative, then these factors may be decomposed into single-degree-of-freedom polynomial effects. Furthermore, interactions of quantitative factors can be partitioned into single-degree-of-freedom components of interaction using a similar process. This is illustrated in the following example.

■**EXAMPLE 6-5. Two Quantitative Factors.** The effective life of a cutting tool installed in a numerically-controlled machine is thought to be affected by the cutting speed and the tool angle. Three speeds and three angles are selected, and a factorial experiment with two replicates is performed. The coded data are shown in Table 6-19. The circled numbers in the cells are the cells totals, $\{y_{ij.}\}$.

Table 6-19 Data for Tool Life Experiment

Tool Angle (degrees)	Cutting Speed (RPM) 125		Cutting Speed (RPM) 150		Cutting Speed (RPM) 175		$y_{i..}$
15	-2 -1	(-3)	-3 0	(-3)	2 3	(5)	-1
20	0 2	(2)	1 3	(4)	4 6	(10)	16
25	-1 0	(-1)	5 6	(11)	0 -1	(-1)	9
$y_{.j.}$	-2		12		14		24 = $y_{...}$

Table 6-20 summarizes the analysis of variance. Both factors are treated as fixed effects. Cutting speed, tool angle, and their interaction are significant at the five percent level.

Table 6-20 Analysis of Variance for Tool Life Data

Source of Variation	Sum of Squares	Degrees of Freedom	Mean Square	F_0
Tool angle (T)	24.33	2	12.17	8.45[a]
Cutting speed (C)	25.33	2	12.67	8.80[a]
TC interaction	61.34	4	~~6.82~~ *15.33* *6.82*	~~4.75[a]~~ *10.61*
Error	13.00	9	1.44	
Total	124.00	17		

[a]Significant at 5 percent.

The linear and quadratic effects of both factors may be determined using the method presented in Example 6-4. The calculations are displayed in Table 6-21.

Table 6-21   Polynomial Effects of Tool Angle (T) and Cutting Speed (C), Example 6-5

Treatment Totals		Orthogonal Contrast Coefficients	
Tool Angle ($y_{i..}$)	Cutting Speed ($y_{.j.}$)	Linear	Quadratic
-1	-2	-1	$+1$
16	12	0	-2
9	14	$+1$	$+1$
Effects		$T_L = 10$	$T_Q = -24$
		$C_L = 16$	$C_Q = -12$

				Totals
Sums of squares		$SS_{T_L} = 8.33$	$SS_{T_Q} = 16.00$	24.33
		$SS_{C_L} = 21.33$	$SS_{C_Q} = 4.00$	25.33

Notice that the sum of squares for a factor is equal to the total of the linear and quadratic component sums of squares for that same factor.

Since both factors in this experiment are quantitative, is is possible to partition the TC interaction into four single-degree-of-freedom components, say

$$TC_{L \times L} \qquad TC_{L \times Q} \qquad TC_{Q \times L} \qquad \text{and} \qquad TC_{Q \times Q}$$

In general, any two-factor interaction between quantitative factors can be partitioned into $(a - 1)(b - 1)$ single-degree-of-freedom components. However, it may be difficult to interpret these interaction components. Often a particular component is significant because of atypical responses in one cell. Thus it is relatively standard practice to compute only the lower-order components. In such a case, the remaining components are considered jointly as higher-order interaction effects. The statistical significance of these higher-order interaction components could indicate a very complex relationship between the factors.

Table 6-22 $TC_{L \times L}$ Interaction Coefficients

T_L	C_L		
	-1	0	$+1$
-1	$+1$	0	-1
0	0	0	0
$+1$	-1	0	$+1$

The technique for computing the single-degree-of-freedom interaction components requires the cell totals (recall that differences between cells define an interaction effect) and the orthogonal contrast coefficients. To illustrate the approach, consider the coefficients for the $TC_{L \times L}$ interaction component shown in Table 6-22. The entries in the body of the table, (e.g., c_{ij}) are obtained by multiplying the coefficients of T_L by the coefficients of C_L.

To obtain the $TC_{L \times L}$ contrast, multiply the coefficient in the ijth cell in Table 6-22 by the ijth cell total (shown in circles in Table 6-19) and add, yielding

$$TC_{L \times L} = \sum_{i=1}^{3} \sum_{j=1}^{3} c_{ij} y_{ij}.$$

$$= +1\,(-3) + (0)\,(-3) + (-1)\,(5) + (0)\,(2) + (0)\,(4)$$

$$+ (0)\,(10) + (-1)\,(-1) + (0)\,(11) + 1\,(-1) = -8$$

The sum of squares for the $TC_{L \times L}$ contrast is

$$SS_{TC_{L \times L}} = \frac{(TC_{L \times L})^2}{n \sum \sum c_{ij}^2} = \frac{(-8)^2}{2[(1)^2 + (0)^2 + (-1)^2 + \cdots + (1)^2]} = \frac{(-8)^2}{2(4)} = 8.00$$

The coefficients for the $TC_{L \times Q}$, $TC_{Q \times L}$, and $TC_{Q \times Q}$ interaction components are derived similarly. They are shown in Table 6-23, along with the contrast and sum of squares for each interaction component.

Table 6-23 Computation of the $TC_{L \times Q}$, $TC_{Q \times L}$, and $TC_{Q \times Q}$ Interactions

T_L	C_Q			T_Q	C_L			T_Q	C_Q		
	+1	-2	+1		-1	0	+1		+1	-2	+1
-1	-1	+2	-1	+1	-1	0	+1	+1	+1	-2	+1
0	0	0	0	-2	+2	0	-2	-2	-2	+4	-2
+1	+1	-2	+1	+1	-1	0	+1	+1	+1	-2	+1

$$TC_{L \times Q} = \sum_{i=1}^{3} \sum_{j=1}^{3} c_{ij} y_{ij}.$$

$$= -32$$

$$SS_{TC_{L \times Q}} = \frac{(TC_{L \times Q})^2}{n \sum \sum c_{ij}^2}$$

$$= \frac{(-32)^2}{2(12)} = 42.67$$

$$TC_{Q \times L} = \sum_{i=1}^{3} \sum_{j=1}^{3} c_{ij} y_{ij}.$$

$$= -8$$

$$SS_{TC_{Q \times L}} = \frac{(TC_{Q \times L})^2}{n \sum \sum c_{ij}^2}$$

$$= \frac{(-8)^2}{2(12)} = 2.67$$

$$TC_{Q \times Q} = \sum_{i=1}^{3} \sum_{j=1}^{3} c_{ij} y_{ij}.$$

$$= -24$$

$$SS_{TC_{Q \times Q}} = \frac{(TC_{Q \times Q})^2}{n \sum \sum c_{ij}^2}$$

$$= \frac{(-24)^2}{2(36)} = 8.00$$

It is easy to see that the sum of the four single-degree-of-freedom interaction components equals the interaction sum of squares; that is,

$$SS_{TC} = SS_{TC_{L \times L}} + SS_{TC_{L \times Q}} + SS_{TC_{Q \times L}} + SS_{TC_{Q \times Q}}$$
$$= 8.00 + 42.67 + 2.67 + 8.00 = 61.34$$

The expanded analysis of variance is summarized in Table 6-24. Both linear and quadratic components of tool angle are significant, while cutting speed has only a linear effect on tool life. The $TC_{L \times L}$, $TC_{L \times Q}$, and $TC_{Q \times Q}$ interactions are also significant, with the $TC_{L \times Q}$ component quite large relative to the others.

Table 6-24 Analysis of Variance for Example 6-5

Source of Variation	Sum of Squares	Degrees of Freedom	Mean Square	F_0
Tool angle, T	24.33	2	12.17	8.45[a]
(T_L)	(8.33)	1	(8.33)	5.78[a]
(T_Q)	(16.00)	1	(16.00)	11.11[a]
Cutting speed, C	25.33	2	12.67	8.80[a]
(C_L)	(21.33)	1	(21.33)	14.81[a]
(C_Q)	(4.00)	1	(4.00)	2.77
TC interaction	61.34	4	~~15.33~~ 6.82	~~4.75[a]~~ 10.61
($TC_{L \times L}$)	(8.00)	1	(8.00)	5.56[a]
($TC_{L \times Q}$)	(42.67)	1	(42.67)	29.63[a]
($TC_{Q \times L}$)	(2.67)	1	(2.67)	1.85
($TC_{Q \times Q}$)	(8.00)	1	(8.00)	5.56[a]
Error	13.00	9	1.44	
Total	124.00			

[a]Significant at 5 percent.

Figure 6-4 plots the cell averages against the tool angle for all three speeds, and illustrates the quadratic effect of tool angle. Figure 6-5 plots cell averages versus cutting speed for the tool angles. Winer (1971) notes that two-dimensional graphs are useful for analyzing interactions between two quantitative factors.

If we let x_1 represent tool angle and x_2 represent cutting speed, then Table 6-24 implies a polynomial approximation for tool life, for example,

$$y = \alpha_0 + \alpha_1 P_1(x_1) + \alpha_{11} P_2(x_1) + \alpha_2 P_1(x_2) + \alpha_{12} P_1(x_1) P_1(x_2)$$
$$+ \alpha_{122} P_1(x_1) P_2(x_2) + \alpha_{1122} P_2(x_1) P_2(x_2) + \epsilon \quad (6\text{-}35)$$

where α_1 denotes the linear coefficient of tool angle, α_{122} denotes the coefficient for linear tool angle and quadratic cutting speed, and so on, and $P_i(x_j)$ are the orthogonal polynomials of Section 3-6. These coefficients and polynomials are evaluated exactly

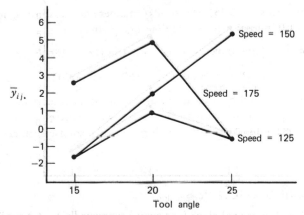

Figure 6-4. Cell averages versus tool angle for example 6-5.

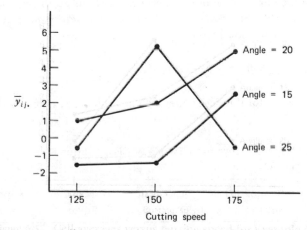

Figure 6-5. Cell averages versus cutting speed for Example 6-6.

as in Section 3-6; for example

$$\frac{\Sigma\Sigma Y_{ij}c_{ij}}{n\Sigma\Sigma c_{ij}^2} = \hat{\alpha}_{12} = \frac{\sum_{i=1}^{3}\sum_{j=1}^{3} y_{ij} P_1(x_{1i}) P_1(x_{2j})}{\sum_{i=1}^{3}\sum_{j=1}^{3} [P_1(x_{1i}) P_1(x_{2j})]^2} = \frac{-8}{8} = -1.00$$

and

$$P_1(x_1) P_1(x_2) = \lambda_1 \left(\frac{x_1 - \bar{x}_1}{d_1}\right) \lambda_1 \left(\frac{x_2 - \bar{x}_2}{d_2}\right)$$

$$= (1) \left(\frac{x_1 - 20}{5}\right) (1) \left(\frac{x_2 - 150}{25}\right)$$

Evaluating all the terms in (6-35) in a similar manner and substituting into (6-35) will yield the following response surface in the factors x_1 and x_2:

$$\hat{y} = 3.40 + 0.544(x_1 - 20) - 0.08(x_1 - 20)^2 + 0.1688(x_2 - 150) - 0.008(x_1 - 20)$$
$$\cdot (x_2 - 150) - 0.0004(x_1 - 20)(x_2 - 150)^2 - 0.00002(x_1 - 20)^2 (x_2 - 150)^2$$

This response surface could be used to predict tool life at intermediate levels of tool angle and cutting speed, or perhaps to find the levels of these factors that maximize tool life. Systematic methods for the exploration of response surfaces are presented in Chapter 14. ■

6-7 ONE OBSERVATION PER CELL

Occasionally one encounters a factorial experiment with only a single replicate; that is, one observation per cell. For example, if there are two factors and only one observation per cell, the linear statistical model is

$$y_{ij} = \mu + \tau_i + \beta_j + (\tau\beta)_{ij} + \epsilon_{ij} \quad \begin{cases} i = 1, 2, \ldots, a \\ j = 1, 2, \ldots, b \end{cases} \tag{6-36}$$

The analysis of variance for this design is shown in Table 6-25, assuming that both factors are fixed.

Table 6-25 Analysis of Variance for a Two-Way Classification, One Observation per Cell

Source of Variation	Sum of Squares	Degrees of Freedom	Mean Square	Expected Mean Square
Rows (A)	$\sum_{i=1}^{a} \dfrac{y_{i.}^2}{b} - \dfrac{y_{..}^2}{ab}$	$a - 1$	MS_A	$\sigma^2 + \dfrac{b \sum \tau_i^2}{a - 1}$
Columns (B)	$\sum_{j=1}^{b} \dfrac{y_{.j}^2}{a} - \dfrac{y_{..}^2}{ab}$	$b - 1$	MS_B	$\sigma^2 + \dfrac{a \sum \beta_j^2}{b - 1}$
Residual or AB	subtraction	$(a-1)(b-1)$	$MS_{Residual}$	$\sigma^2 + \dfrac{\sum\sum (\tau\beta)_{ij}^2}{(a-1)(b-1)}$
Total	$\sum_{i=1}^{a} \sum_{j=1}^{b} y_{ij}^2 - \dfrac{y_{..}^2}{ab}$	$ab - 1$		

From examining the expected mean squares, we see that the error variance σ^2 is not estimable; that is, the two-factor interaction effect $(\tau\beta)_{ij}$ and the experimental error cannot be separated in any obvious manner. Consequently, there are no tests on main effects unless the interaction effect is zero. If there is no interaction present, then $(\tau\beta)_{ij} = 0$ for all i and j, and a plausible model is

$$y_{ij} = \mu + \tau_i + \beta_j + \epsilon_{ij} \quad \begin{cases} i = 1, 2, \ldots, a \\ j = 1, 2, \ldots, b \end{cases} \tag{6-37}$$

If the model (6-37) is appropriate, then the residual mean square in Table 6-25 is an unbiased estimator of σ^2, and main effects may be tested by comparing MS_A and MS_B to $MS_{Residual}$.

To help determine whether or not interaction is present, we may use a procedure developed by Tukey (1949a). The technique partitions the residual sum of squares into a single-degree-of-freedom component due to nonadditivity (interaction), and a component for error with $(a - 1)(b - 1) - 1$ degrees of freedom. Computationally, we have

$$SS_N = \frac{\left[\displaystyle\sum_{i=1}^{a} \sum_{j=1}^{b} y_{ij}\, y_{i.}\, y_{.j} - y_{..}\left(SS_A + SS_B + \frac{y_{..}^2}{ab} \right) \right]^2}{ab\, SS_A\, SS_B} \tag{6-38}$$

with one degree of freedom, and

$$SS_{Error} = SS_{Residual} - SS_N \tag{6-39}$$

with $(a - 1)(b - 1) - 1$ degrees of freedom. To test for the presence of interaction, compute

$$F_0 = \frac{SS_N}{SS_{Error}/[(a - 1)(b - 1) - 1]} \tag{6-40}$$

If $F_0 > F_{\alpha, 1, (a-1)(b-1)-1}$, the hypothesis of no interaction must be rejected.

■ **EXAMPLE 6-6.** The impurity present in a chemical product is affected by two factors, pressure and temperature. The data from a single replicate of a factorial experiment is shown in Table 6-26.

Table 6-26   Impurity Data for Example 6-6

Temperature (°F)	Pressure					$y_{i.}$
	25	30	35	40	45	
100	5	4	6	3	5	23
125	3	1	4	2	3	13
150	1	1	3	1	2	8
$y_{.j}$	9	6	13	6	10	$44 = y_{..}$

We may compute

$$SS_A = \sum_{i=1}^{a} \frac{y_{i.}^2}{b} - \frac{y_{..}^2}{ab}$$

$$= \frac{23^2 + 13^2 + 8^2}{5} - \frac{44^2}{(3)(5)} = 23.33$$

$$SS_B = \sum_{j=1}^{b} \frac{y_{.j}^2}{a} - \frac{y_{..}^2}{ab}$$

$$= \frac{9^2 + 6^2 + 13^2 + 6^2 + 10^2}{3} - \frac{44^2}{(3)(5)} = 11.60$$

$$SS_T = \sum_{i=1}^{a} \sum_{j=1}^{b} y_{ij}^2 - \frac{y_{..}^2}{ab}$$

$$= 166 - 129.07 = 36.93$$

and

$$SS_{Residual} = SS_T - SS_A - SS_B$$

$$= 36.93 - 23.33 - 11.60 = 2.00$$

The sum of squares for nonadditivity is computed from (6-38) as follows:

$$\sum_{i=1}^{a} \sum_{j=1}^{b} y_{ij} y_{i.} y_{.j} = (5)(23)(9) + (4)(23)(6) + \cdots + (2)(8)(10) = 7236$$

$$SS_N = \frac{\left[\sum_{i=1}^{a} \sum_{j=1}^{b} y_{ij} y_{i.} y_{.j} - y_{..} \left(SS_A + SS_B + \frac{y_{..}^2}{ab} \right) \right]^2}{ab SS_A SS_B}$$

$$= \frac{[7236 - (44)(23.33 + 11.60 + 129.07)]^2}{(3)(5)(23.33)(11.60)}$$

$$= \frac{[20.00]^2}{4059.42} = 0.0985$$

and the error sum of squares is, from (6-39),

$$SS_{Error} = SS_{Residual} - SS_N$$

$$= 2.00 - 0.0985 = 1.9015$$

The complete analysis of variance is summarized in Table 6-27. The test statistic for nonadditivity is $F_0 = 0.0985/0.2716 = 0.36$, so we conclude that there is no evidence of interaction in these data. The main effects of temperature and pressure are significant. ∎

Table 6-27 Analysis of Variance for Example 6-6

Source of Variation	Sum of Squares	Degrees of Freedom	Mean Square	F_0
Temperature	23.33	2	11.67	42.97^a
Pressure	11.60	4	2.90	10.68^a
Nonadditivity	0.0985	1	0.0985	0.36
Error	1.9015	7	0.2716	
Total	36.93	14		

[a]Significant at 1 percent.

In concluding this section, we note that the two-factor factorial model with one observation per cell (6-37) looks exactly like the randomized complete block model (4-1). However, the *experimental situations* that lead to these two models are very different. Equation 6-37 is the linear model for a *factorial*, that is, all *ab* runs have been made in *random order;* while in (4-1) the *a* levels of the factor have only been randomized *within* the block. The blocks are a randomization restriction, and hence the manner in which the data are collected and the interpretation of the two models is quite different.

6-8 PROBLEMS

6-1. The yield of a chemical process is being studied. The two most important variables are thought to be the pressure and the temperature. Three levels of each factor are selected, and a factorial experiment with two replicates performed. the yield data are:

Pressure

Temperature	200	215	230
Low	90.4 90.2	90.7 90.6	90.2 90.4
Medium	90.1 90.3	90.5 90.6	89.9 90.1
High	90.5 90.7	90.8 90.9	90.4 90.1

Analyze the data and draw conclusions. Estimate the components in the appropriate statistical model. Under which conditions would you operate this process?

6-2. An engineer suspects that the surface finish of a metal part is influenced by the feed rate and the depth of cut. He selects three feed rates and randomly chooses four depths of cut. He conducts a factorial experiment and obtains the following data:

	Depth of Cut (in.)			
Feed Rate (in./min)	0.15	0.18	0.20	0.25
0.20	74	79	82	99
	64	68	88	104
	60	73	92	96
0.25	92	98	99	104
	86	104	108	110
	88	88	95	99
0.30	99	104	108	114
	98	99	110	111
	102	95	99	107

Analyze the data and state your conclusions. Estimate the components of the appropriate model.

6-3. For the data in Problem 6-2 compute a 95 percent interval estimate of the mean difference in response for feed rates of 0.20 and 0.25 in./min.

6-4. The factors that influence the breaking strength of synthetic fiber are being studied. Four machines and three operators are chosen at random and a factorial experiment is run using fiber from the same production batch. The results are as follows:

	Machine			
Operator	1	2	3	4
1	109	110	108	110
	110	115	109	108
2	110	110	111	114
	112	111	109	112
3	116	112	114	120
	114	115	119	117

Analyze the data and draw conclusions. Estimate the components of variance.

6-5. Suppose that in Problem 6-4 the operators were chosen at random, but only four machines were avilable for the test. Test for interaction and main effects at the 5 percent level. Does the new experimental situation influence either the analysis or your conclusions?

6-6. A mechanical engineer is studying the thrust force developed by a drill press. He suspects that the drilling speed and the feed rate of the material are the most important factors. He randomly selects four feed rates, and uses a high and low drill speed chosen to represent the extreme operating conditions. He obtains the results shown below. Analyze the data and draw conclusions.

Feed Rate

Drill Speed	.015	.030	.045	.060
125	2.70	2.45	2.60	2.75
	2.78	2.49	2.72	2.86
200	2.83	2.85	2.86	2.94
	2.86	2.80	2.87	2.88

6-7. An experiment is conducted to study the influence of operating temperature and three types of face-plate glass in the light output of an oscilloscope tube. The following data are collected.

Temperature

Glass Type	100	125	150
1	580	1090	1392
	568	1087	1380
	570	1085	1386
2	550	1070	1328
	530	1035	1312
	579	1000	1299
3	546	1045	867
	575	1053	904
	599	1066	889

Assume that both factors are fixed. Is there a significant interaction effect? Does glass type or temperature affect the response? What conclusions can you draw? Use the method discussed in the text to partition the temperature effect into its linear and quadratic components. Break the interaction down into appropriate components.

6-8. Consider the data in Problem 6-1. Use the method described in the text to extract the linear and quadratic effect of pressure.

6-9. Use Duncan's multiple range test to determine which levels of the pressure factor are significantly different for the data in Problem 6-1.

6-10. Consider the following data:

Column Factor

Row Factor	1	2	3
1	570	1063	565
	565	1080	510
	583	1043	590
2	528	988	526
	547	1026	538
	521	1004	532

Suppose it is known in advance that no interaction exists. Write down the statistical model. Conduct the analysis of variance assuming both treatments to be fixed effects. What conclusions can be drawn? Derive the expected mean squares. Estimate the components of the statistical model.

6-11. Derive the expected mean squares for a two-way classification analysis of variance with one observation per cell, assuming that both factors are fixed.

6-12. Consider a two-way classification analysis of variance with one observation per cell. What are the expected mean squares, assuming both row and column factors are random? Under what conditions do we have exact tests for main effects? Write down the analysis of variance table.

6-13. Consider a two-way classification analysis of variance with one observation per cell. If the row factor is fixed and the column factor is random, what are the expected mean squares? Under what conditions do we have exact tests for main effects? Write down the analysis of variance table.

6-14. Consider the following data, where row and column factors are both fixed. Analyze the data and draw conclusions. Perform a test for nonadditivity.

Column Factor

Row Factor	1	2	3	4
1	36	39	36	32
2	18	20	22	20
3	30	37	33	34

6-15. The shear strength of an adhesive is thought to be affected by the application pressure and temperature. A factorial experiment is performed, assuming both factors to be fixed. Analyze the data and draw conclusions. Perform a test for nonadditivity.

Temperature (°F)

Pressure (lb/in.^{2})	250	260	270
120	9.60	11.28	9.00
130	9.69	10.10	9.57
140	8.43	11.01	9.03
150	9.98	10.44	9.80

6-16. For the data in Problem 6-15, suppose the temperature and pressure levels are chosen at random. Assuming no interaction, conduct the analysis of variance. How do your conclusions differ, if at all, from the conclusions of Problem 6-15?

6-17. Consider the three-way analysis of variance model

$$y_{ijk} = \mu + \tau_i + \beta_j + \gamma_k + (\tau\beta)_{ij} + (\beta\gamma)_{jk} + \epsilon_{ijk} \begin{cases} i = 1, 2, \ldots, a \\ j = 1, 2, \ldots, b \\ k = 1, 2, \ldots, c \end{cases}$$

Notice that there is only one replicate. Assuming all factors as fixed, write down the analysis of variance table, including the expected mean squares. What would you use as the "experimental error" in order to test hypotheses?

6-18. The percentage of hardwood concentration in raw pulp, the vat pressure, and the cooking time of pulp are being investigated for their effects on the strength of paper. Three levels of hardwood concentration, three levels of pressure, and two cooking times are selected. All factors may be regarded as fixed effects. A factorial experiment with two replicates is conducted, and the following data are obtained:

Percentage of Hardwood Concentration	Cooking Time 3.0 hours Pressure			Cooking Time 4.0 hours Pressure		
	400	500	650	400	500	650
2	196.6 196.0	197.7 196.0	199.8 199.4	198.4 198.6	199.6 200.4	200.6 200.9
4	198.5 197.2	196.0 196.9	198.4 197.6	197.5 198.1	198.7 198.0	199.6 199.0
8	197.5 196.6	195.6 196.2	197.4 198.1	197.6 198.4	197.0 197.8	198.5 199.8

Analyze the data and draw conclusions. Under what set of conditions would you operate this process? Why?

6-19. The quality control department of a fabric finishing plant is studying the effect of several factors on the dyeing of cotton-synthetic cloth used to manufacture men's shirts. Three operators, three cycle times, and two temperatures were selected and three small specimens of cloth were dyed under each set of conditions. The finished cloth was compared to a standard and a numerical score was assigned. The results are shown below. Analyze the data and draw conclusions.

	Temperature					
	300°			350°		
	Operator			Operator		
Cycle Time	1	2	3	1	2	3
40	23	27	31	24	38	34
	24	28	32	23	36	36
	25	26	29	28	35	39
50	36	34	33	37	34	34
	35	38	34	39	38	36
	36	39	35	35	36	31
60	28	35	26	26	36	28
	24	35	27	29	37	26
	27	34	25	25	34	24

6-20. In Problem 6-1, determine the power of the test for detecting a true difference in pressure such that $\sum_{j=1}^{3} \beta_j^2 = 0.08$, where β_j is the jth pressure effect.

6-21. In Problem 6-4, determine the power of the test for detecting a machine effect such that $\sigma_\beta^2 = \sigma^2$, where σ_β^2 is the variance component for the machine factor. Are two replicates sufficient?

6-22. In the two-way classification mixed model analysis of variance show that Cov $[(\tau\beta)_{ij}, (\tau\beta)_{i'j}] = -(1/a)\,\sigma_{\tau\beta}^2$ for $i \neq i'$.

6-23. Show that the method of analysis of variance always produces unbiased point estimates of the variance components in any random or mixed model.

6-24. Invoking the usual normality assumptions, find an expression for the probability that a negative estimate of a variance component will be obtained by the analysis of variance method. Using this result, write a statement giving the probability that $\hat{\sigma}_\tau^2 < 0$ in a one-way analysis of variance. Comment on the usefulness of this probability statement.

6-25. Analyze the data in Problem 6-2 using both the mixed models discussed in the text. Compare the results obtained from the two models.

6-26. Consider the two-factor mixed model. Show that the standard error of the fixed factor mean (e.g., A) for Duncan's multiple range test is $[MS_{AB}/bn]^{1/2}$.

Chapter Seven
Rules for Sums of Squares and Expected Mean Squares

An important part of any experimental design problem is conducting the analysis of variance. This involves determining the sum of squares for each component in the model and the number of degrees of freedom associated with each sum of squares. Then, to construct appropriate test statistics, the expected mean squares must be determined. In complex design situations, particularly those involving random or mixed models, it is frequently helpful to have a formal procedure for this process.

We will present a set of rules for writing down the sums of squares for the effects in model (the expression *effect* now includes both main effects and interactions), the number of degrees of freedom, and the expected mean squares for any balanced factorial, nested,[1] or nested-factorial experiment (note that partially balanced arrangements, such as Latin squares and incomplete block designs are specifically excluded). These rules are discussed by several authors, including Scheffé (1959), Bennett and Franklin (1954), Hicks (1973), and Searle (1971). The two-factor factorial model will be used to illustrate the application of the rules.

[1]Nested designs are discussed in Chapter 11.

7-1   RULES FOR SUMS
OF SQUARES

RULE 1.   The error term in the model, $\epsilon_{ij\dots m}$, is written as $\epsilon_{(ij\dots)m}$, where the subscript m denotes the replication subscript. For the two-way model, this rule implies that ϵ_{ijk} becomes $\epsilon_{(ij)k}$.

RULE 2.   In addition to an overall mean (μ) and an error term $[\epsilon_{(ij\dots)m}]$, the model contains all main effects and any interactions which the experimenter assumes exist. If all possible interactions between k factors exist, then there are $\binom{k}{2}$ two-factor interactions, $\binom{k}{3}$ three-factor interactions, $\dots$, 1 k-factor interaction. If one of the factors in a term appears in parenthesis, then there is no interaction between that factor and the other factors in that term.

indices of see p. 267

α

RULE 3.   For each term in the model, divide the subscripts into three classes: (a) live—those subscripts that are present in the term and not in parentheses; (b) dead—those subscripts that are present in the term and in parentheses; and (c) absent—those subscripts that are present in the model, but not in that particular term.

Thus, in $(\tau\beta)_{ij}$, i and j are live and k is missing, and in $\epsilon_{(ij)k}$, k is live and i and j are dead.

RULE 4.   Degrees of Freedom.   The number of degrees of freedom for any term in the model is the product of the number of levels associated with each dead subscript and the number of levels minus one associated with each live subscript.

For example, the number of degrees of freedom associated with $(\tau\beta)_{ij}$ is $(a-1)(b-1)$, and the number of degrees of freedom associated with $\epsilon_{(ij)k}$ is $ab(n-1)$.

RULE 5.   Sums of Squares.   To obtain a computing formula for the sum of squares for any effect, first expand the number of degrees of freedom for that effect. For example, for β_j the expanded degrees of freedom are simply $b-1$. Each term in this quantity is a *symbolic form* of an uncorrected sum of squares. Then

(a)   Let 1 represent the correction factor; that is,

$$1 = \frac{\left(\sum_{i=1}^{a} \sum_{j=1}^{b} \cdots \sum_{m=1}^{n} y_{ij\dots m} \right)^2}{ab \cdots n}$$

For each remaining symbolic form, write down the total of all observations in summation form. Thus, for b write down $\sum_{i=1}^{a} \sum_{j=1}^{b} \sum_{k=1}^{c} y_{ijk}$.

(b)   Reorder the summation signs so that those relative to the letters in the symbolic form of interest (in this case b) come first, and enclose the remaining elements of the term in parentheses. Therefore, for b we obtain

$$\sum_{j=1}^{b} \left(\sum_{i=1}^{a} \sum_{k=1}^{n} y_{ijk} \right)$$

(c) Convert the quantity within parentheses to the standard "dot subscript" notation, where a dot replacing a subscript indicates summation over that subscript. Thus

$$\sum_{j=1}^{b} \left(\sum_{i=1}^{a} \sum_{k=1}^{n} y_{ijk} \right) = \sum_{j=1}^{b} (y_{.j.})$$

(d) Square the term within parenthesis and divide by the product of the number of levels of the "dot" subscripts. For our example, the b term becomes $\sum_{j=1}^{b} y_{.j.}^2 /an$, which is an *uncorrected sum of squares*. The *corrected* sum of squares of the effect results from replacing each symbolic form by the corresponding uncorrected sum of squares. Thus, to obtain the sum of squares for β_j, replace the symbolic forms in $b - 1$ by the uncorrected sum of squares formed above, and we see that

$$SS_B = \sum_{j=1}^{b} \frac{y_{.j.}^2}{an} - \frac{y_{...}^2}{abn}$$

This is clearly the sum of squares for the main effect B in a two-way classification.

As another illustration of Rule 5, consider finding the sum of squares for $(\tau\beta)_{ij}$; that is, SS_{AB}. The expanded number of degrees of freedom is $(a - 1)(b - 1) = ab - a - b + 1$. Therefore the sum of squares is determined as follows:

ab	$-a$	$-b$	$+1$	
$\displaystyle\sum_{i=1}^{a}\sum_{j=1}^{b}\sum_{k=1}^{n} y_{ijk}$	$\displaystyle\sum_{i=1}^{a}\sum_{j=1}^{b}\sum_{k=1}^{n} y_{ijk}$	$\displaystyle\sum_{i=1}^{a}\sum_{j=1}^{b}\sum_{k=1}^{n} y_{ijk}$	$\dfrac{y_{...}^2}{abn}$	part (a)
$\displaystyle\sum_{i=1}^{a}\sum_{j=1}^{b}\left(\sum_{k=1}^{n} y_{ijk}\right)$	$\displaystyle\sum_{i=1}^{a}\left(\sum_{j=1}^{b}\sum_{k=1}^{n} y_{ijk}\right)$	$\displaystyle\sum_{j=1}^{b}\left(\sum_{i=1}^{a}\sum_{k=1}^{n} y_{ijk}\right)$	$\dfrac{y_{...}^2}{abn}$	part (b)
$\displaystyle\sum_{i=1}^{a}\sum_{j=1}^{b} (y_{ij.})$	$\displaystyle\sum_{i=1}^{a} (y_{i..})$	$\displaystyle\sum_{j=1}^{b} (y_{.j.})$	$\dfrac{y_{...}^2}{abn}$	part (c)
$\displaystyle\sum_{i=1}^{a}\sum_{j=1}^{b} \frac{y_{ij.}^2}{n}$	$\displaystyle\sum_{i=1}^{a} \frac{y_{i..}^2}{bn}$	$\displaystyle\sum_{j=1}^{b} \frac{y_{.j.}^2}{an}$	$\dfrac{y_{...}^2}{abn}$	part (d)

Now, combining the uncorrected sums of squares in the last row according to the signs at the top of each column, we obtain

$$SS_{AB} = \sum_{i=1}^{a}\sum_{j=1}^{b} \frac{y_{ij.}^2}{n} - \sum_{i=1}^{a} \frac{y_{i..}^2}{bn} - \sum_{j=1}^{b} \frac{y_{.j.}^2}{an} + \frac{y_{...}^2}{abn}$$

This last equation is equivalent to $SS_{AB} = SS_{\text{subtotals}} - SS_A - SS_B$, which we obtained as Equation 6-9.

7-2 RULES FOR EXPECTED MEAN SQUARES

We have seen that the expected mean squares play an important role in the analysis of variance. By examining the expected mean squares, one may develop the appropriate statistic for testing hypotheses about any model parameter. The test statistic is a ratio of mean squares, chosen so that the expected value of the *numerator* mean square differs from the expected value of the *denominator* mean square only by the variance component or fixed factor in which we are interested.

It is always possible to determine the expected mean squares in any model as we did in Chapter 3; that is, by direct application of the expectation operator. This "brute force" method, as it is often called, can be very tedious. The rules given below will always produce the expected mean squares for any design without resort to the brute force approach, and with practice, they become relatively simple to use. When applied to a mixed model, these rules produce expected mean squares consistent with the assumptions for the standard mixed model of Section 6-4. We shall illustrate the rules for the two-factor fixed effects factorial model.

RULE 1. Each effect has either a variance component (random effect) or a fixed factor (fixed effect) associated with it. If an interaction contains at least one random effect, the entire interaction is considered as random. A variance component has Greek letters as subscripts to identify the particular random effect. Thus, in a two-way mixed model with factor A fixed and factor B random, the variance component for B is σ_β^2, and the variance component for AB is $\sigma_{\tau\beta}^2$. A fixed effect is always represented by the sum of squares of the model components associated with that factor, divided by its degrees of freedom. In our example, the effect of A is

$$\frac{\sum_{i=1}^{a} \tau_i^2}{a-1} .$$

RULE 2. Expected Mean Squares. To obtain the expected mean squares, prepare the following table. There is a row for each model component (mean square), and a column for each subscript. Over each subscript write the number of levels of the factor associated with that subscript, and whether the factor is fixed (F) or random (R). Replicates are always considered random.

(a) In each row, write 1 if one of the dead subscripts in the row component matches the subscript in the column.

	F	F	R
	a	b	n
Factor	i	j	k
τ_i			
β_j			
$(\tau\beta)_{ij}$			
$\epsilon_{(ij)k}$	1	1	

(b) In each row, if any of the subscripts on the row component match the subscript in the column, write 0 if the column is headed by a fixed factor and 1 if the column is headed by a random factor.

Factor	$\begin{matrix}F\\a\\i\end{matrix}$	$\begin{matrix}F\\b\\j\end{matrix}$	$\begin{matrix}R\\n\\k\end{matrix}$
τ_i	0		
β_j		0	
$(\tau\beta)_{ij}$	0	0	
$\epsilon_{(ij)k}$	1	1	1

(c) In the remaining empty row positions, write the number of levels shown above the column heading.

Factor	$\begin{matrix}F\\a\\i\end{matrix}$	$\begin{matrix}F\\b\\j\end{matrix}$	$\begin{matrix}R\\n\\k\end{matrix}$
τ_i	0	b	n
β_j	a	0	n
$(\tau\beta)_{ij}$	0	0	n
$\epsilon_{(ij)k}$	1	1	1

(d) To obtain the expected mean square for any model component, first cover all columns headed by live subscripts on that component. Then, in each row that contains *at least* the *same* subscripts as those on the component being considered, take the product of the visible numbers and multiply by the appropriate fixed or random factor from Rule 1. The sum of these quantities is the expected mean square of the model component being considered. To find $E(MS_A)$, for example, cover column i. The product of the visible numbers in the rows that contain at least subscript i are bn (row 1), 0 (row 3), and 1 (row 4). Note that i is missing in row 2. Therefore, the expected mean square is

$$E(MS_A) = \sigma^2 + \frac{bn\sum_{i=1}^{a}\tau_i^2}{a-1}$$

The complete table of expected mean squares for this design is shown in Table 7-1.

Table 7-1 Expected Mean Square Derivation, Two-Factor
Fixed Effects Model

Factor	F a i	F b j	R n k	Expected Mean Square
τ_i	0	b	n	$\sigma^2 + \dfrac{bn \sum \tau_i^2}{a-1}$
β_j	a	0	n	$\sigma^2 + \dfrac{an \sum \beta_j^2}{b-1}$
$(\tau\beta)_{ij}$	0	0	n	$\sigma^2 + \dfrac{n \sum \sum (\tau\beta)_{ij}^2}{(a-1)(b-1)}$
$\epsilon_{(ij)k}$	1	1	1	σ^2

Tables 7-2 and 7-3 display the expected mean square derivations for the two-factor random and mixed models, respectively.

Table 7-2 Expected Mean Square Derivation, Two-Factor
Random Effects Model

Factor	R a i	R b j	R n k	Expected Mean Square
τ_i	1	b	n	$\sigma^2 + n\sigma_{\tau\beta}^2 + bn\sigma_\tau^2$
β_j	a	1	n	$\sigma^2 + n\sigma_{\tau\beta}^2 + an\sigma_\beta^2$
$(\tau\beta)_{ij}$	1	1	n	$\sigma^2 + n\sigma_{\tau\beta}^2$
$\epsilon_{(ij)k}$	1	1	1	σ^2

A three-factor factorial design is treated in the following example.

■**EXAMPLE 7-1.** Consider a three-factor factorial experiment with a levels of factor A, b levels of factor B, c levels of factor C and n replicates. The analysis of this design assuming that all factors are fixed effects is given in Section 6-5. We will now determine the expected mean squares assuming that all factors are *random*. The appropriate statistical model is

$$y_{ijkl} = \mu + \tau_i + \beta_j + \gamma_k + (\tau\beta)_{ij} + (\tau\gamma)_{ik} + (\beta\gamma)_{jk} + (\tau\beta\gamma)_{ijk} + \epsilon_{ijkl}$$

Using the rules described above, the expected mean squares are derived in Table 7-4.

Table 7-3 Expected Mean Square Derivation, Two-Factor Mixed Model

Factor	F a i	R b j	R n k	Expected Mean Square
τ_i	0	b	n	$\sigma^2 + n\sigma^2_{\tau\beta} + \dfrac{bn\sum \tau_i^2}{a-1}$
β_j	a	1	n	$\sigma^2 + an\sigma^2_\beta$
$(\tau\beta)_{ij}$	0	1	n	$\sigma^2 + n\sigma^2_{\tau\beta}$
$\epsilon_{(ij)k}$	1	1	1	σ^2

We notice, by examining the expected mean squares in Table 7-4, that if A, B, and C are all random factors, then no exact test exists for the main effects. That is, if we wish to test the hypothesis that $\sigma^2_\tau = 0$, we cannot form a ratio of two expected mean squares such that the only term in the numerator which is not in the denominator is $bcn\sigma^2_\tau$. The same phenomenon occurs for the main effects of B and C. Notice that proper tests do exist for the two-factor and three-factor interactions. However, it is likely that tests on the main effects are of primary importance to the experimenter. Therefore, how should the main effects be tested? This problem is considered in the next section. ∎

Table 7-4 Expected Mean Square Derivation for the Three-Way Random Model

Factor	R a i	R b j	R c k	R n l	Expected Mean Squares
τ_i	1	b	c	n	$\sigma^2 + cn\sigma^2_{\tau\beta} + bn\sigma^2_{\tau\gamma} + n\sigma^2_{\tau\beta\gamma} + bcn\sigma^2_\tau$
β_j	a	1	c	n	$\sigma^2 + cn\sigma^2_{\tau\beta} + an\sigma^2_{\beta\gamma} + n\sigma^2_{\tau\beta\gamma} + acn\sigma^2_\beta$
γ_k	a	b	1	n	$\sigma^2 + bn\sigma^2_{\tau\gamma} + an\sigma^2_{\beta\gamma} + n\sigma^2_{\tau\beta\gamma} + abn\sigma^2_\gamma$
$(\tau\beta)_{ij}$	1	1	c	n	$\sigma^2 + n\sigma^2_{\tau\beta\gamma} + cn\sigma^2_{\tau\beta}$
$(\tau\gamma)_{ik}$	1	b	1	n	$\sigma^2 + n\sigma^2_{\tau\beta\gamma} + bn\sigma^2_{\tau\gamma}$
$(\beta\gamma)_{jk}$	a	1	1	n	$\sigma^2 + n\sigma^2_{\tau\beta\gamma} + an\sigma^2_{\beta\gamma}$
$(\tau\beta\gamma)_{ijk}$	1	1	1	n	$\sigma^2 + n\sigma^2_{\tau\beta\gamma}$
$\epsilon_{(ijk)l}$	1	1	1	1	σ^2

7-3 APPROXIMATE *F* TESTS

In factorial experiments with three or more factors involving a random or mixed model, and certain other more complex designs, there is frequently no exact test statistic for at least one effect in the model. One possible solution to this dilemma is to assume that certain interactions are negligible. To illustrate, in Example 7-1, if we could reasonably assume that all two-factor interactions are negligible, then we could put $\sigma_{\tau\beta}^2 = \sigma_{\tau\gamma}^2 = \sigma_{\beta\gamma}^2 = 0$, and tests for main effects could be conducted.

While this seems to be an attractive possibility, we must point out that there must be something in the nature of the process—or some strong prior knowledge—in order for us to assume that one or more interactions are negligible. In general, this assumption is not easily made, nor should it be taken lightly. We should not eliminate certain interactions from the model without conclusive evidence that it is appropriate to do so. A procedure advocated by some experimenters is to test interactions first, then set to zero those interactions found insignificant, and then assume these interactions to be zero when testing other effects in the same experiment. While sometimes done in practice, this procedure can be dangerous because any decision regarding the interaction is subject to both types of errors.

A variation of this idea is pooling mean squares in the analysis of variance to obtain an estimate of error with more degrees of freedom. For instance, suppose that in Example 7-1 the test statistic $F_0 = MS_{ABC}/MS_E$ was not significant. Thus, $H_0: \sigma_{\tau\beta\gamma}^2 = 0$ is not rejected, and both MS_{ABC} and MS_E estimate the error variance σ^2. The experimenter might consider pooling or combining MS_{ABC} and MS_E according to

$$MS_{E'} = \frac{abc(n-1)MS_E + (a-1)(b-1)(c-1)MS_{ABC}}{abc(n-1) + (a-1)(b-1)(c-1)}$$

so that $E(MS_{E'}) = \sigma^2$. Note that $MS_{E'}$ has $abc(n-1) + (a-1)(b-1)(c-1)$ degrees of freedom, compared to $abc(n-1)$ degrees of freedom for the original MS_E. The danger of pooling is that one may make a type II error and combine the mean square for a factor that really *is* significant with error, thus obtaining a new residual mean square $(MS_{E'})$ that is too large. This will make other significant effects more difficult to detect. On the other hand, if the original error mean square has a very small number of degrees of freedom (e.g., less than 6), then the experimenter may have much to gain by pooling as it could potentially increase the precision of further tests considerably. A reasonably practical procedure is as follows. If the original error mean square has more than 6 degrees of freedom, do not pool. If the original error mean square has fewer than 6 degrees of freedom, pool only if the *F* statistic for the mean square to be pooled is not significant at a large value of α, such as $\alpha = .25$.

If we cannot assume certain interactions to be negligible and still need to make inferences about those effects for which exact tests do not exist, a procedure due to Satterthwaite (1946) can be employed. Satterthwaite's method uses linear combinations of mean squares, for example

$$MS' = MS_r + \cdots + MS_s \tag{7-1}$$

and

$$MS'' = MS_u + \cdots + MS_v \tag{7-2}$$

where the mean squares in (7-1) and (7-2) are chosen so that $E(MS') - E(MS'')$ is equal to the effect (model parameter or variance component) considered in the null hypothesis. Then the test statistic would be

$$F = \frac{MS'}{MS''} \sim F_{p,q} \tag{7-3}$$

where

$$p = \frac{(MS_r + \cdots + MS_s)^2}{MS_r^2/f_r + \cdots + MS_s^2/f_s} \tag{7-4}$$

and

$$q = \frac{(MS_u + \cdots + MS_v)^2}{MS_u^2/f_u + \cdots + MS_v^2/f_v} \tag{7-5}$$

In p and q, f_i is the number of degrees of freedom associated with the mean square MS_i. There is no assurance that p and q will be integers, and so it will be necessary to interpolate in the tables of the F distribution. For example, in the three-way classification random effects model (Table 7-4), it is relatively easy to see that an appropriate test statistic for H_0: $\sigma_T^2 = 0$ would be $F = MS'/MS''$ with

$$MS' = MS_A + MS_{ABC}$$

and

$$MS'' = MS_{AB} + MS_{AC}$$

The degrees of freedom for F would be computed from (7-4) and (7-5).

The theory underlying this test is that both numerator and denominator of the test statistic (7-3) are distributed approximately as multiples of chi-square random variables, and as no mean square appears both in the numerator and denominator of (7-3), numerator and denominator are independent. Thus F in (7-3) is distributed approximately as $F_{p,q}$. Satterthwaite has remarked that caution should be used in applying the procedure when some of the mean squares in MS' and MS'' are involved negatively. Gaylor and Hopper (1969) report that if $MS' = MS_1 - MS_2$, then Satterthwaite's approximation holds reasonably well if

$$\frac{MS_1}{MS_2} > F_{.025, f_2, f_1} \times F_{.50, f_1, f_2}$$

if $f_1 \leqslant 100$ and $f_2 \geqslant f_1/2$.

Table 7-5   Coded Pressure Drop Data for the Turbine Experiment

Pressure (C)	60°F Speed (B) 150	200	225	300	75°F Speed (B) 150	200	225	300	90°F Speed (B) 150	200	225	300
50 psi	-2	0	-1	4	14	6	1	-7	-8	2	-1	-2
	-3	-9	-8	4	14	0	2	6	-8	20	-2	1
75 psi	-6	-5	-8	-3	22	8	6	-5	-8	1	-9	-8
	4	-1	-2	-7	24	6	2	2	3	-7	-8	3
85 psi	-1	-4	0	-2	20	2	3	-5	-2	-1	-4	1
	-2	-8	-7	4	16	0	0	-1	-1	-2	-7	3

The column headers for this table are: Gas Temperature (A) spanning 60°F, 75°F, 90°F, each with Speed (B) at 150, 200, 225, 300.

■**EXAMPLE 7-2.** The pressure drop experienced across an expansion valve in a turbine is being studied. The design engineer considers the important variables that influence pressure drop to be gas temperature on the inlet side (A), turbine speed (B), and inlet gas pressure (C). These three factors are arranged in a factorial design, with gas temperature fixed, and turbine speed and pressure random. The coded data for two replicates are shown in Table 7-5. The linear model for this design is

$$y_{ijkl} = \mu + \tau_i + \beta_j + \gamma_k + (\tau\beta)_{ij} + (\tau\gamma)_{ik} + (\beta\gamma)_{jk} + (\tau\beta\gamma)_{ijk} + \epsilon_{ijkl}$$

where τ_i is the effect of the gas temperature (A), β_j is the effect of the turbine speed (B), and γ_k is the effect of the gas pressure (C).

The analysis of variance is shown in Table 7-6. A column headed "expected mean squares" has been added to this table, and the entries in this column are derived by the methods of Section 7-2. From the expected mean square column we observe that exact tests exist for all effects except the main effect A. Results for these tests are shown in Table 7-6. To test $H_0: \tau_i = 0$ we could use the statistic

$$F = \frac{MS'}{MS''}$$

where

$$MS' = MS_A + MS_{ABC}$$

and

$$MS'' = MS_{AB} + MS_{AC}$$

since

$$E(MS') - E(MS'') = \frac{bcn \sum \tau_i^2}{a - 1}$$

Table 7-6 Analysis of Variance for the Pressure Drop Data

Source of Variation	Sum of Squares	Degrees of Freedom	Expected Mean Squares	Mean Square	F_0
Temperature, A	1005.03	2	$\sigma^2 + bn\sigma_{\tau\gamma}^2 + cn\sigma_{\tau\beta}^2 + n\sigma_{\tau\beta\gamma}^2 + \dfrac{bcn\sum \tau_i^2}{a-1}$	502.51	2.12
Speed, B	423.48	3	$\sigma^2 + an\sigma_{\beta\gamma}^2 + acn\sigma_\beta^2$	141.16	3.76[b]
Pressure, C	11.19	2	$\sigma^2 + an\sigma_{\beta\gamma}^2 + abn\sigma_\gamma^2$	5.60	0.15
AB	1242.31	6	$\sigma^2 + n\sigma_{\tau\beta\gamma}^2 + cn\sigma_{\tau\beta}^2$	207.05	12.42[a]
AC	149.22	4	$\sigma^2 + n\sigma_{\tau\beta\gamma}^2 + bn\sigma_{\tau\gamma}^2$	37.31	2.24[c]
BC	225.47	6	$\sigma^2 + an\sigma_{\beta\gamma}^2$	37.58	1.96
ABC	200.11	12	$\sigma^2 + n\sigma_{\tau\beta\gamma}^2$	16.68	0.87
Error	690.50	36	σ^2	19.18	
Total	3947.31	71			

[a] Significant at 1 percent.
[b] Significant at 10 percent.
[c] Significant at 25 percent.

To determine the test statistic for $H_0: \tau_i = 0$ we compute

$$MS' = MS_A + MS_{ABC}$$

$$= 502.51 + 16.68 = 519.15$$

$$MS'' = MS_{AB} + MS_{AC}$$

$$= 207.05 + 37.31 = 244.36$$

Figure 7-1. Interactions in pressure drop experiment.

and

$$F = \frac{MS'}{MS''} = \frac{519.15}{244.36} = 2.12$$

The degrees of freedom for this statistic are

$$p = \frac{(MS_A + MS_{ABC})^2}{MS_A^2/2 + MS_{ABC}^2/12}$$

$$= \frac{(519.15)^2}{(502.51)^2/2 + (16.68)^2/12} = 2.13$$

and

$$q = \frac{(MS_{AB} + MS_{AC})^2}{MS_{AB}^2/6 + MS_{AC}^2/4}$$

$$= \frac{(244.36)^2}{(207.05)^2/6 + (37.31)^2/4} = 7.97$$

Comparing $F = 2.12$ to $F_{.05,2.13,7.97}$ we cannot reject H_0.

The main effect of B and the AB interaction are large, and there is some indication of an AC interaction. The graphical analysis of the AB and AC interactions, shown in Figure 7-1, indicates that the effect of A may be large at low levels of B and at $C = 75$. Thus, it seems possible that the main effect of A is masked by the large AB interaction.   ∎

7-4 PROBLEMS

7-1. Use the procedure discussed in Section 7-2 to determine the expected mean squares for the three-factor factorial, fixed effects model.

7-2. By application of the expectation operator, develop the expected mean squares for the two-factor factorial, mixed model. Check your results with the expected mean squares given in Table 7-3 to see that they agree.

7-3. Consider the three-factor factorial design in Example 7-1. Propose appropriate test statistics for all main effects and interactions. Repeat for the case where A and B are fixed and C is random.

7-4. Consider the data in Example 7-2. Analyze the data for the case where A, B, and C are random.

7-5. Derive the expected mean squares shown in Table 7-6.

7-6. Consider a four-factor factorial experiment where factor A is at a levels, factor B is at b levels, factor C at c levels, factor D at d levels, and there are n replicates. Write down the sums of squares, degrees of freedom, and the expected mean squares for the following cases:
(a) A, B, C, and D are fixed factors.

(b) $A, B, C,$ and D are random factors.

(c) A is fixed, and $B, C,$ and D are random.

(d) A and B are fixed, and C and D are random.

(e) $A, B,$ and C are fixed, and D is random.

Do exact tests exist for all effects? If not, propose test statistics for those effects which cannot be directly tested.

7-7. Consider the data in Problem 6-18. Suppose that hardwood concentrations and cooking times are fixed while pressure is random. Analyze the data under these conditions and draw appropriate conclusions.

7-8. In Problem 6-19, assume that operators, cycle times, and temperatures were selected at random. Analyze the data under these conditions and draw conclusions. Estimate the variance components.

7-9. Consider the three-way factorial model

$$y_{ijk} = \mu + \tau_i + \beta_j + \gamma_k + (\tau\beta)_{ij} + (\beta\gamma)_{jk} + \epsilon_{ijk} \begin{cases} i = 1, 2, \dots, a \\ j = 1, 2, \dots, b \\ k = 1, 2, \dots, c \end{cases}$$

Assuming all factors random, outline the analysis of variance table, including the expected mean squares. Propose appropriate test statistics for all effects.

7-10. The three-way factorial model for a single replicate is

$$y_{ijk} = \mu + \tau_i + \beta_j + \gamma_k + (\tau\beta)_{ij} + (\beta\gamma)_{jk} + (\tau\gamma)_{ik} + (\tau\beta\gamma)_{ijk} + \epsilon_{ijk}$$

If all factors are random, can any effects be tested? If the three-way interaction and the $(\tau\beta)_{ij}$ interaction do not exist, can all remaining effects be tested?

Chapter Eight
2^k and 3^k Factorial Designs

8-1 INTRODUCTION

Factorial designs are widely used in experiments involving several factors where it is necessary to study the joint effect of these factors on a response. Chapter 6 presented general methods for the analysis of factorial designs. However, there are several special cases of the general factorial design that are important because they are widely used in research work, and also because they form the basis of other designs of considerable practical value.

The first of these special cases is that of k factors, each at only two levels. These levels may be quantitative, such as two values of temperature, pressure, or time; or they may be qualitative, such as two machines, two operators, the "high" and "low" levels of a factor, or perhaps the presence and absence of a factor. A complete replicate of such a design requires $2 \times 2 \times \cdots \times 2 = 2^k$ observations, and is called a 2^k factorial design. The second special case is that of k factors, each at three levels, which is called a 3^k factorial design.

This chapter presents special methods for the analysis of these two useful series of designs. Throughout this chapter we will assume that (1) the factors are fixed, (2) the designs are completely randomized, and (3) the usual normality assumptions are satisfied.

180

8-2 ANALYSIS OF THE 2^k FACTORIAL DESIGN

The 2^k design is particularly useful in the early stages of experimental work, when there are likely to be many factors to be investigated. It provides the smallest number of treatment combinations with which k factors can be studied in a complete factorial arrangement. As there are only two levels for each factor, we must assume that the response is approximately linear over the range of the factor levels chosen.

8-2.1 The 2^2 Design

The first design in the 2^k series is one with only two factors, say A and B, each run at two levels. This design is called a 2^2 factorial design. The levels of the factors may be arbitrarily called "low" and "high." As an example, consider an investigation into the effect of concentration of reactant and the presence of catalyst on the reaction time of a chemical process. Let reactant concentration be factor A, and let the two levels of interest be 15 percent and 25 percent. Catalyst is factor B, with the high level denoting the presence of the catalyst and the low level denoting its absence. Assuming three replicates, the data from this experiment are displayed below.

Treatment Combination		Replicate I	II	III	Total
A low, B low	(1)	28	25	27	80
A high, B low	a	36	32	32	100
A low, B high	b	18	19	23	60
A high, B high	ab	31	30	29	90

The treatment combinations in this design are shown graphically in Figure 8-1. By convention, we denote the effect of a factor by a capital Latin letter. Thus "A" refers to the effect of factor A, "B" refers to the effect of factor B, and "AB" refers to the AB interaction. In the 2^2 design the low and high levels of A and B are denoted by 0 and 1, respectively, on the A and B axes. Thus, 0 on the A axis represents the low level of concentration (15%) while 1 represents the high level (25%), and 0 on the B axis represents the absence of catalyst while 1 denotes its presence.

The coordinates of the vertices of the square also represent the four treatment combinations as follows: 00 represents both factors at the low level, 10 represents A at the high level and B at the low level, 01 represents B at the high level and A at the low level, and 11 represents both factors at the high level. These treatment combinations are usually represented by lower case letters, as shown in Figure 8-1. We see from the figure that the high level of any factor in the treatment combination is denoted by the corresponding lower case letter, and the low level of a factor in the treatment combination is denoted by the absence of the corresponding letter. Thus, a represents the

Figure 8-1.   Treatment combinations in the 2^2 design.

treatment combination of A at the high level and B at the low level, b represents A at low level and B at the high level, and ab represents both factors at the high level. By convention, (1) is used to denote both factors at the low level. This notation is used throughout the 2^k series.

Define the average effect of a factor to be the change in response produced by a change in the level of that factor, averaged over the levels of the other factor. Also, the lower-case letters (1), a, b, and ab now represent the *total* of all n replicates taken at that treatment combination, as illustrated in Figure 8-1. Now the effect of A at the low level of B is $[a - (1)]/n$ and the effect of A at the high level of B is $[ab - b]/n$. Averaging these two quantities yields

$$A = \frac{1}{2n} \{[ab - b] + [a - (1)]\}$$

$$= \frac{1}{2n}[ab + a - b - (1)] \tag{8-1}$$

The average B effect is found from the effect of B at the low level of A (i.e., $[b - (1)]/n$) and at the high level of A (i.e., $[ab - a]/n$), as

$$B = \frac{1}{2n} \{[ab - a] + [b - (1)]\}$$

$$= \frac{1}{2n}[ab + b - a - (1)] \tag{8-2}$$

Define the interaction effect AB as the average difference between the effect of A at the high level of B and the effect of A at the low level of B. Thus

$$AB = \frac{1}{2n} \{[ab - b] - [a - (1)]\}$$

$$= \frac{1}{2n} [ab + (1) - a - b] \qquad (8\text{-}3)$$

Alternatively, we may define AB as the average difference between the effect of B at the high level of A and the effect of B at the low level of A. This will also lead to Equation 8-3.

These formulas for the effects of A, B, and AB may be derived by another method. Note that the effect of A, Equation 8-1, is found by comparing the two treatment combinations on the right-hand side of the square in Figure 8-1 with the two treatment combinations on the left-hand side. Also, the effect of B, Equation 8-2, if found by comparing the two observations on top of the square with the two observations on the bottom. Finally, the interaction effect AB is the sum of the right-to-left diagonal treatment combinations in the square [ab and (1)] minus the sum of the left-to-right diagonal treatment combinations (b and a).

Using the example data in Figure 8-1, we may estimate the average effects as

$$A = \frac{1}{2(3)} (90 + 100 - 60 - 80) = 8.33$$

$$B = \frac{1}{2(3)} (90 + 60 - 100 - 80) = -5.00$$

$$AB = \frac{1}{2(3)} (90 + 80 - 100 - 60) = 1.67$$

Now consider finding the sums of squares for A, B, and AB. Note from Equation 8-1 that a contrast is used in estimating A, namely

$$\text{Contrast}_A = ab + a - b - (1) \qquad (8\text{-}4)$$

Kempthorne (1952) and Anderson and McLean (1974) call this contrast the *total effect* of A. From (8-2) and (8-3) we see that contrasts are also used to estimate B and AB. Furthermore, these three contrasts are orthogonal. The sum of squares for any contrast can be computed from Equation 3-17, which states that the contrast sum of squares is equal to the contrast squared divided by the number of observations in each total in the contrast times the sum of the squares of the contrast coefficients. Consequently, we have

$$SS_A = \frac{[ab + a - b - (1)]^2}{n \cdot 4} \qquad (8\text{-}5)$$

$$SS_B = \frac{[ab + b - a - (1)]^2}{n \cdot 4} \qquad (8\text{-}6)$$

and

$$SS_{AB} = \frac{[ab + (1) - a - b]^2}{n \cdot 4} \tag{8-7}$$

as the sums of squares for A, B, and AB.

Using the data in Figure 8-1, we may find the sums of squares from Equations 8-5, 8-6, and 8-7 as

$$SS_A = \frac{(50)^2}{4(3)} = 208.33$$

$$SS_B = \frac{(-30)^2}{4(3)} = 75.00 \tag{8-8}$$

and

$$SS_{AB} = \frac{(10)^2}{4(3)} = 8.33$$

The total sum of squares is found in the usual way, that is, by

$$SS_T = \sum_{i=1}^{2} \sum_{j=1}^{2} \sum_{k=1}^{n} y_{ijk}^2 - \frac{y_{...}^2}{4n} \tag{8-9}$$

In general, SS_T has $4n - 1$ degrees of freedom. The error sum of squares, with $4(n-1)$ degrees of freedom, is usually computed by subtraction as

$$SS_E = SS_T - SS_A - SS_B - SS_{AB} \tag{8-10}$$

For the data in Figure 8-1, we obtain

$$SS_T = \sum_{i=1}^{2} \sum_{j=1}^{2} \sum_{k=1}^{3} y_{ijk}^2 - \frac{y_{...}^2}{4(3)}$$

$$= 9398.00 - 9075.00 = 323.00$$

and

$$SS_E = SS_T - SS_A - SS_B - SS_{AB}$$

$$= 323.00 - 208.33 - 75.00 - 8.33$$

$$= 31.34$$

using SS_A, SS_B, and SS_{AB} from (8-8). The complete analysis of variance is summarized in Table 8-1. Both main effects are statistically significant at one percent.

Table 8-1 Analysis of Variance for Data in Figure 8-1

Source of Variation	Sum of Squares	Degrees of Freedom	Mean Square	F_0
A	208.33	1	208.33	53.15[a]
B	75.00	1	75.00	19.13[a]
AB	8.33	1	8.33	2.13
Error	31.34	8	3.92	
Total	323.00	11		

[a]Significant at 1 percent.

It is often convenient to write down the treatment combinations in the order (1), a, b, ab. This is referred to as *standard order*. Using this standard order, we see that the contrast coefficients used in estimating the effects are

	(1)	a	b	ab
A:	-1	$+1$	-1	$+1$
B:	-1	-1	$+1$	$+1$
AB:	$+1$	-1	-1	$+1$

Note that the contrast coefficients for estimating the interaction effect are just the product of the corresponding coefficients for the two main effects. The contrast coefficient is always either +1 or -1, and a table of plus and minus signs such as in Table 8-2 can be used to determine the proper sign for each treatment combination. The column headings in Table 8-2 are the main effects (A and B), the AB interaction, and I, which represents the total or average of the entire experiment. Notice that the column corresponding to I has only plus signs. The row designators are the treatment combinations. To find the contrast for estimating any effect, simply multiply the signs in the appropriate column of the table by the corresponding treatment combination and add. For example, to estimate A, the contrast is $- (1) + a - b + ab$, which agrees with Equation 8-1.

Table 8-2 Algebraic Signs for Calculating Effects in the 2^2 Design

Treatment Combination	Factorial Effect			
	I	A	B	AB
(1)	+	−	−	+
a	+	+	−	−
b	+	−	+	−
ab	+	+	+	+

8-2.2 The 2^3 Design

Suppose that three factors, A, B, and C, each at two levels, are under study. The design is called a 2^3 factorial, and the eight treatment combinations can now be displayed graphically as a cube, as shown in Figure 8-2. Extending the notation of Section 8-2.1, we write the treatment combinations in standard order as (1), a, b, ab, c, ac, bc, and abc. Remember that these lower-case letters also represent the *total* of all n observations taken at that particular treatment combination.

Figure 8-2.

The seven degrees of freedom between the eight treatment combinations are associated with the main effects of A, B, and C; the two-factor interactions AB, AC, and BC; and the three-factor interaction ABC. Consider estimating the main effect A. The effect of A when B and C are at the low level is $[a - (1)]$ /n. Similarly, the effect of A when B is at the high level and C is at the low level is $[ab - b]$ /n. The effect of A when C is at the high level and B is at the low level is $[ac - c]$ /n. Finally, the effect of A when both B and C are at the high level is $[abc - bc]$ /n. Thus, the average effect of A is just the average of these four, or

$$A = \frac{1}{4n} [a - (1) + ab - b + ac - c + abc - bc] \tag{8-11}$$

This equation can be developed as a contrast between the four treatment combinations in the right face of the cube (where A is at the high level) and the four in the left face (where A is at the low level). Combining these gives

$$A = \frac{1}{4n} [a + ab + ac + abc - (1) - b - c - bc]$$

which is identical to Equation 8-11.

In a similar manner, the effect of B is a contrast between the four treatment combinations in the front face of the cube and the four in the back. This yields

$$B = \frac{1}{4n} [b + ab + bc + abc - (1) - a - c - ac] \tag{8-12}$$

The effect of C is a contrast between the four treatment combinations in the top face of the cube and the four in the bottom, that is,

$$C = \frac{1}{4n} [c + ac + bc + abc - (1) - a - b - ab] \tag{8-13}$$

When C is at the low level the effect of the AB interaction is the average difference in the A effect at the two levels of B, that is, $(1/2n) \{[ab - b] - [a - (1)]\}$. When C is at the high level, the AB interaction is $(1/2n) [(abc - bc) - (ac - c)]$. The average AB effect is just the average of these two, or

$$AB = \frac{1}{4n} [ab - b - a + (1) + abc - bc - ac + c] \tag{8-14}$$

Similarly, we find the average effects of AC and BC as

$$AC = \frac{1}{4n} [(1) - a + b - ab - c + ac - bc + abc] \tag{8-15}$$

and

$$BC = \frac{1}{4n} [(1) + a - b - ab - c - ac + bc + abc] \tag{8-16}$$

The ABC effect is defined as the average difference between the AB interaction for the two different levels of C. Thus

$$ABC = \frac{1}{4n} \{[abc - bc] - [ac - c] - [ab - b] + [a - (1)]\}$$

$$= \frac{1}{4n} [abc - bc - ac + c - ab + b + a - (1)] \tag{8-17}$$

In Equations 8-11 through 8-17 the quantities in brackets are contrasts in the treatment combinations. A table of plus and minus signs can be developed from the contrast constants and is shown in Table 8-3. Signs for the main effects are determined by associating a plus with the high level and a minus with the low level. Once the signs for the main effects have been established, the signs for the remaining columns can be obtained by multiplying the appropriate preceding columns, row by row. For example, the signs in the AB column are the product of the A and B column signs in each row. The contrast for any effect can be obtained easily from this table.

Table 8-3   Algebraic Signs for Calculating Effects in the 2^3 Design

Treatment Combination	Factorial Effect

Treatment Combination	I	A	B	AB	C	AC	BC	ABC
(1)	+	−	−	+	−	+	+	−
a	+	+	−	−	−	−	+	+
b	+	−	+	−	−	+	−	+
ab	+	+	+	+	−	−	−	−
c	+	−	−	+	+	−	−	+
ac	+	+	−	−	+	+	−	−
bc	+	−	+	−	+	−	+	−
abc	+	+	+	+	+	+	+	+

Table 8-3 has several interesting properties. (1) Except for column I, every column has an equal number of plus and minus signs. (2) The sum of products of signs in any two columns is zero. (3) Column I multiplied times any column leaves that column unchanged. That is, I is an *identity element*. (4) The product of any two columns yields a column in the table. For example, $A \times B = AB$, and

$$AB \times B = AB^2 = A$$

We see that the products are formed *modulus* 2 (the exponent can only be zero or one; if it is greater than one it is reduced by multiples of two until it is either zero or one). All of these properties are implied by orthogonality of the contrasts used to estimate the effects.

Sums of squares for the effects are easily determined, since each effect has a corresponding single-degree-of-freedom contrast. In the 2^3 design with n replicates, the

Table 8-4   Data for the Fill Volume Problem, Example 6-3

	Operating Pressure (B)

	25 psi		30 psi	
	Line Speed (C)		Line Speed (C)	
Percent Carbonation (A)	200	250	200	250
10	−3 −1 −4 = (1)	−1 0 −1 = c	−1 0 −1 = b	1 1 2 = bc
12	0 1 1 = a	2 1 3 = ac	2 3 5 = ab	6 5 11 = abc

sum of squares for any effect is

$$SS = \frac{(\text{Contrast})^2}{8n} \tag{8-18}$$

■EXAMPLE 8-1. Recall Example 6-3, which presented a study on the effect of percent carbonation, operating pressure, and line speed on the fill volume of a carbonated beverage. Suppose that only two levels of carbonation are used so that the experiment is a 2^3 factorial design with two replicates. The coded data are shown in Table 8-4.

Using the totals under the treatment combinations shown in Table 8-4, we may compute the average effects as follows:

$$A = \frac{1}{4n} [a - (1) + ab - b + ac - c + abc - bc]$$

$$= \frac{1}{8} [1 - (-4) + 5 - (-1) + 3 - (-1) + 11 - 2]$$

$$= \frac{1}{8} [24] = 3.00$$

$$B = \frac{1}{4n} [b + ab + bc + abc - (1) - a - c - ac]$$

$$= \frac{1}{8} [-1 + 5 + 2 + 11 - (-4) - 1 - (-1) - 3]$$

$$= \frac{1}{8} [18] = 2.25$$

$$C = \frac{1}{4n} [c + ac + bc + abc - (1) - a - b - ab]$$

$$= \frac{1}{8} [-1 + 3 + 2 + 11 - (-4) - 1 - (-1) - 5]$$

$$= \frac{1}{8} [14] = 1.75$$

$$AB = \frac{1}{4n} [ab - a - b + (1) + abc - bc - ac + c]$$

$$= \frac{1}{8} [5 - 1 - (-1) + (-4) + 11 - 2 - 3 + (-1)]$$

$$= \frac{1}{8} [6] = 0.75$$

$$AC = \frac{1}{4n} [(1) - a + b - ab - c + ac - bc + abc]$$

$$= \frac{1}{8} [-4 - 1 + (-1) - 5 - (-1) + 3 - 2 + 11]$$

$$= \frac{1}{8} [2] = 0.25$$

$$BC = \frac{1}{4n} [(1) + a - b - ab - c - ac + bc + abc]$$

$$= \frac{1}{8} [-4 + 1 - (-1) - 5 - (-1) - 3 + 2 + 11]$$

$$= \frac{1}{8} [4] = 0.50$$

and

$$ABC = \frac{1}{4n} [abc - bc - ac + c - ab + b + a - (1)]$$

$$= \frac{1}{8} [11 - 2 - 3 + (-1) - 5 + (-1) + 1 - (-4)]$$

$$= \frac{1}{8} [4] = 0.50$$

From Equation 8-18, the sums of squares are

$$SS_A = \frac{(24)^2}{16} = 36.00$$

$$SS_B = \frac{(18)^2}{16} = 20.2\llcorner$$

$$SS_C = \frac{(4)^2}{16} = 12.25$$

$$SS_{AB} = \frac{(6)}{16} = 2.25$$

$$SS_{AC} = \frac{(2)^2}{16} = 0.25$$

$$SS_{BC} = \frac{(4)^2}{16} = 1.00$$

and

$$SS_{ABC} = \frac{(4)^2}{16} = 1.00$$

It is easy to verify that the usual methods for the analysis of factorials will yield the same values for these sums of squares. From the individual observations we may compute the total sum of squares as 78.00, and by subtraction we find $SS_E = 5.00$. The analysis of variance is summarized in Table 8-5. All three main effects are significant. ∎

Table 8-5 Analysis of Variance for the Fill Volume Data

Source of Variation	Sum of Squares	Degrees of Freedom	Mean Square	F_0
Percent Carbonation (A) ·	36.00	1	36.00	57.14[a]
Pressure (B)	20.25	1	20.25	32.14[a]
Line speed (C)	12.25	1	12.25	19.44[a]
AB	2.25	1	2.25	3.57
AC	0.25	1	0.25	0.40
BC	1.00	1	1.00	1.59
ABC	1.00	1	1.00	1.59
Error	5.00	8	0.63	
Total	78.00	15		

[a]Significant at 1 percent.

8-2.3 The General 2^k Design

The methods of analysis that we have presented thus far may be generalized to the case of a 2^k factorial design, that is, a design with k factors each at two levels. The statistical model for a 2^k design would include k main effects, $\binom{k}{2}$ two-factor interactions, $\binom{k}{3}$ three-factor interactions, . . . , one k-factor interaction. That is, for a 2^k design the complete model would contain $2^k - 1$ effects. The notation introduced earlier for treatment combinations is also used here. For example, in a 2^5 design abd denotes the treatment combination with factors A, B, and D at the high level and factors C and E at the low level. The treatment combinations may be written in standard order by introducing the factors one at a time; each new factor being successively combined with those above it. For example, the standard order for a 2^4 design is (1), a, b, ab, c, ac, bc, abc, d, ad, bd, abd, cd, acd, bcd, and $abcd$.

To estimate an effect or to compute the sum of squares for the effect, we must first determine the contrast associated with that effect. This can always be done by using a table of plus and minus signs, such as Table 8-2 or Table 8-3. However, for values of k larger than 2 or 3 this is awkward, and we prefer a more efficient method. In general we determine the contrast for effect $AB \ldots K$ by expanding the right-hand side of

$$\text{Contrast}_{AB\ldots K} = (a \pm 1)(b \pm 1) \cdots (k \pm 1) \qquad (8\text{-}19)$$

In expanding (8-19), ordinary algebra is used with "1" being replaced by (1) in the final expression. The sign in each set of parentheses is negative if the factor is included in the effect, and positive if the factor is not included.

To illustrate the use of (8-19), consider a 2^3 factorial. The contrast for AB would be

$$\text{Contrast}_{AB} = (a - 1)(b - 1)(c + 1)$$

$$= abc + ab + c + (1) - ac - bc - a - b$$

Table 8-6 Analysis of Variance for a 2^k Design

Source of Variation	Sum of Squares	Degrees of Freedom
k Main Effects		
A	SS_A	1
B	SS_B	1
.	.	.
.	.	.
.	.	.
K	SS_K	1
$\binom{k}{2}$ Two-Factor Interactions		
AB	SS_{AB}	1
AC	SS_{AC}	1
.	.	.
.	.	.
JK	SS_{JK}	1
$\binom{k}{3}$ Three-Factor Interactions		
ABC	SS_{ABC}	1
ABD	SS_{ABD}	1
.	.	.
.	.	.
IJK	SS_{IJK}	1
.	.	.
.	.	.
$\binom{k}{k} = 1$ k-Factor Interaction		
$ABC \dots K$	$SS_{ABC \dots K}$	1
Error	SS_E	$2^k(n - 1)$
Total	SS_T	$n2^k - 1$

As a further example, in a 2^5, the contrast for $ABCD$ would be

$$\text{Contrast}_{ABCD} = (a - 1)(b - 1)(c - 1)(d - 1)(e + 1)$$

$$= abcde + cde + bde + ade + bce + ace + abe + e + abcd + cd + bd$$

$$+ ad + bc + ac + ab + (1) - a - b - c - abc - d - abd - acd$$

$$- bcd - ae - be - ce - abce - de - abde - acde - bcde$$

Once the contrasts for the effects have been computed, we may estimate the effects and compute the sums of squares according to

$$AB \ldots K = \frac{1}{n2^{k-1}}(\text{Contrast}_{AB \ldots K}) \tag{8-20}$$

and

$$SS_{AB \ldots K} = \frac{1}{n2^k}(\text{Contrast}_{AB \ldots K})^2 \tag{8-21}$$

respectively, where n denotes the number of replicates. The analysis of variance for the 2^k is summarized in Table 8-6.

8-2.4 A Single Replicate of the 2^k Design

For even a moderate number of factors the total number of treatment combinations in a 2^k factorial design is large. For example, a 2^5 has 32 treatment combinations, a 2^6 has 64 treatment combinations, and so on. Since resources are usually limited, the number of replicates that the experimenter can employ may be restricted. Frequently, available resources will only allow a single replicate of the design to be run, unless the experimenter is willing to omit some of the original factors.

With only a single replicate of the 2^k it is impossible to compute an estimate of experimental error, that is, a mean square for error. Thus, it seems that hypotheses concerning main effects and interactions cannot be tested. However, the usual approach to the analysis of a single replicate of the 2^k is to assume that certain higher-order interactions are negligible, and then since their mean squares will all have expectation σ^2, they may be combined to estimate the experimental error. As discussed in Chapter 7, we should select the interactions that will make up the error mean square *before* the data are analyzed, since testing high-order interactions and taking those that appear small as error can lead to substantial underestimation of the true experimental error. Usually, the smallest design for which this procedure is recommended is the 2^4.

The practice of combining higher-order interaction mean squares to estimate the error is subject to criticism on statistical grounds. If some of these interactions are significant, then the estimate of error will be inflated. As a result, other significant effects may not be detected and the significant interactions taken as error will be discovered. As a general rule, it is probably unwise to assume two-factor interactions to be zero without prior information. If most two-factor interactions are small, then it seems likely that all higher-order interactions will be insignificant also. (A word of caution here—one does not have to look very far for counterexamples to these rules.)

The experimenter must use both his knowledge of the phenomena under study and common sense in the analysis of such a design. For example, if in the analysis of a 2^5 the main effects of A, B, and C as well as AB and AC are very large, then it is likely that ABC may be large. Thus ABC should not be included in the interactions used as error. Such a decision is best made before ABC is examined.

■**EXAMPLE 8-2. A Single Replicate of the 2^4 Design.** A chemical product is produced in a pressure vessel. A factorial experiment was carried out in the pilot plant to study the factors thought to influence the filtration rate of this product. The four factors are temperature (A), pressure (B), concentration of reactant (C), and stirring

Table 8-7   Filtration Data from the Pilot Plant Experiment

	A_0				A_1			
	B_0		B_1		B_0		B_1	
	C_0	C_1	C_0	C_1	C_0	C_1	C_0	C_1
D_0	45	68	48	80	71	60	65	65
D_1	43	75	45	70	100	86	104	96

Table 8-8   Analysis of Variance for the Pilot Plant Data

Source of Variation	Sum of Squares	Degrees of Freedom	Mean Square	F_0
A	1870.56	1	1870.56	73.15[b]
B	39.06	1	39.06	1.53
C	390.06	1	390.06	15.25[a]
D	855.56	1	855.56	33.46[b]
AB	0.06	1	0.06	<1
AC	1314.06	1	1314.06	51.39[b]
AD	1105.56	1	1105.56	43.24[b]
BC	22.56	1	22.56	<1
BD	0.56	1	0.56	<1
CD	5.06	1	5.06	<1
Error	127.84	5	25.57	
Total	5730.94	15		

[a] Significant at the 5 percent level.
[b] Significant at the 1 percent level.

rate (D). Each factor is present at two levels, and the data obtained from a single replicate of the 2^4 experiment are shown in Table 8-7. The 16 runs are made in random order.

We assume that the three-factor and four-factor interactions are negligible, and may be used to provide an estimate of error. The analysis of variance for the pilot plant data is shown in Table 8-8. The main effects of A (temperature), C (concentration), and D (stirring rate) are significant, as are the AC and AD interactions.   ∎

Another approach to the analysis of unreplicated 2^k factorial designs, due to Daniel (1959), is often useful. If the data are normally and independently distributed, then the $2^k - 1$ estimates of the effects obtained from a 2^k design are normally distributed. To help identify significant effects, the estimates of the effects may be plotted on normal probability paper.[1]   Insignificant effects, that is, those due primarily to random error, should have the appearance of a normal distribution.

Arrange the estimates of the effects in ascending order, and plot the jth of these ordered values against $P_j = (j - .5)/(2^k - 1), j = 1, 2, \ldots, 2^k - 1$, on normal probability paper. The effects that are negligible will tend to fall along a straight line on this graph, while significant effects will be far from the line. The negligible effects can then be combined to form an estimate of error.

We will demonstrate this method of analysis using the data in Example 8-2. The table of plus and minus signs for the contrast constants for the 2^4 design are shown in Table 8-9. From these contrasts, we may estimate the 15 factorial effects shown below:

Order (j)	Effect	Estimate	(j - .5)/15
15	A	21.67	.9667
14	AD	16.63	.9000
13	D	14.63	.8333
12	C	9.88	.7667
11	ABD	4.13	.7000
10	B	3.13	.6333
9	BC	2.38	.5667
8	ABC	1.88	.5000
7	$ABCD$	1.38	.4333
6	AB	0.13	.3667
5	BD	-0.38	.3000
4	CD	-0.13 ~~-1.125~~	.2333
3	ACD	~~=0.63~~ -1.625	.1667
2	BCD	-2.63	.1000
1	AC	-18.13	.0333

[1] Normal probability paper is simply graph paper with the ordinate scaled so that the graph of the cumulative $N(\mu, \sigma^2)$ distribution is a straight line.

Figure 8-3.   Ordered effects for the 2^4 factorial in Example 8-2.

Table 8-9   Contrast Constants for the 2^4 Design

	A	B	AB	C	AC	BC	ABC	D	AD	BD	ABD	CD	ACD	BCD	ABCD
(1)	−	−	+	−	+	+	−	−	+	+	−	+	−	−	+
a	+	−	−	−	−	+	+	−	−	+	+	+	+	−	−
b	−	+	−	−	+	−	+	−	+	−	+	+	−	+	−
ab	+	+	+	−	−	−	−	−	−	−	−	+	+	+	+
c	−	−	+	+	−	−	+	−	+	+	−	−	+	+	−
ac	+	−	−	+	+	−	−	−	−	+	+	−	−	+	+
bc	−	+	−	+	−	+	−	−	+	−	+	−	+	−	+
abc	+	+	+	+	+	+	+	−	−	−	−	−	−	−	−
d	−	−	+	−	+	+	−	+	−	−	+	−	+	+	−
ad	+	−	−	−	−	+	+	+	+	−	−	−	−	+	+
bd	−	+	−	−	+	−	+	+	−	+	−	−	+	−	+
abd	+	+	+	−	−	−	−	+	+	+	+	−	−	−	−
cd	−	−	+	+	−	−	+	+	−	−	+	+	−	−	+
acd	+	−	−	+	+	−	−	+	+	−	−	+	+	−	−
bcd	−	+	−	+	−	+	−	+	−	+	−	+	−	+	−
abcd	+	+	+	+	+	+	+	+	+	+	+	+	+	+	+

Table 8-10   Analysis of Variance for 2^3 Design in A, C, and D

Source of Variation	Sum of Squares	Degrees of Freedom	Mean Square	F_0
A	1870.56	1	1870.56	83.36[a]
C	390.06	1	390.06	17.38[a]
D	855.56	1	855.56	38.13[a]
AC	1314.06	1	1314.06	58.56[a]
AD	1105.56	1	1105.56	49.27[a]
CD	5.06	1	5.06	<1
ACD	10.56	1	10.56	<1
Error	179.52	8	22.44	
Total	5730.94	15		

[a]Significant at the 1 percent level.

The ordered effects plotted on normal probability paper are shown in Figure 8-3. From examining this display, we see that all of the small effects lie approximately along a straight line while the large effects are far from the line. As none of the large effects are three-factor or four-factor interactions, we may combine these effects as an estimate of error, as was done in Example 8-2.

On viewing Figure 8-3, another interpretation of the data is possible. Since B (pressure) is not significant, and all interactions involving B are negligible, we may discard B from the experiment and the design becomes a 2^3 factorial in A, C, and D with two replicates. This is easily seen from Table 8-9, as the columns for A, C, and D are now the levels in the 2^3 design. The analysis of variance for the data using this simplifying assumption is summarized in Table 8-10. The conclusions that we would draw from this analysis are essentially unchanged from those of Example 8-2. In general, if we have a single replicate of a 2^k and $h(h < k)$ factors can be dropped, then the original data will correspond to a full two-level factorial in the remaining $k - h$ factors with 2^h replicates.

8-2.5   Yates' Algorithm for the 2^k Design

There is a very simple technique devised by Yates (1937) for estimating the effects and determining the sums of squares in a 2^k factorial design. The procedure is best learned through study of a numerical example.

Consider the data for the 2^3 design in Example 8-1. These data have been entered in Table 8-11. The treatment combinations are always written down in standard order, and the column labeled "Response" contains the corresponding observation (or total of all observations) at that treatment combination. The first half of column (1) is obtained by adding the responses in adjacent pairs. The second half of column (1) is

Table 8-11   Yates' Algorithm for the Data in Example 8-1

Treatment Combination	Response	(1)	(2)	(3)	Effect	Estimate of Effect $(3) \div n2^{k-1}$	Sum of Squares $(3)^2 \div n2^k$
(1)	-4	-3	1	16	I	$-$	$-$
a	1	4	15	24	A	3.00	36.00
b	-1	2	11	18	B	2.25	20.25
ab	5	13	13	6	AB	0.75	2.25
c	-1	5	7	14	C	1.75	12.25
ac	3	6	11	2	AC	0.25	0.25
bc	2	4	1	4	BC	0.50	1.00
abc	11	9	5	4	ABC	0.50	1.00

obtained by changing the sign of the first entry in each of the pairs in the Response column and adding the adjacent pairs.   For example, in column (1) we obtain $5 = -(-4), + 1, 6 = -(-1) + 5$, and so on.

Column (2) is obtained from column (1) just as column (1) is obtained from the Response column. Column (3) is obtained from column (2) similarly. In general, for a 2^k design we would construct k columns of this type. Column (3) [in general, column (k)] is the contrast for the effect designated in small letters at the beginning of the row.   To obtain the estimate of the effect, we divide the entries in column (3) by $n2^{k-1}$ (in our example, $n2^{k-1} = 8$).   Finally, the sums of squares for the effects are obtained by squaring the entries in column (3) and dividing by $n2^k$ (in our example, $n2^k = 16$).

The estimates of the effects and sums of squares obtained by Yates' algorithm for the data in Example 8-1 are in agreement with the results found there by the usual methods.   Note that the entry in column (3) [in general, column (k)] for the row corresponding to (1) is always equal to the grand total of the observations.

In spite of its apparent simplicity, it is notoriously easy to make numerical errors in Yates' algorithm, and we should be extremely careful in executing the procedure. As a partial check on the computations, we may use the fact that the sum of the squares of the elements in the jth column is 2^j times the sum of the squares of the elements in the response column.   Note, however, that this check is subject to errors in sign in column j.   See Good (1958, 1960), Rayner (1967), Kempthorne (1952), and Davies (1956) for other error-checking techniques.

8-3   ANALYSIS OF THE 3^k FACTORIAL DESIGN

8-3.1   Notation for the 3^k Series

We now discuss the analysis of the 3^k factorial design; that is, a factorial arrangement with k factors each at three levels. Factors and interactions will be denoted by capital letters. Without loss of generality, we may refer to the three levels of the factors as low, intermediate, and high. These levels will be designated by the digits 0(low), 1(intermediate), and 2(high). Each treatment combination in the 3^k design will be denoted by k digits, where the first digit indicates the level of factor A, the second digit indicates the level of factor $B, \ldots$, and the kth digit indicates the level of factor K. For example, in a 3^2 design, 00 denotes the treatment combination corresponding to A and B both at the low level, and 01 denotes the treatment combination corresponding to A at the low level and B at the intermediate level.

This system of notation could be used for the 2^k series of designs. The primary reason for not using it is that most of the literature has adopted the $(1), a, b, ab, \ldots$ notation. However, the digital notation employed in the 3^k series has the advantage that it can be easily extended to other factorial design systems.

8-3.2 The 3^2 Design

The simplest design in the 3^k system is the 3^2, that is, two factors each at three levels. The treatment combinations for this design are shown in Figure 8-4. Since there are $3^2 = 9$ treatment combinations, there are eight degrees of freedom between these treatment combinations. The main effects of A and B each have two degrees of freedom, and the AB interaction has four degrees of freedom. If there are n replicates, there will be $n3^2 - 1$ total degrees of freedom and $3^2(n-1)$ degrees of freedom for error.

The sums of squares for A, B, and AB may be computed by the usual methods for factorial designs discussed in Chapter 6. Alternatively, in Section 8-3.5 we give an extension of Yates' algorithm to the 3^k series. The sum of squares for any main effect may be partitioned into a linear and a quadratic component, each with a single degree of freedom, using the orthogonal contrast constants, as demonstrated in Chapter 6. Of

Figure 8-4. Treatment combinations in a 3^2 design.

course, this is only meaningful if the factor is quantitative and if the three levels are equally-spaced.

As an illustration, consider the tool-life problem in Example 6-5. The objective of this experiment was to investigate the effect of cutting speed and tool angle on tool life in a numerically-controlled machine. Three cutting speeds and three tool angles were selected for study. This is a 3^2 design with $n = 2$ replicates. The analysis of variance is repeated for convenience in Table 8-12. For details of the sum of squares calculations, and partitioning of the main effects into linear and quadratic components, refer to Chapter 6, Section 6-6.

Table 8-12   Analysis of Variance for Example 6-5

Source of Variation	Sum of Squares	Degrees of Freedom	Mean Square	F_0
Tool Angle (A)	24.33	2	12.17	8.45[a]
(A_L)	(8.33)	1	(8.33)	5.78[a]
(A_Q)	(16.00)	1	(16.00)	11.11[a]
Cutting Speed (B)	25.33	2	12.67	8.80[a]
(B_L)	(21.33)	1	(21.33)	14.81[a]
(B_Q)	(4.00)	1	(4.00)	2.77
AB interaction	61.34	4	6.82	4.75[a]
Error	13.00	9	1.44	
Total	124.00			

[a]Significant at 5 percent.

The two-factor interaction, AB, may be partitioned in two ways. The first method consists of subdividing AB into the four single-degree-of-freedom components corresponding to $AB_{L \times L}, AB_{L \times Q}, AB_{Q \times L}$, and $AB_{Q \times Q}$. This is done using orthogonal contrast constants, as demonstrated in Example 6-5. For the tool life data, this yields $SS_{AB_{L \times L}} = 8.00$, $SS_{AB_{L \times Q}} = 42.67$, $SS_{AB_{Q \times L}} = 2.67$, and $SS_{AB_{Q \times Q}} = 8.00$. Since this is an orthogonal partitioning of AB, note that $SS_{AB} = SS_{AB_{L \times L}} + SS_{AB_{L \times Q}} + SS_{AB_{Q \times L}} + SS_{AB_{Q \times Q}}$.

The second method is based on orthogonal Latin squares. Consider the totals of the treatment combinations for the data in Example 6-5. These totals are shown in Figure 8-5, as the circled numbers in the squares. The two factors, A and B, correspond to the rows and columns, respectively, of a 3 $\times$ 3 Latin square. In Figure 8-5 (a) and (b) two particular 3 $\times$ 3 Latin squares are shown, superimposed on the cell totals.

These two Latin square are *orthogonal*, that is, if one square is superimposed on the other, each letter in the first square will appear exactly once with each letter in the second square. The totals for the letters in the (a) square are $Q = 18$, $R = -2$, and $S = 8$, and the sum of squares between these totals is $[18^2 + (-2)^2 + 8^2]/(3)(2) -$

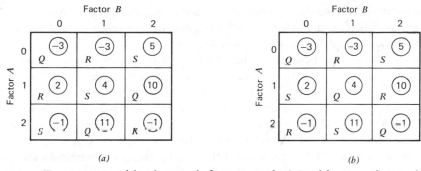

(a)   *(b)*

Figure 8-5. Treatment combination totals from example 6-5, with two orthogonal latin squares superimposed.

$[24^2/(9)(2)] = 33.34$, with two degrees of freedom. Similarly, the letter totals in the (b) square are $Q = 0$, $R = 6$, and $S = 18$, and the sum of squares between these totals is $[0^2 + 6^2 + 18^2]/(3)(2) - [24^2/(9)(2)] = 28.00$, with two degrees of freedom. Note that the sum of these two components is

$$33.34 + 28.00 = 61.34 = SS_{AB}$$

with $2 + 2 = 4$ degrees of freedom.

In general, the sum of squares computed from square (a) is called the AB component of interaction, and the sum of squares computed from square (b) is called the AB^2 component of interaction. The components AB and AB^2 each have two degrees of freedom. This terminology is used because if we denote the levels (0, 1, 2) for A and B by x_1 and x_2, respectively, then we find that the letters occupy cells according to the following pattern:

	Square (a)		Square (b)
J_0	$Q: x_1 + x_2 = 0 \pmod 3$	I_0	$Q: x_1 + 2x_2 = 0 \pmod 3$
J_1	$R: x_1 + x_2 = 1 \pmod 3$	$I_1 \ S$	$R: x_1 + 2x_2 = 1 \pmod 3$
J_2	$S: x_1 + x_2 = 2 \pmod 3$	$I_2 \ R$	$S: x_1 + 2x_2 = 2 \pmod 3$

For example, in square (b), note that the middle cell corresponds to $x_1 = 1$ and $x_2 = 1$; thus $x_1 + 2x_2 = 1 + (2)(1) = 3 = 0 \pmod 3$, and Q would occupy the middle cell. When considering expressions of the form $A^p B^q$ we establish the convention that the only exponent allowed on the first letter is one. If the first letter is not one, the entire expression is squared and the exponents reduced modulus 3. For example, $A^2 B$ is the same as AB^2, since

$$A^2 B = (A^2 B)^2 = A^4 B^2 = AB^2$$

The AB and AB^2 components of the AB interaction have no actual meaning, and are usually not displayed in the analysis of variance table. However, this rather arbitrary partitioning of the AB interaction into two orthogonal two-degree-of-freedom components is very useful in constructing more complex designs. Also, there is no connection between the AB and AB^2 components of interaction and the sums of squares for $AB_{L \times L}$, $AB_{L \times Q}$, $AB_{Q \times L}$, and $AB_{Q \times Q}$.

The AB and AB^2 components of interaction may be computed another way. Consider the treatment combination totals in either square in Figure 8-5. If we add the data by diagonals downward from left to right we obtain the totals $-3 + 4 - 1 = 0$, $-3 + 10 - 1 = 6$ and $5 + 11 + 2 = 18$. The sum of squares between these totals is 28.00 (AB^2). Similarly, the diagonal totals downward from right to left are $5 + 4 - 1 = 8$, $-3 + 2 - 1 = -2$ and $-3 + 11 + 10 = 18$. The sum of squares between these totals is 33.34 (AB). Yates called these components of interaction the I and J components of interaction respectively. We will use both notations interchangeably, that is

$$I(AB) = AB^2$$
$$J(AB) = AB$$

8-3.3 The 3^3 Design

Now suppose there are three factors (A, B, and C) under study, and each factor is at three levels arranged in a factorial experiment. This is a 3^3 factorial design, and the experimental layout and treatment combination notation are shown in Figure 8-6.

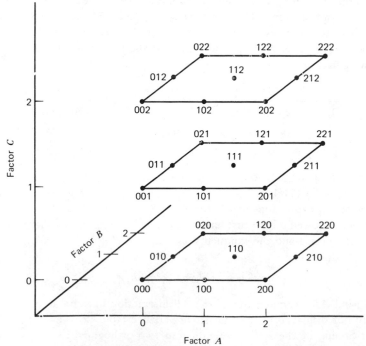

Figure 8-6. Treatment combinations in a 3^3 design.

The 27 treatment combinations have 26 degrees of freedom. Each main effect has two degrees of freedom, each two-factor interaction has 4 degrees of freedom, and the three-factor interaction has 8 degrees of freedom. If there are n replicates, there are $n3^3 - 1$ total degrees of freedom and $3^3(n - 1)$ degrees of freedom for error.

The sums of squares may be calculated using the standard methods for factorial designs. Additionally, if the factors are quantitative and equally-spaced the main effects may be partitioned into linear and quadratic components, each with a single degree of freedom. The two-factor interactions may be decomposed into linear $\times$ linear, linear $\times$ quadratic, quadratic $\times$ linear, and quadratic $\times$ quadratic effects. Finally, the three-factor interaction ABC can be partitioned into eight single-degree-of-freedom components corresponding to linear $\times$ linear $\times$ linear, linear $\times$ linear $\times$ quadratic, and so on. Such a breakdown for the three-factor interaction is generally not very useful.

It is also possible to partition the two-factor interactions into their I and J components. These would be designated AB, AB^2, AC, AC^2, BC, and BC^2, and each component would have two degrees of freedom. As in the 3^2 design, these components have no physical significance.

The three-factor interaction ABC may be partitioned into four orthogonal two-degree-of-freedom components, which are usually called the W, X, Y, and Z components of interactions. They are also referred to as the AB^2C^2, AB^2C, ABC^2, and ABC components of the ABC interaction, respectively. The two notations are used interchangably, that is

$$W(ABC) = AB^2C^2$$
$$X(ABC) = AB^2C$$
$$Y(ABC) = ABC^2$$
$$Z(ABC) = ABC$$

Note that no first letter can have an exponent other than 1. Like the I and J components, the W, X, Y, and Z components have no practical interpretation. They are, however, useful in constructing more complex designs.

Table 8-13 Syrup Loss Data for Example 8-3
Units are (cubic centimeters -70)

C, Pressure	A, Nozzle								
	0			1			2		
	B, Operator								
	0	1	2	0	1	2	0	1	2
0	-35	-45	-40	17	-65	20	-39	-55	15
	-25	-60	15	24	-58	4	-35	-67	-30
1	100	-10	80	55	-55	110	90	-28	110
	75	30	54	120	-44	44	113	-26	135
2	4	-40	31	-23	-64	-20	-30	-61	54
	5	-30	36	-5	-62	-31	-55	-52	4

Figure 8-7. Two-factor interactions for Example 8-3.

■**EXAMPLE 8-3.** A machine is used to fill five-gallon metal containers with soft drink syrup. The variable of interest is the amount of syrup loss due to frothing. Three factors are thought to influence frothing, the nozzle configuration (A), the operator (B), and the operating pressure (C). Three nozzles, three operators, and three pressures are chosen and two replicates of a 3^3 factorial experiment are run. The coded data are shown in Table 8-13.

The analysis of variance for the syrup loss data is shown in Table 8-14. The sums of squares have been computed by the usual methods. We see that operators and operating pressure are significant at the one percent level. All three two-factor interactions are also significant. The two-factor interactions are analyzed graphically in Figure 8-7.

Table 8-14   Analysis of Variance for Syrup Loss Data

Source of Variation	Sum of Squares	Degrees of Freedom	Mean Square	F_0
A, nozzle	993.77	2	496.89	1.17
B, operator	61,190.33	2	30,595.17	71.74[b]
C, pressure	69,105.33	2	34,552.67	81.01[b]
AB	6,300.90	4	1,575.22	3.69[a]
AC	7,513.90	4	1,878.47	4.40[b]
BC	12,854.34	4	3,213.58	7.53[b]
ABC	4,628.76	8	578.60	1.36
Error	11,515.50	27	426.50	
Total	174,102.83	53		

[a]Significant at 5 percent.
[b]Significant at 1 percent.

Operator 1 is superior to operator 0 and 2, and nozzle type 1 and the high pressure (2) seems most effective in reducing syrup loss.   ∎

We will demonstrate numerically the partitioning of the ABC interaction into its four orthogonal two-degree-of-freedom components using the data in Example 8-3. The general procedure has been described by Cochran and Cox (1957), Davies (1956), and Hicks (1973). First, select any two of the three factors, say AB, and compute the I and J totals of the AB interaction at each level of the third factor C. These calculations are displayed below.

			A			Totals	
C	B	0	1	2	I	J	
	0	− 60	41	− 74	I_0 −198	−222	
0	1	−105	−123	−122	I_1 −106	− 79	
	2	− 25	24	− 15	I_2 −155	−158	
	0	185	175	203	I_0 331	238	
1	1	20	− 99	− 54	I_1 255	440	
	2	134	154	245	I_2 377	285	
	0	9	− 28	− 85	I_0 − 59	−144	
2	1	− 70	−126	−113	I_1 − 74	− 40	
	2	67	− 51	58	I_2 −206	−155	

The $I(AB)$ and $J(AB)$ totals are now arranged in a two-way table with factor C, and the I and J diagonal totals of this new display computed.

C	I(AB)			Totals I( ) J(C)		C	J(AB)			Totals I( ) J(C)	
0	-198	-106	-155	-149	41	0	-222	-79	-158	63	138
1	331	255	377	212	19	1	238	440	285	62	4
2	-59	-74	-206	102	105	2	-144	-40	-155	40	23

The I and J diagonal totals computed above are actually the totals representing the quantities $I[I(AB) \times C] = AB^2 C^2$, $J[I(AB) \times C] = AB^2 C$, $I[J(AB) \times C] = ABC^2$, and $J[J(AB) \times C] = ABC$, or the W, X, Y, and Z components of ABC. The sums of squares are found in the usual way, that is

$$I[I(AB) \times C] = AB^2 C^2 = W(ABC)$$

$$= \frac{(-149)^2 + (212)^2 + (102)^2}{18} - \frac{(165)^2}{54} = 3804.11$$

$$J[I(AB) \times C] = AB^2 C = X(ABC)$$

$$= \frac{(41)^2 + (19)^2 + (105)^2}{18} - \frac{(165)^2}{54} = 221.77$$

$$I[J(AB) \times C] = ABC^2 = Y(ABC)$$

$$= \frac{(63)^2 + (62)^2 + (40)^2}{18} - \frac{(165)^2}{54} = 18.77$$

$$J[J(AB) \times C] = ABC = Z(ABC)$$

$$= \frac{(138)^2 + (4)^2 + (23)^2}{18} - \frac{(165)^2}{54} = 584.11$$

While this is an orthogonal partitioning of SS_{ABC}, we point out again that it is not customarily displayed in the analysis of variance table. In Chapter 9, we will discuss the occasional need for computation of one or more of these components.

8-3.4 The General 3^k Design

The concepts utilized in the 3^2 and 3^3 designs can be readily extended to the case of k factors each at three levels, that is, a 3^k factorial design. The usual digital notation is employed for the treatment combinations, so that 0120 represents a treatment combination in a 3^4 design with A and D at the low levels, B at the intermediate level, and C at the high level. There are 3^k treatment combinations, with $3^k - 1$ degrees of freedom between them. These treatment combinations allow sums of squares to be determined for k main effects, each with two degrees of freedom, $\binom{k}{2}$ two-factor interactions each with four degrees of freedom, . . . , one k-factor interaction with 2^k degrees of freedom. In general, an b-factor interaction has 2^b degrees of freedom. If there are n replicates, there are $n3^k - 1$ total degrees of freedom and $3^k(n - 1)$ degrees of freedom for error. The analysis of variance is outlined in Table 8-15.

Sums of squares for effects and interactions are computed by the usual methods for factorial designs. As a general rule, three-factor and higher interactions are not broken down any further. However, any b-factor interaction has 2^{b-1} orthogonal two-degree-of-freedom components. For example, the four-factor interaction $ABCD$ has $2^{4-1} = 8$

Table 8-15 Analysis of Variance for a 3^k Design

Source of Variation	Sum of Squares	Degrees of Freedom
k Main Effects		
A	SS_A	2
B	SS_B	2
.	.	.
.	.	.
K	SS_K	2
$\binom{k}{2}$ Two-Factor Interactions		
AB	SS_{AB}	4
AC	SS_{AC}	4
.	.	.
.	.	.
JK	SS_{JK}	4
$\binom{k}{3}$ Three-Factor Interactions		
ABC	SS_{ABC}	8
ABD	SS_{ABD}	8
.	.	.
.	.	.
IJK	SS_{IJK}	8
.	.	.
.	.	.
$\binom{k}{k} = 1$ k-Factor Interaction		
$ABC \ldots K$	$SS_{ABC\ldots K}$	2^k
Error	SS_E	$3^k(n-1)$
Total	SS_T	$n3^k - 1$

orthogonal two-degree-of-freedom components, denoted by $ABCD^2$, ABC^2D, AB^2CD, $ABCD$, ABC^2D^2, AB^2C^2D, AB^2CD^2, and $AB^2C^2D^2$. In writing these components, note that the only exponent allowed on the first letter is 1. If the first letter is not 1, then the entire expression must be squared and the exponents reduced modulus 3. To demonstrate, consider

$$A^2BCD = (A^2BCD)^2 = A^4B^2C^2D^2 = AB^2C^2D^2$$

These interaction components have no physical interpretation, but are useful in constructing more complex designs.

The size of the design increases rapidly with k. For example, a 3^3 has 27 treatment combinations per replication, the 3^4 has 81, the 3^5 has 243, and so on. Therefore, frequently only a single replicate of the 3^k design is used, and higher-order interactions are combined to provide an estimate of error. As an illustration, if three-factor and higher interactions are negligible, then a single replicate of the 3^3 provides 8 degrees of freedom for error, and a single replicate of the 3^4 provides 48 degrees of freedom for error.

8-3.5 Yates' Algorithm for the 3^k Design

Yates' algorithm can be modified for use in the 3^k factorial design. We illustrate the procedure using the data in Example 6-1. The data for this example are given in Table 6-6. This is a 3^2 design used to investigate the effect of material type (A) and temperature (B) on the output voltage of a battery. There are $n = 4$ replicates.

The procedure is displayed in Table 8-16. The treatment combinations are written down in standard order, that is, the factors are introduced one at a time, each level being combined successively with every set of factor levels above it in the table. (The standard order for a 3^3 would be 000, 100, 200, 010, 110, 210, 020, 120, 220, 001, . . .). The Response column contains the total of all observations taken under the corresponding treatment combination. The entries in column (1) are computed as

Table 8-16   Yates' Algorithm for the 3^2 Design in Example 6-1

Treatment Combination	Response	(1)	(2)	Effect	Divisor	Sum of Squares
00	539	1738	3799	—	—	—
10	623	1291	503	A_L	$2^1 \times 3^1 \times 4$	10,542.04
20	576	770	-101	A_Q	$2^1 \times 3^2 \times 4$	141.68
01	229	37	-968	B_L	$2^1 \times 3^1 \times 4$	39,042.66
11	479	354	75	$AB_{L \times L}$	$2^2 \times 3^0 \times 4$	351.56
21	583	112	307	$AB_{Q \times L}$	$2^2 \times 3^1 \times 4$	1,963.52
02	230	-131	-74	B_Q	$2^1 \times 3^2 \times 4$	76.06
12	198	-146	-559	$AB_{L \times Q}$	$2^2 \times 3^1 \times 4$	6,510.02
22	342	176	337	$AB_{Q \times Q}$	$2^2 \times 3^2 \times 4$	788.67

follows: The first third of the column consists of the sums of each of the three sets of three values in the response column. The second third of the column is the third minus the first observation in the same set of three. This operation computes the linear component of the effect. The last third of the column is obtained by taking the sum of the first and third value minus twice the second in each set of three observations. This computes the quadratic component. For example, in column (1), the second, fifth, and eighth entries are $229 + 479 + 583 = 1291$, $-229 + 583 = 354$, and $229 - (2)(479) + 583 = -146$, respectively. Column (2) is obtained similarly from column (1). In general, k columns must be constructed.

The effect column is determined by converting the treatment combinations at the left of the row into corresponding effects. That is, 10 represents the linear effect of A, A_L, and 11 represents the $AB_{L \times L}$ component of the AB interaction. The entries in the divisor column are found from

$$2^r 3^t n$$

where r is the number of factors in the effect considered, t is the number of factors in the experiment minus the number of linear terms in this effect, and n is the number of replicates. For example, B_L has the divisor $2^1 \times 3^1 \times 4 = 24$.

The sums of squares are obtained by squaring the element in column (2) and dividing by the corresponding entry in the divisor column. The sum of squares column now contains all of the required quantities to construct an analysis of variance table, if both factors A and B are quantitative. However, in this example, factor A, material type, is qualitative and thus the linear and quadratic partitioning of A is not appropriate. Individual observations are required to compute the total sum of squares, and the error sum of squares is obtained by subtraction.

The analysis of variance is summarized in Table 8-17. Comparison with Table 6-18

Table 8-17 Analysis of Variance for the 3^2 Design in Example 6-1

Source of Variation	Sum of Squares	Degrees of Freedom	Mean Square	F_0
$A = \text{"}A_L\text{"} + \text{"}A_Q\text{"}$	10,683.72	2	5,341.86	7.91[b]
B, Temperature	39,118.72	2	19,558.36	28.97[b]
(B_L)	(39,042.67)	1	39,042.67	57.82[b]
(B_Q)	(76.05)	1	76.05	0.12
AB	9,613.77	4	2,403.44	3.56[a]
$(A \times B_L = \text{"}AB_{L \times L}\text{"} + \text{"}AB_{Q \times L}\text{"})$	(2,315.07)	2	1,157.54	1.71
$(A \times B_Q = \text{"}AB_{L \times Q}\text{"} + \text{"}AB_{Q \times Q}\text{"})$	(7,298.70)	2	3,649.75	5.41[a]
Error	18,230.75	27	675.21	
Total	77,646.96	35		

[a] Significant at 5 percent.
[b] Significant at 1 percent.

shows that substantially the same results were obtained by conventional analysis of variance methods. Examples 6-1 and 6-4 provide interpretation of this experiment.

8-4 PROBLEMS

8-1. An engineer is interested in the effect of cutting speed (A), tool geometry (B), and cutting angle (C) on the life of a machine tool. Two levels of each factor are chosen, and three replicates of a 2^3 factorial design are run. The results are shown below. Analyze the data from this experiment.

Treatment Combination	Replicate		
	I	II	III
(1)	22	31	25
a	32	43	29
b	35	34	50
ab	55	47	46
c	44	45	38
ac	40	37	36
bc	60	50	54
abc	39	41	47

8-2. An experiment was performed to improve the yield of a chemical process. Four factors were selected, and two replicates of a completely randomized experiment were run. The results are shown below. Analyze the data and draw appropriate conclusions.

Treatment Combination	Replicate		Treatment Combination	Replicate	
	I	II		I	II
(1)	90	93	d	98	95
a	74	78	ad	72	76
b	81	85	bd	87	83
ab	83	80	abd	85	86
c	77	78	cd	99	90
ac	81	80	acd	79	75
bc	88	82	bcd	87	84
abc	73	70	abcd	80	80

8-3. A bacteriologist is interested in the effect of two different culture mediums and two different times on the growth of a particular virus. He performs six replicates of a 2^2 design, making the runs in random order. Analyze the data shown below and draw appropriate conclusions.

Culture Medium

Time	1		2	
	21	22	25	26
12 hr	23	28	24	25
	20	26	29	27
	37	39	31	34
18 hr	38	38	29	33
	35	36	30	35

8-4. An industrial engineer employed by a beverage bottler is interested in the effect of two different types of 32-ounce bottles on the time to deliver 12-bottle cases of the product. The two bottle types, glass and plastic, and two delivery men are used to perform a task consisting of moving 40 cases of product 50 feet on a standard type of hand truck and stacking the cases in a display. Four replicates of a 2^2 factorial design are performed, and the times observed are shown below. Analyze the data and draw appropriate conclusions.

Man

Bottle Type	1		2	
Glass	5.12	4.89	6.65	6.24
	4.98	5.00	5.49	5.55
Plastic	4.95	4.43	5.28	4.91
	4.27	4.25	4.75	4.71

8-5. In Problem 8-4, the engineer was also interested in potential fatigue differences between the two types of bottles. As a measure of the amount of effort required, he measured the elevation of the heart rate (pulse) induced by the task. The results are shown below. Analyze the data and draw conclusions.

Man

Bottle Type	1		2	
Glass	39	45	20	13
	58	35	16	11
Plastic	44	35	13	10
	42	21	16	15

8-6. A pilot plant experiment is conducted with five factors, each at two levels. Two replicates of a 2^5 design are run, and the results are shown below. Analyze the data and draw conclusions.

Treatment Combination	Replicate I	Replicate II	Treatment Combination	Replicate I	Replicate II
(1)	2	2	e	-3	-2
a	3	1	ae	-3	0
b	3	1	be	-2	1
ab	4	2	abe	1	1
c	-3	0	ce	1	-4
ac	2	0	ace	-1	-1
bc	-1	1	bce	-2	1
abc	-1	1	abce	-2	1
d	0	1	de	-1	-1
ad	-1	-1	ade	1	1
bd	1	0	bde	0	-1
abd	5	3	abde	3	2
cd	-3	-1	cde	-2	-1
acd	-1	-1	acde	1	1
bcd	-1	1	bcde	0	-3
abcd	-1	-1	abcde	0	-2

8-7. The data shown below represent a single replicate of a 2^5 design used in an experiment to study the tensile strength of rubber. The factors are mix (A), time (B), laboratory (C), temperature (D), and pressure (E). Analyze the data assuming that three-factor and higher interactions are negligible.

(1) = 7	d = 10	e = 8	de = 10
a = 9	ad = 11	ae = 12	ade = 15
b = 34	bd = 30	be = 35	bde = 40
ab = 55	abd = 61	abe = 62	abde = 65
c = 6	cd = 8	ce = 5	cde = 15
ac = 10	acd = 11	ace = 12	acde = 20
bc = 30	bcd = 34	bce = 25	bcde = 34
abc = 50	abcd = 60	abce = 55	abcde = 65

8-8. Plot the estimates of the effects from Problem 8-7 on normal probability paper. Does the assumption that three-factor and higher interactions are negligible seem justified? Does this change your interpretation of the data?

8-9. In an experiment on yield, four factors were studied, each at two levels. A single replicate of a 2^4 design in the factors time (A), concentration (B), pressure (C) and temperature (D) was run, and the resulting data are shown at the top of page 213. Analyze the data assuming that three-factor and higher interactions are negligible.

	A_0				A_1			
	B_0		B_1		B_0		B_1	
	C_0	C_1	C_0	C_1	C_0	C_1	C_0	C_1
D_0	12	17	13	20	18	15	16	15
D_1	10	19	13	17	25	21	24	23

8-10. Plot the estimates of the effects from Problem 8-9 on normal probability paper. Is the assumption of negligible three-factor and higher interactions reasonable?

8-11. One of the main effects in Problem 8-9 is probably not significant. Collapse the single replicate of the 2^4 design in that problem into two replicates of a 2^3 in the three appropriate factors.

8-12. A factor that appears at 4 levels can be considered as two pseudo-factors, each at two levels. In general, a factor with 2^k levels can be thought of as k pseudo-factors at two levels each. Yates' method can then be used on these designs, and the sums of squares associated with the pseudo-factors can be combined to yield the sums of squares for the main effects and interactions for the original factors. Use the data in Problem 6-6 to demonstrate this technique.

8-13. Analyze the data in Example 8-2 by Yates' algorithm.

8-14. A procedure that is occasionally used when an observation is missing in a 2^k is to replace it by the value that makes the k-factor interaction zero. Use this technique on the data of Problem 8-9, assuming that the run bd is missing. Compare the analysis with the results of Problem 8-9.

8-15. The effects of developer strength (A) and development time (B) on the density of photographic plate film are being studied. Three strengths and three times are used, and four replicates of a 3^2 factorial experiment are run. The data from this experiment are shown below. Analyze this data, using the standard methods for factorial experiments.

Development Time (Minutes)

Developer Strength	10		15		18	
1	0	2	1	3	2	5
	5	4	4	2	4	6
2	4	6	6	8	9	10
	7	5	7	7	8	5
3	7	10	10	10	12	10
	8	7	8	7	9	8

8-16. Analyze the data in Problem 8-15 using Yates' algorithm.

8-17. Compute the I and J components of the two factor interaction in Problem 8-15.

8-18. An experiment was performed to study the effect of three different types of 32-ounce bottles (A), and three different shelf types (smooth permanent shelves, end-aisle displays with grilled shelves, and beverage coolers) (B) on the time to stock ten 12-bottle cases on the shelves. Three men (factor C) were employed in the experiment, and two replicates of a 3^3 factorial design were run. The observed time data are shown below. Analyze these data and draw conclusions.

Man	Bottle Type	Replicate I Permanent	End Aisle	Cooler	Replicate II Permanent	End Aisle	Cooler
	Plastic	3.45	4.14	5.80	3.36	4.19	5.23
1	28 mm glass	4.07	4.38	5.48	3.52	4.26	4.85
	38 mm glass	4.20	4.26	5.67	3.68	4.37	5.58
	Plastic	4.80	5.22	6.21	4.40	4.70	5.88
2	28 mm glass	4.52	5.15	6.25	4.44	4.65	6.20
	38 mm glass	4.96	5.17	6.03	4.39	4.75	6.38
	Plastic	4.08	3.94	5.14	3.65	4.08	4.49
3	28 mm glass	4.30	4.53	4.99	4.04	4.08	4.59
	38 mm glass	4.17	4.86	4.85	3.88	4.48	4.90

8-19. A medical researcher is studying the effect of lidocaine on the enzyme level in the heart muscle of beagle dogs. Three different commercial brands of lidocaine (A), three dosage levels (B), and three dogs (C) are used in the experiment, and two replicates of a 3^3 factorial design are run. The observed enzyme levels are shown below. Analyze the data from this experiment.

Lidocaine Brand	Dosage Strength	Replicate I Dog 1	2	3	Replicate II Dog 1	2	3
	1	86	84	85	84	85	86
1	2	94	99	98	95	97	90
	3	101	106	98	105	104	103
	1	85	84	86	80	82	84
2	2	95	98	97	93	99	95
	3	108	114	109	110	102	100
	1	84	83	81	83	80	79
3	2	95	97	93	92	96	93
	3	105	100	106	102	111	108

8-20. Compute the I and J components of the two-factor interactions for Example 8-3.

Chapter Nine
Confounding

9-1 INTRODUCTION

There are many problems in which it is impossible to perform a complete replicate of a factorial design in one block, where the block might be one day, or one homogeneous batch of raw material, or one laboratory, and so forth. *Confounding* is a design technique for arranging a complete factorial experiment in blocks, where the block size is smaller than the number of treatment combinations in one replicate. The technique causes information about certain treatment effects (usually high-order interactions) to be *indistinguishable from* or *confounded* with blocks. In this chapter we concentrate on confounding systems for the 2^k and 3^k factorial designs. Note that even though the designs presented are incomplete block designs because each block does not contain all treatments or treatment combinations, the special structure of the 2^k and 3^k factorial system allows a simplified method of analysis. As in Chapter 8, we continue to assume that all factors are fixed effects.

9-2 CONFOUNDING IN THE 2^k FACTORIAL DESIGN

The simplest confounding schemes are those for the 2^k factorial series. In this section we consider the construction and analysis of the 2^k factorial design in 2^p incomplete blocks, where $p < k$. Consequently, these designs can be run in two blocks, four blocks, eight blocks, and so on.

9-2.1 The 2^k Factorial Design in Two Blocks

Suppose that we wish to run a single replicate of the 2^2 design. Each of the $2^2 = 4$ treatment combinations requires a quantity of raw material, for example, and each batch of raw material is only large enough for two treatment combinations to be tested. Thus, two batches of raw material are required. If batches of raw material are considered as blocks, then we must assign two of the four treatment combinations to each block.

Consider the design shown in Figure 9-1. Notice that block 1 contains the treatment combinations (1) and ab, and that block two contains a and b. Of course, the *order*

Figure 9-1. A 2^2 design in two blocks.

in which the treatment combinations are run within a block is randomly determined. We would also randomly decide which block to run first. Suppose we estimate the main effects of A and B just as if no blocking had occurred. From Equations 8-1 and 8-2, we obtain

$$A = \tfrac{1}{2} \ [ab + a - b - (1)]$$

$$B = \tfrac{1}{2} \ [ab + b - a - (1)]$$

Note that both A and B are unaffected by blocking since in each estimate there is one plus and one minus treatment combination from each block. That is, any difference between block 1 and block 2 will cancel out.

Now consider the AB interaction:

$$AB = \tfrac{1}{2} \ [ab + (1) - a - b]$$

Since the two treatment combinations with the plus sign [ab and (1)] are in block 1 and the two with the minus sign [a and b] are in block 2, the block effect and the AB interaction are identical. That is, AB is *confounded* with blocks.

The reason for this is apparent from the table of plus and minus signs for the 2^2 design. This was originally given as Table 8-2, but for convenience it is reproduced as Table 9-1. From this table, we see that all treatment combinations that have a plus on AB are assigned to block 1, while all treatment combinations which have a minus sign on AB are assigned to block 2. Furthermore, notice that if (1) and b had been assigned to block 1 and a and ab to block 2, then the main effect A would have been con-

Table 9-1 Table of Plus and Minus
Signs for the 2^2 Design

Treatment Combination	Factorial Effect			
	I	A	B	AB
(1)	+	−	−	+
a	+	+	−	−
b	+	−	+	−
ab	+	+	+	+

founded with blocks. The usual practice is to confound the highest-order interaction with blocks.

This scheme can be used to confound any 2^k design in two blocks. As a second example, consider a 2^3 design, run in two blocks. Suppose we wish to confound the three-factor interaction ABC with blocks. From the table of plus and minus signs, shown in Table 9-2, we assign the treatment combinations that are minus on ABC to

Table 9-2 Table of Plus and Minus Signs for the 2^3 Design

Treatment Combination	Factorial Effect							
	I	A	B	AB	C	AC	BC	ABC
(1)	I			+	−	+	+	−
a	+	+	−	−	−	−	+	+
b	+	−	+	−	−	+	−	+
ab	+	+	+	+	−	−	−	−
c	+	−	−	+	+	−	−	+
ac	+	+	−	−	+	+	−	−
bc	+	−	+	−	+	−	+	−
abc	+	+	+	+	+	+	+	+

block 1 and those that are plus on ABC to block 2. The resulting design is shown in Figure 9-2. Once again, we emphasize that the treatment combinations *within* a block are run in random order.

Block 1	Block 2
(1)	a
ab	b
ac	c
bc	abc

Figure 9-2. The 2^3 design in two blocks, ABC confounded.

Kempthorne (1952), Hicks (1973), and Anderson and McLean (1974) have described another method for constructing these designs. The method uses the linear combination

$$L = \alpha_1 x_1 + \alpha_2 x_2 + \cdots + \alpha_k x_k \qquad (9\text{-}1)$$

where x_i is the level of the ith factor appearing in a particular treatment combination and α_i is the exponent appearing on the ith factor in the effect to be confounded. For the 2^k system, we have either $\alpha_i = 0$ or 1, and either $x_i = 0$ (low level) or $x_i = 1$ (high level). Equation 9-1 is called a *defining contrast*. Treatment combinations that produce the same value of L modulus (2) will be placed in the same block. Since the only possible values of L (mod 2) are 0 and 1, this will assign the 2^k treatment combinations to exactly two blocks.

To illustrate the approach, consider a 2^3 design with ABC confounded with blocks. Here x_1 corresponds to A, x_2, to B, x_3 to C, and $\alpha_1 = \alpha_2 = \alpha_3 = 1$. Thus the defining contrast corresponding to ABC is

$$L = x_1 + x_2 + x_3$$

The treatment combination (1) is written 000 in the (0, 1) notation, therefore

$$L = 1(0) + 1(0) + 1(0) = 0 = 0 (\text{mod } 2)$$

Similarly, the treatment combination a is 100, yielding

$$L = 1(1) + 1(0) + 1(0) = 1 = 1 (\text{mod } 2)$$

Thus (1) and a would be run in different blocks. For the remaining treatment combinations we have:

$$b: \; L = 1(0) + 1(1) + 1(0) = 1 = 1 (\text{mod } 2)$$

$$ab: \; L = 1(1) + 1(1) + 1(0) = 2 = 0 (\text{mod } 2)$$

$$c: \; L = 1(0) + 1(0) + 1(1) = 1 = 1 (\text{mod } 2)$$

$$ac: \; L = 1(1) + 1(0) + 1(1) = 2 = 0 (\text{mod } 2)$$

$$bc: \; L = 1(0) + 1(1) + 1(1) = 2 = 0 (\text{mod } 2)$$

$$abc: \; L = 1(1) + 1(1) + 1(1) = 3 = 1 (\text{mod } 2)$$

Thus (1), ab, ac, and bc are run in block 1 and a, b, c, and abc are run in block 2. This is the same design shown in Figure 9-2 that was generated from the table of plus and minus signs.

Another method may be used to construct these designs. The block containing the treatment combination (1) is called the *principal block*. The treatment combinations in this block have a useful group-theoretic property; namely, they form a group with respect to multiplication modulus 2. This implies that any element [except (1)] in the principal block may be generated by multiplying two other elements in the principal block modulus 2. For example, consider the principal block of the 2^3 design with

ABC confounded, shown in Figure 9-2. Note that

$$ab \cdot ac = a^2 bc = bc$$

$$ab \cdot bc = ab^2 c = ac$$

$$ac \cdot bc = abc^2 = ab$$

Treatment combinations in the other block (or blocks) may be generated by multiplying one element in the new block by each element in the principal block modulus 2. For the 2^3 with ABC confounded, since the principal block is (1), ab, ac, and bc, we know that b is in the other block. Thus, the elements of this second block are

$$b \cdot (1) \qquad = b$$

$$b \cdot ab = ab^2 = a$$

$$b \cdot ac \qquad = abc$$

$$b \cdot bc = b^2 c = c$$

This agrees with the results obtained previously.

Unless the experimenter has an independent estimate of experimental error, or can assume certain interactions to be negligible, replication of the experiment will be necessary to test hypotheses. Suppose that a 2^3 factorial must be run in two incomplete blocks with ABC confounded, and the experimenter decides to replicate the design four times. The resulting design might appear as in Figure 9-3. Note that ABC is confounded in each replicate.

Replicate I		Replicate II		Replicate III		Replicate IV	
Block 1	Block 2	Block 1	Block 2	Block 1	Block 2	Block 1	Block 2
(1)	abc	(1)	abc	(1)	abc	(1)	abc
ac	a	ac	a	ac	a	ac	a
ab	b	ab	b	ab	b	ab	b
bc	c	bc	c	bc	c	bc	c

Figure 9-3. Four replicates of the 2^3 design, ABC confounded.

The analysis of variance for this design is shown in Table 9-3. There are 32 observations and 31 total degrees of freedom. Furthermore, since there are eight blocks, 7 degrees of freedom must be associated with these blocks. The breakdown of those 7 degrees of freedom suggested by Cochran and Cox (1957) is shown in Table 9-3. The error sum of squares actually consists of the two-factor interactions between replicates and each of the effects (A, B, C, AB, AC, BC). It is usually safe to consider these interactions zero and treat the resulting mean square as an estimate of error. Main

Table 9-3   Analysis of Variance for Four Replicates
of a 2^3 Design with ABC Confounded

Source of Variation	Degrees of Freedom
Replicates	3
Blocks (ABC)	1
Error for ABC (replicates X blocks)	3
A	1
B	1
C	1
AB	1
AC	1
BC	1
Error (or replicates X effects)	18
Total	31

effects and two-factor interactions are tested against the mean square error. Cochran and Cox (1957) state that the block or ABC mean square could be compared to the error for ABC mean square, which is really replicates X blocks. This test usually is very insensitive.

   If resources are sufficient to allow replication of confounded designs, it is generally better to use a slightly different method of designing the blocks in each replicate. This approach consists of confounding a different effect in each replicate so that some information on all effects is obtained. Such a procedure is called partial confounding, and is discussed in Section 9-4. If k is moderately large, say $k \geqslant 4$, we frequently can only afford a single replicate. The experimenter usually assumes certain higher-order interactions to be negligible and combines their sums of squares as error. This is illustrated in the following example.

■**EXAMPLE 9-1.**   Consider the situation described in Example 8-2. Recall that four factors—temperature (A), pressure (B), concentration of reactant (C), and stirring rate (D)—are studied in a pilot plant to determine their effect on product filtration rate. Suppose now that the 2^4 = 16 treatment combinations cannot all be run on the same day. The experimenter can run 8 treatment combinations in one day, so a 2^4 design confounded in two blocks seems appropriate. It is logical to confound the highest-order interaction $ABCD$ with blocks. The defining contrast is

$$L = x_1 + x_2 + x_3 + x_4$$

and it is easy to verify that the blocks are

Block 1	Block 2

Block 1	Block 2
(1) = 45	a = 71
ab = 65	b = 48
ac = 60	c = 68
bc = 80	d = 43
ad = 100	abc = 65
bd = 45	bcd = 70
cd = 75	acd = 86
abcd = 96	abd = 104

The data may be analyzed by Yates' algorithm, as shown in Table 9-4. Since the block totals are 566, and 555, we have the sum of squares for blocks as

$$SS_{\text{Blocks}} = \frac{(566)^2 + (555)^2}{8} - \frac{(1121)^2}{16} = 7.5625$$

which is identical to the sum of squares for $ABCD$ in Table 9-4.

The sums of squares for the main effects and two-factor interactions are taken directly from the Yates' algorithm. The experimenter has decided that three-factor

Table 9-4 Yates' Algorithm for Example 9-1

Treatment Combination	Response	(1)	(2)	(3)	(4)	Effects	Sum of Squares
(1)	45	116	229	502	1121		
a	71	113	273	619	173	A	1870.5625
b	48	128	292	20	25	B	39.0625
ab	65	145	327	153	1	AB	0.0625
c	68	143	43	14	79	C	390.0625
ac	60	149	-23	11	-145	AC	1314.0625
bc	80	161	116	-16	19	BC	22.5625
abc	65	166	37	17	15	ABC	14.0625
d	43	26	-3	44	117	D	855.5625
ad	100	17	17	35	133	AD	1105.5625
bd	45	-8	6	-66	-3	BD	0.5625
abd	104	-15	5	-79	33	ABD	68.0625
cd	75	57	-9	20	-9	CD	5.0625
acd	86	59	-7	-1	-13	ACD	10.5625
bcd	70	11	2	2	-21	BCD	27.5625
abcd	96	26	15	13	11	ABCD	7.5625

Table 9-5   Analysis of Variance for Example 9-1

Source of Variation	Sum of Squares	Degrees of Freedom	Mean Square	F_0
Blocks (*ABCD*)	7.5625	1		
A	1870.5625	1	1870.5625	62.22[b]
B	39.0625	1	39.0625	1.30
C	390.0625	1	390.0625	12.98[a]
D	855.5625	1	855.5625	28.46[b]
AB	0.0625	1	0.0625	<1
AC	1314.0625	1	1314.0625	43.71[b]
AD	1105.5625	1	1105.5625	36.78[b]
BC	22.5625	1	22.5625	<1
BD	0.5625	1	0.5625	<1
CD	5.0625	1	5.0625	<1
Error (or *ABC* + *ABD* + *ACD* + *BCD*)	120.2500	4	30.0625	
Total		15		

[a]Significant at 5 percent.
[b]Significant at 1 percent.

interactions are negligible, so the error sum of squares is

$$SS_E = SS_{ABC} + SS_{ABD} + SS_{ACD} + SS_{BCD}$$

$$= 14.0625 + 68.0625 + 10.5625 + 27.5625 = 120.2500$$

with four degrees of freedom. The complete analysis of variance is summarized in Table 9-5. For further details of the interpretation of this experiment, refer to Example 8-2. ∎

9-2.2   The 2^k Factorial Design in Four Blocks

It is possible to construct 2^k factorial designs confounded in four blocks of 2^{k-2} observations each. These designs are particularly useful in situations where the number of factors is moderately large, say $k \geqslant 4$, and block sizes are relatively small.

As an example, consider the 2^5 design. If each block will hold only eight treatment combinations, then four blocks must be used. The construction of this design is relatively straightforward. Select *two* effects to be confounded with blocks, say *ADE* and *BCE*. These effects have the two defining contrasts

$$L_1 = x_1 + x_4 + x_5$$

$$L_2 = x_2 + x_3 + x_5$$

associated with them. Now every treatment combination will yield a particular pair of values of L_1 (mod 2) and L_2 (mod 2); that is, either $(L_1, L_2) = (0, 0), (0, 1), (1, 0)$, or $(1, 1)$. Treatment combinations yielding the same values of (L_1, L_2) are assigned to the same block. In our example we find

$$L_1 = 0, L_2 = 0 \quad \text{for} \quad (1), ad, bc, abcd, abe, ace, cde, bde$$

$$L_1 = 1, L_2 = 0 \quad \text{for} \quad a, d, abc, bcd, be, abde, ce, acde$$

$$L_1 = 0, L_2 = 1 \quad \text{for} \quad b, abd, c, acd, ae, de, abce, bcde$$

$$L_1 = 1, L_2 = 1 \quad \text{for} \quad e, ade, bce, abcde, ab, bd, ac, cd$$

These treatment combinations would be assigned to different blocks. The complete design is as shown in Figure 9-4.

Block 1		Block 2		Block 3		Block 4	
$L_1 = 0$		$L_1 = 1$		$L_1 = 0$		$L_1 = 1$	
$L_2 = 0$		$L_2 = 0$		$L_2 = 1$		$L_2 = 1$	
(1)	abe	a	bc	b	abce	e	abcde
ad	ace	d	abde	abd	ae	ade	bd
bc	cde	abc	ce	c	bcde	bce	ac
abcd	bde	bcd	acde	acd	de	ab	cd

Figure 9-4. The 2^5 design in four blocks with ADE, BCE, and $ABCD$ confounded.

With a little reflection we will realize that another effect in addition to ADE and BCE must be confounded with blocks. Since there are four blocks, with three degrees of freedom between them, and since ADE and BCE have only one degree of freedom each, clearly an additional effect with one degree of freedom must be confounded. This effect is the *generalized interaction* of ADE and BCE, defined as the product of ADE and BCE modulus 2. Thus in our example the generalized interaction (ADE) $(BCE) = ABCDE^2 = ABCD$ is also confounded with blocks. It is easy to verify this by referring to a table of plus and minus signs for the 2^5 design, such as in Davies (1956). Inspection of such a table reveals that the treatment combinations are assigned to the blocks as follows:

Treatment Combinations in	Sign on ADE	Sign on BCE	Sign on $ABCD$
Block 1	−	−	+
Block 2	+	−	−
Block 3	−	+	−
Block 4	+	+	+

Notice that the product of signs of any two effects for a particular block (e.g., ADE and BCE) yields the sign of the other effect for that block (in this case $ABCD$). Thus ADE, BCE, and $ABCD$ are all confounded with blocks.

The group-theoretic properties of the principal block mentioned in Section 9-2.1 still hold. For example, we see that the product of two treatment combinations in the principal block yields another element of the principal block. That is,

$$ad \cdot bc = abcd, \quad \text{and} \quad abe \cdot bde = ab^2 de^2 = ad$$

and so forth. To construct another block select a treatment combination which is not in the principal block (e.g., b) and multiply b by all treatment combinations in the principal block. This yields

$$b \cdot (1) = b \quad b \cdot ad = abd \quad b \cdot bc = c \quad b \cdot abcd = ab^2 cd = acd$$

and so forth, which will produce the eight treatment combinations in block 3. In practice, the principal block can be obtained from the defining contrasts and the group-theoretic property, and the remaining blocks determined from these treatment combinations by the method shown above.

The general procedure for constructing a 2^k design confounded in four blocks is to choose two effects, automatically confounding a third effect that is the generalized interaction of the first two. Then the design is constructed by using the two defining contrasts (L_1, L_2) and the group-theoretic properties of the principal block. In selecting effects to be confounded with blocks care must be exercised to obtain a design that does not confound effects that may be of interest. For example, in a 2^5 we might choose to confound $ABCDE$ and ABD, which automatically confounds CE, an effect which is probably of interest. A better choice is to confound ADE and BCE, which automatically confounds $ABCD$. It is preferable to sacrifice information on the three-factor interactions ADE and BCE instead of the two-factor interaction CE.

9-2.3 The 2^k Factorial Design in 2^p Blocks

The methods described above may be extended to the construction of a 2^k factorial design confounded in 2^p blocks $(p < k)$, where each block contains exactly 2^{k-p} runs. We select p independent effects to be confounded, where by "independent" we mean that no effect chosen is the generalized interaction of the others. The blocks may be formed by use of the p defining contrasts $L_1, L_2, \ldots, L_p$ associated with these effects. In addition, exactly $2^p - p - 1$ other effects will be confounded with blocks, these being the generalized interactions of those p independent effects initially chosen. Care should be exercised in selecting effects to be confounded so that information on effects that may be of potential interest is not sacrificed.

The statistical analysis of these designs is straightforward. Sums of squares for all effects are computed as if no blocking had occurred. Then the block sum of squares is found by adding the sums of squares for all effects confounded with blocks.

As an example, suppose we wish to construct a 2^6 design confounded in $2^3 = 8$ blocks of $2^{6-3} = 8$ runs each. Suppose we choose $ABCD$, AEF, and $ACDE$ as the $p = 3$ independent effects to be confounded. The remaining $2^p - p - 1 = 2^3 - 3 - 1 = 4$ effects that are confounded are the generalized interactions of these three, that is,

$$(ABCD)(AEF) = A^2 BCDEF = BCDEF$$

$$(ABCD)(ACDE) = A^2 BC^2 D^2 E = BE$$

$$(AEF)(ACDE) = A^2 CDF^2 F = CDF$$

$$(ABCD)(AEF)(ACDE) = A^3 BC^2 D^2 E^2 F = ABF$$

The reader is asked to generate the eight blocks for this design in Problem 9-8.

9-3 CONFOUNDING IN THE 3^k FACTORIAL DESIGN

In this section we discuss confounding the 3^k factorial design in 3^p incomplete blocks, where $p < k$. Thus these designs may be confounded in three blocks, nine blocks, and so on.

9-3.1 The 3^k Factorial Design in Three Blocks

Suppose that we wish to confound the 3^k design in three incomplete blocks. These three blocks have two degrees of freedom between them, thus there must be two degrees of freedom confounded with blocks. Recall that in the 3^k factorial series each main effect has two degrees of freedom. Furthermore, every two-factor interaction has four degrees of freedom and can be decomposed into two components of interaction (e.g., AB and AB^2) each with two degrees of freedom, every three-factor interaction has eight degrees of freedom and can be decomposed into four components of interaction (e.g., ABC, ABC^2, AB^2C, and AB^2C^2) each with two degrees of freedom, and so on. Therefore, it is convenient to confound a component of interaction with blocks.

The general procedure is to construct a defining contrast

$$L = \alpha_1 x_1 + \alpha_2 x_2 + \cdots + \alpha_k x_k \tag{9-2}$$

where α_i represents the exponent on the ith factor in the effect to be confounded and x_i is the level of the ith factor in a particular treatment combination. For the 3^k series, we have $\alpha_1 = 0$, 1, or 2 with the first nonzero α_i unity, and $x_i = 0$ (low level), 1 (intermediate level), or 2(high level). The treatment combinations in the 3^k design are assigned to blocks based on the value of L modulus 3. Since L (mod 3) can only take on values 0, 1, or 2, three blocks are uniquely defined. The treatment combinations satisfying $L = 0$ (mod 3) constitute the *principal block*. This block will always contain the treatment combination $00 \ldots 0$.

For example, suppose we wish to construct a 3^2 factorial design in three blocks.

Either component of the AB interaction, AB or AB^2, may be confounded with blocks. Arbitrarily choosing AB^2, we obtain the defining contrast

$$L = x_1 + 2x_2$$

The value of L (mod 3) of each treatment combination may be found as follows:

00: $L = 1(0) + 2(0) = 0 = 0(\text{mod } 3)$ 11: $L = 1(1) + 2(1) = 3 = 0(\text{mod } 3)$

01: $L = 1(0) + 2(1) = 2 = 2(\text{mod } 3)$ 21: $L = 1(2) + 2(1) = 4 = 1(\text{mod } 3)$

02: $L = 1(0) + 2(2) = 4 = 1(\text{mod } 3)$ 12: $L = 1(1) + 2(2) = 5 = 2(\text{mod } 3)$

10: $L = 1(1) + 2(0) = 1 = 1(\text{mod } 3)$ 22: $L = 1(2) + 2(2) = 6 = 0(\text{mod } 3)$

20: $L = 1(2) + 2(0) = 2 = 2(\text{mod } 3)$.

The blocks are shown in Figure 9-5.

Block 1	Block 2	Block 3
00	10	01
11	21	12
22	02	20

Figure 9-5. The 3^2 design in three blocks with AB^2 confounded.

The elements in the principal block form a group with respect to addition modulus 3. Referring to Figure 9-5 we see that $11 + 11 = 22$, and $11 + 22 = 00$. Treatment combinations in the other two blocks may be generated by adding modulus 3, any element in the new block, to the elements of the principal block. Thus, for block 2 we use 10 and obtain

$$10 + 00 = 10 \qquad 10 + 11 = 21 \qquad \text{and} \qquad 10 + 22 = 02$$

To generate block 3, using 01, we find

$$01 + 00 = 01 \qquad 01 + 11 = 12 \qquad \text{and} \qquad 01 + 22 = 20$$

■**EXAMPLE 9-2.** We illustrate the statistical analysis of the 3^2 design confounded in three blocks by using the data shown below, which represent a single replicate of the

Block 1	Block 2	Block 3
00 = 4	10 = -2	01 = 5
11 = -4	21 = 1	12 = -5
22 = 0	02 = 8	20 = 0

Block Totals =   0   7   0

3^2 design shown in Figure 9-5. Using conventional methods for the analysis of factorials, we find that $SS_A = 131.56$ and $SS_B = 0.22$. We also find

$$SS_{Blocks} = \frac{(0)^2 + (7)^2 + (0)^2}{3} - \frac{(7)^2}{9} = 10.89$$

However, SS_{Blocks} is exactly equal to the AB^2 component of interaction. To see this, write the observations as follows:

		Factor B		
		0	1	2
	0	4	5	8
Factor A	1	-2	-4	-5
	2	0	1	0

Recall from Section 8-3.2 that the I or AB^2 component of the AB interaction may be found by computing the sum of squares between the left-to-right diagonal totals in the above layout. This yields

$$SS_{AB^2} = \frac{(0)^2 + (0)^2 + (7)^2}{3} - \frac{(7)^2}{9} = 10.89$$

which is identical to SS_{Blocks}.

The analysis of variance is shown in Table 9-6. As there is only one replicate, no formal tests can be performed. Note that it is unwise to use the AB component of interaction as an estimate of error, since it has an insufficient number of degrees of freedom. ∎

Table 9-6 Analysis of Variance for Data in Example 9-2

Source of Variation	Sum of Squares	Degrees of Freedom
Blocks (AB^2)	10.89	2
A	131.56	2
B	0.22	2
AB	2.89	2
Total	145.56	8

We now give a slightly more complicated design, a 3^3 factorial confounded in three blocks of nine runs each. A component of the three-factor interaction (e.g., AB^2C^2) is to be confounded with blocks. The defining contrast is

$$L = x_1 + 2x_2 + 2x_3$$

It is easy to verify that the treatment combinations 000, 012, and 101 belong in the principal block. The remaining elements in the principal block are generated as follows:

(1) 000   (4) 101 + 101 = 202   (7) 101 + 021 = 122

(2) 012   (5) 012 + 012 = 021   (8) 012 + 202 = 211

(3) 101   (6) 101 + 012 = 110   (9) 021 + 202 = 220

To find the elements in another block, note that the treatment combination 200 is not in the principal block. Thus, the elements of block 2 are

(1) 200 + 000 = 200   (4) 200 + 202 = 102   (7) 200 + 122 = 022

(2) 200 + 012 = 212   (5) 200 + 021 = 221   (8) 200 + 211 = 111

(3) 200 + 101 = 001   (6) 200 + 110 = 010   (9) 200 + 220 = 120

Notice that these elements all satisfy $L = 2 \pmod 3$. The final block is found by observing that 100 does not belong in block 1 or 2. Using 100 as above yields

(1) 100 + 000 = 100   (4) 100 + 202 = 002   (7) 100 + 122 = 222

(2) 100 + 012 = 112   (5) 100 + 021 = 121   (8) 100 + 211 = 011

(3) 100 + 101 = 201   (6) 100 + 110 = 210   (9) 100 + 220 = 020

The blocks are shown in Fig. 9-6.

Block 1	Block 2	Block 3
000	200	100
012	212	112
101	001	201
202	102	002
021	221	121
110	010	210
122	022	222
211	111	011
220	120	020

Figure 9-6. The 3^3 design in three blocks with AB^2C^2 confounded.

The analysis of variance for this design is shown in Table 9-7. Using this confounding scheme information on all the main effects and two-factor interactions is available. The remaining components of the three-factor interaction (ABC, AB^2C, and ABC^2) are combined as an estimate of error. These components could be obtained jointly by subtraction. In general, for the 3^k design in three blocks, we would always select a component of the highest-order interaction to confound with blocks. The remaining

Table 9-7 Analysis of Variance for a 3^3 Design with $AB^2 C^2$ Confounded

Source of Variation	Degrees of Freedom
Blocks $(AB^2 C^2)$	2
A	2
B	2
C	2
AB	4
AC	4
BC	4
Error $(ABC + AB^2 C + ABC^2)$	6
Total	26

unconfounded components of this interaction could be obtained by computing the k-factor interaction in the usual way and subtracting from this quantity the sum of squares for blocks.

9-3.2 The 3^k Factorial Design in Nine Blocks

In some experimental situations it may be necessary to confound the 3^k design in nine blocks. Thus, eight degrees of freedom will be confounded with blocks. To construct these designs, we choose *two* components of interaction, and as a result, two more will be confounded automatically yielding the required eight degrees of freedom. These two are the generalized interactions of the two effects originally chosen. In the 3^k system, the *generalized interactions* of two effects (e.g., P and Q), are defined as PQ and PQ^2 (or $P^2 Q$).

The two components of interaction initially chosen yield *two* defining contrasts

$$L_1 = \alpha_1 x_1 + \alpha_2 x_2 + \cdots + \alpha_k x_k = u \pmod 3, \quad u = 0, 1, 2$$

$$L_2 = \beta_1 x_1 + \beta_2 x_2 + \cdots + \beta_k x_k = h \pmod 3, \quad h = 0, 1, 2$$

(9-3)

where $\{\alpha_i\}$ and $\{\beta_j\}$ are the exponents in the first and second generalized interactions, respectively, with the convention that the first nonzero α_i and β_j is unity. The defining contrasts in (9-3) imply nine simultaneous equations, specified by the pair of values for L_1 and L_2. Treatment combinations having the same pair of values for (L_1, L_2) are assigned to the same block.

The principal block consists of treatment combinations satisfying $L_1 = L_2 = 0$ (mod 3). The elements of this block form a group with respect to addition modulus 3; thus, the scheme given in Section 9-3.1 can be used to generate the blocks.

Block 1	Block 2	Block 3	Block 4	Block 5	Block 6	Block 7	Block 8	Block 9
0000	0001	2000	0200	0020	0010	1000	0100	0002
0122	0120	2122	0022	0112	0102	1122	0222	0121
0211	0212	2211	0111	0201	0221	1211	0011	0210
1021	1022	0021	0221	1011	1001	2021	1121	1020
1110	1111	0110	1010	1100	1120	2110	1210	1112
1201	1200	0202	1102	1222	1212	2202	1002	1201
2012	2010	1012	2212	2002	2022	0012	2112	2011
2101	2012	1101	2001	2121	2111	0101	2201	2100
2220	2221	1220	2120	2210	2200	0220	2020	2222

$(L_1, L_2) = $ (0,0) (0,1) (2,2) (2,0) (2,1) (1,2) (1,1) (1,0) (0,2)

Figure 9-7. The 3^4 design in nine blocks with ABC, AB^2D^2, AC^2D, and BC^2D^2 confounded.

As an example, consider the 3^4 factorial design confounded in nine blocks of nine runs each. Suppose we choose to confound ABC and AB^2D^2. Their generalized interactions

$$(ABC)(AB^2D^2) = A^2B^3CD^2 = (A^2B^3CD^2)^2 = AC^2D$$

$$(ABC)(AB^2D^2)^2 = A^3B^5CD^4 = B^2CD = (B^2CD)^2 = BC^2D^2$$

are also confounded with blocks. The defining contrasts for ABC and AB^2D^2 are

$$L_1 = x_1 + x_2 + x_3$$
$$L_2 = x_1 + 2x_2 + 2x_4$$

$$(9\text{-}4)$$

The nine blocks may be constructed by use of the defining contrasts (9-4) and the group-theoretic property of the principal block. The design is shown in Figure 9-7.

For the 3^k design in nine blocks there will be four components of interaction confounded. The remaining unconfounded components of these interactions can be determined by subtracting the sum of squares for the confounded component from the sum of squares for the entire interaction. The method described in Section 8-3.3 may be useful in computing components of interaction.

9-3.3 The 3^k Factorial Design in 3^p Blocks

The 3^k factorial design may be confounded in 3^p blocks of 3^{k-p} observations each, where $p < k$. The procedure is to select p independent effects to be confounded with blocks. As a result, exactly $(3^p - 2p - 1)/2$ other effects are automatically confounded. These effects are the generalized interactions of those effects originally chosen.

As an illustration, consider a 3^7 design to be confounded in 27 blocks. Since $p = 3$, we would select three independent components of interaction and automatically confound $[3^3 - 2(3) - 1]/2 = 10$ others. Suppose we choose $ABC^2 DG$, $BCE^2 F^2 G$, and $BDEFG$. Three defining contrasts can be constructed from these effects and the 27 blocks generated by the methods previously described. The other 10 effects confounded with blocks are:

$$(ABC^2 DG)(BCE^2 F^2 G) = AB^2 C^2 DE^2 F^2 G$$

$$(ABC^2 DG)(BCE^2 F^2 G)^2 = AB^3 C^4 DE^4 F^4 G^2 = ACDEFG$$

$$(ABC^2 DG)(BDEFG) = AB^2 CD^2 EFG^2$$

$$(ABC^2 DG)(BDEFG)^2 = A^2 B^3 C^2 D^3 E^2 F^2 G^2 = ACEFG$$

$$(BCE^2 F^2 G)(BDEFG) = B^2 CDE^3 F^3 G^2 = B^2 CDG$$

$$(BCE^2 F^2 G)(BDEFG)^2 = B^3 CD^2 E^4 F^4 G = CD^2 EFG$$

$$(ABC^2 DG)(BCE^2 F^2 G)(BDEFG) = AB^3 C^2 D^2 E^3 F^3 G^2 = AC^2 D^2 G^2$$

$$(ABC^2 DG)^2 (BCE^2 F^2 G)(BDEFG) = A^2 B^4 C^5 D^3 G^4 = AB^2 CG^2$$

$$(ABC^2 DG)(BCE^2 F^2 G)^2 (BDEFG) = ABCD^2 E^2 F^2 G^2$$

$$(ABC^2 DG)(BCE^2 F^2 G)(BDEFG)^2 = ABC^4 D^2 E^2 F^2 G^3 = ABCD^2 E^2 F^2.$$

This is a very large design requiring $3^7 = 2187$ observations, arranged in 27 blocks of 81 observations each. In practice, we would usually resort to some alternative experiment which, hopefully, could be run more economically.

9-4 PARTIAL CONFOUNDING

We have remarked in Section 9-2.1 that unless the experimenter has a prior estimate of error or is willing to assume certain interactions to be negligible, he must replicate the design to obtain an estimate of error. Figure 9-3 shows a 2^3 factorial in two blocks with ABC confounded, replicated four times. From the analysis of variance for this design, shown in Table 9-3, we note that information on the ABC interaction cannot be retrieved, since ABC is confounded with blocks *in each replicate*. This design is said to be *completely confounded*.

Consider an alternative design, shown in Figure 9-8. Once again, there are four replicates of the 2^3 design, but a *different* interaction has been confounded in each replicate. That is, ABC is confounded in replicate I, AB is confounded in replicate II, BC is confounded in replicate III, and AC is confounded in replicate IV. As a result information on ABC can be obtained from the data in replicates II, III, and IV, information on AB can be obtained from replicates I, III, and IV, information on AC

Figure 9-8. Partial confounding in the 2^3 design.

can be obtained from replicates I, II, and III, while information on *BC* can be obtained from replicates I, II, and IV. We say that three-quarters information is obtained on the interactions, because they are unconfounded in only three replicates. Yates (1937) calls the ratio 3/4 the relative information for the confounded effects. This design is said to be *partially confounded*.

The analysis of variance for this design is shown in Table 9-8. In calculating interaction sums of squares, only data from the replicates in which an interaction is unconfounded is used. The error sum of squares consists of replicates × main effect sums of squares, plus replicates × interaction sums of squares for each replicate in which that interaction is unconfounded (e.g., replicates × *ABC* for replicates II, III, and IV). Furthermore, there are seven degrees of freedom between the eight blocks. This is usually partitioned into three degrees of freedom for replicates and four degrees of freedom for blocks within replicates. The composition of the sum of squares for blocks is shown in Table 9-8, and follows directly from the choice of effect confounded in each replicate.

Table 9-8　Analysis of Variance for a Partially Confounded 2^3 Design

Source of Variation	Degrees of Freedom
Replicates	3
Blocks within replicates [or *ABC* (rep. I) + *AB* (rep. II) + *BC* (rep. III) + *AC* (rep. IV)]	4
A	1
B	1
C	1
AB (from replicates I, III, and IV)	1
AC (from replicates I, II, and III)	1
BC (from replicates I, II, and IV)	1
ABC (from replicates II, III, and IV)	1
Error	17
Total	31

■EXAMPLE 9-3. Consider Example 8-1, in which a study was performed to determine the effect of percent carbonation (A), operating pressure (B), and line speed (C) on the fill volume of a carbonated beverage. Suppose that each batch of syrup is only large enough to test four treatment combinations. Thus each replicate of the 2^3 design must be run in two blocks. Two replicates are run, with ABC confounded in replicate I and AB confounded in replicate II. The data are as follows:

Replicate I		Replicate II	
ABC Confounded		AB Confounded	

$(1) = -3$	$a = \ \ \ 0$	$(1) = -1$	$a = 1$
$ab = \ \ \ 2$	$b = -1$	$c = \ \ \ 0$	$b = 0$
$ac = \ \ \ 2$	$c = -1$	$ab = \ \ \ 3$	$ac = 1$
$bc = \ \ \ 1$	$abc = \ \ \ 6$	$abc = \ \ \ 5$	$bc = 1$

The sums of squares for A, B, C, AC, and BC may be calculated in the usual manner. Yates' algorithm may be employed, as shown in Table 9-9. Note that in this table, the calculation of sums of squares for AB and ABC is not completed. We must find SS_{ABC} using only the data in replicate II and SS_{AB} using only the data in replicate I as follows:

$$SS_{ABC} = \frac{[a + b + c + abc - ab - ac - bc - (1)]^2}{n2^k}$$

$$= \frac{[1 + 0 + 0 + 5 - 3 - 1 - 1 - (-1)]^2}{(1)(8)} = 0.50$$

$$SS_{AB} = \frac{[(1) + abc - ac + c - a - b + ab - bc]^2}{n2^k}$$

$$= \frac{[-3 + 6 - 2 + (-1) - 0 - (-1) + 2 - 1]^2}{(1)(8)} = 0.50$$

The replicates sum of squares is, in general,

$$SS_{Rep} = \sum_{b=1}^{n} \frac{R_b^2}{2^k} - \frac{y_{...}^2}{N}$$

$$= \frac{(6)^2 + (10)^2}{8} - \frac{(16)^2}{16} = 1.00$$

where R_b is the total of the observations in the bth replicate. The block sum of squares is the sum of SS_{ABC} from replicate I and SS_{AB} from replicate II, or $SS_{Blocks} = 2.50$.

Table 9-9 Yates' Algorithm for the Data in Example 9-3

Treatment Combination	Response	(1)	(2)	(3)	Effect	Sum of Squares
(1)	-4	-3	1	16	I	—
a	1	4	15	24	A	36.00
b	-1	2	11	18	B	20.25
ab	5	13	13	6	AB	Not direct
c	-1	5	7	14	C	12.25
ac	3	6	11	2	AC	0.25
bc	2	4	1	4	BC	1.00
abc	11	9	5	4	ABC	Not direct

The analysis of variance is shown in Table 9-10. All three main effects are significant at one percent. ∎

Table 9-10 Analysis of Variance for Example 9-3

Source of Variation	Sum of Squares	Degrees of Freedom	Mean Square	F_0
Replicates	1.00	1	1.00	
Blocks within replicates	2.50	2	1.25	
A	36.00	1	36.00	48.00^a
B	20.25	1	20.25	27.00^a
C	12.25	1	12.25	16.33^a
AB (rep. I only)	0.50	1	0.50	0.67
AC	0.25	1	0.25	0.33
BC	1.00	1	1.00	1.33
ABC (rep. II only)	0.50	1	0.50	0.67
Error	3.75	5	0.75	
Total	78.00	15		

[a]Significant at 1 percent.

Partial confounding may also be employed in the 3^k factorial design. The general procedure is to confound a different component of interaction in each replicate. The statistical analysis of these designs is similar to that for partially confounded 2^k factorials.

For example, consider a 3^2 design to be run in three blocks and replicated twice. If we confound AB^2 in replicate I and AB in replicate II, we obtain the design shown in Figure 9-9. The analysis of variance for this design is shown in Table 9-11.

Replicate I
AB² Confounded

00		10		12
11		21		20
22		02		01

Replicate II
AB Confounded

00		01		02
12		10		11
21		22		20

Figure 9-9. Partial confounding in the 3^2 design.

Table 9-11   Analysis of Variance for a Partially Confounded 3^2 Design

Source of Variation	Degrees of Freedom
Replicates	1
Blocks within replicates $[AB^2$ (rep. I) + AB (rep. II)]	4
A	2
B	2
AB (rep. I only)	2
AB^2 (rep. II only)	2
Error	4
Total	17

In partially confounded 3^k designs it is usually necessary to compute components of interactions from the replicates in which they are not confounded. The method of Section 8-3.3 may be used for this.

9-5 OTHER CONFOUNDING SYSTEMS

The combinatorial properties of the 3^k factorial design [and the 2^k factorial also, if treatment combinations are represented as 00, 10, . . . , instead of (1), a . . .] can be extended to confounding the general p^k factorial design, where p is a prime number or a power of a prime.   The treatment combinations are represented by numbers $x_1 x_2 \ldots x_k$, where x_i is the level of the ith factor and $0 \leqslant x_i \leqslant p - 1$. All calculated numbers are reduced modulus p. The $p^k - 1$ degrees of freedom between the p^k treatment combinations may be partitioned into $(p^k - 1)/(p - 1)$ sets containing $p - 1$ degrees of freedom.   Each set of $p - 1$ degrees of freedom is given by the contrasts between the p sets of p^{k-1} treatment combinations specified by the equations

$$L = \alpha_1 x_1 + \alpha_2 x_2 + \cdots + \alpha_k x_k = 0, 1, \ldots, (p - 1)(\text{mod } p) \tag{9-5}$$

In (9-5), the $\{\alpha_i\}$ are positive integers between 0 and $p - 1$, not all zero, and the first nonzero α_i is unity. Any of these p sets may be used to confound the experiment in p blocks, each block containing p^{k-1} observations.

We present two examples. Consider a 3^2 design with nine treatment combinations. The $3^2 - 1 = 8$ degrees of freedom between these treatment combinations are partitioned into $(3^2 - 1)/(3 - 1) = 4$ sets of $3 - 1 = 2$ degrees of freedom each, that is, A, B, AB, and AB^2. As a second example, consider the 5^2 design, with 25 treatment combinations. The 24 degrees of freedom between these treatment combinations are partitioned into $(5^2 - 1)/(5 - 1) = 6$ sets of $5 - 1 = 4$ degrees of freedom each; these being the effects A, B, AB, AB^2, AB^3, and AB^4. Any one of the four components of the AB interaction may be used to confound the 5^2 design in five blocks of five observations each. If we select AB^3, then $(9 - 5)$ becomes

$$L = x_1 + 3x_2 = 0, 1, 2, 3, 4 \ (\text{mod } 5)$$

which uniquely defines the treatment combinations that belong in any specific block. These ideas can be extended to confounding the general p^k design in p^s blocks, where p is prime and $s < k$.

If p is a power of a prime number then it is still possible to generate special incomplete block designs. We will illustrate for the 4^k series; that is, k factors each at four levels. The approach consists of splitting every four-level factor into two pseudo-factors, each with two levels. For example, let A be a four-level factor with levels a_0, a_1, a_2, and a_3. Then the arrangement of the two-level factors (e.g., P and Q) is as in Table 9-12.

Table 9-12   Four-Level
Factor Expressed as Two
Factors, Each at Two Levels

A	P	Q
a_0	0	$0 = (1)$
a_1	0	$1 = q$
a_2	1	$0 = p$
a_3	1	$1 = pq$

The effects of P, Q, and PQ are mutually orthogonal with one degree of freedom each, and correspond to the three-degree-of-freedom A effect.

For two factors, A and B, each at four levels, we would replace A with two-level pseudo-factors P and Q and B with two-level pseudo-factors R and S. Thus P, Q, and PQ correspond to A; R, S, and RS correspond to B; and the other nine interactions between the pseudo-factors correspond to the AB interaction. Therefore, the 4^2 design can be confounded in four blocks of four runs each by using the pseudo-factors P, Q, R, and S in a 2^4 design in four blocks with PQR, QRS, and PS confounded.

The general principles of confounding can be applied to factorial experiments in which all of the factors do not have the same number of levels. For example, consider the 4×2^k design; that is, a factorial experiment with k factors at two levels and one factor at four levels. By the use of pseudo-factors, this design could be analyzed as a

2^{k+2} factorial. In general, a $4^r 2^k$ could be confounded by this approach. Other useful designs that may be confounded include the $2^k 3^r$ factorial series. Since both 2 and 3 are prime numbers no pseudo-factors are required to construct these designs. Margolin (1967) has presented systematic methods for the analysis of $2^k 3^r$ factorial designs.

9-6 PROBLEMS

9-1. Consider the data from the first replicate of Problem 8-1. Suppose that these observations could not all be run using the same bar stock. Set up a design to run these observations in two blocks of four observations each with ABC confounded. Analyze the data.

9-2. Consider the data from the first replicate of Problem 8-2. Construct a design with two blocks of eight observations each, with $ABCD$ confounded. Analyze the data.

9-3. Repeat Problem 9-2, assuming that four blocks are required. Confound ABD and ABC (and consequently CD) with blocks.

9-4. Using the data from the first replicate of the 2^5 design in Problem 8-6, construct and analyze a design in four blocks with $ABCD$ and CDE (and consequently ABE) confounded with blocks.

9-5. Repeat Problem 9-4, assuming that eight blocks are necessary. Suggest a reasonable confounding scheme.

9-6. Consider the data from the 2^5 design in Problem 8-7. Suppose that it was necessary to run this design in four blocks, with $ACDE$ and BCD (and consequently ABE) confounded. Analyze the data from this design.

9-7. Design an experiment for confounding a 2^6 factorial in four blocks. Suggest an appropriate confounding scheme.

9-8. Consider the 2^6 design in eight blocks of eight runs each, with $ABCD$, AEF, and $ACDE$ as the independent effects chosen to be confounded with blocks. Generate the design. Find the other effects confounded with blocks.

9-9. Consider the 2^2 design in two blocks with AB confounded. Prove algebraically that $SS_{AB} = SS_{Blocks}$.

9-10. Confound a 3^3 design in three blocks, using the ABC^2 component of the three-factor interaction. Compare your results with the design in Figure 9-6.

9-11. Confound a 3^4 design in three blocks, using the AB^2CD component of the four-factor interaction.

9-12. Consider the data from the first replicate of Problem 8-18. Assuming that all 27 observations could not be run on the same day, set up a design for conducting the experiment over three days, with AB^2C confounded with blocks. Analyze the data.

9-13. Outline the analysis of variance table for the 3^4 design in nine blocks shown in Figure 9-7. Is this a practical design?

9-14. Suppose that in Problem 8-1, we had confounded ABC in replicate I, AB in replicate II, and BC in replicate III. Construct the analysis of variance table.

9-15. Repeat Problem 8-1, assuming that ABC was confounded with blocks in each replicate.

9-16. Suppose that in Problem 8-2, $ABCD$ was confounded in replicate I and ABC was confounded in replicate II. Perform the statistical analysis of this design.

9-17. Consider the data in Problem 8-18. If ABC is confounded in replicate I and ABC^2 is confounded in replicate II, perform the analysis of variance.

9-18. Construct a 2^3 design with ABC confounded in the first two replicates and BC confounded in the third. Outline the analysis of variance and comment on the information obtained.

9-19. Consider the 5^2 factorial design. Using the AB^4 component of the two-factor interaction, confound the design in five blocks. Outline the analysis of variance table.

9-20. Construct a 4×2^3 design confounded in two blocks of 16 observations each. How could Yates' algorithm be employed to analyze this design?

9-21. Outline the analysis of variance table for a $2^2 3^2$ factorial design. Discuss how this design may be confounded in blocks.

Chapter Ten
Fractional Replication

10-1 INTRODUCTION

As the number of factors in a 2^k or 3^k factorial design increases, the number of runs required for a complete replicate of the design rapidly outgrows the resources of most experimenters. A complete replicate of the 2^6 design requires 64 runs. In this design only 6 of the 63 degrees of freedom correspond to main effects, and only 15 degrees of freedom correspond to two-factor interactions. The remaining 42 degrees of freedom are associated with three-factor and higher interactions. In the 3^k series, the situation is worse; for example, the 3^6 factorial requires 243 runs, and only 12 of the 242 degrees of freedom correspond to main effects.

If the experimenter can reasonably assume that certain high-order interactions are negligible, then information on main effects and low-order interactions may be obtained by running only a fraction of the complete factorial experiment. These *fractional factorial designs* are widely used in industrial research, and are the subject of this chapter. A major use of fractional factorials is in *screening experiments*. These are experiments in which many factors are considered with the purpose of identifying those factors (if any) that have large effects. Screening experiments are usually performed in the early stages of a project, and most of the factors initially considered frequently have little or no effect on the response. The factors that are identified as important are then investigated more thoroughly in subsequent experiments.

10-2   FRACTIONAL REPLICATION OF THE 2^k FACTORIAL DESIGN

The simplest fractional factorial designs are those in the 2^k series. A complete replicate of the 2^k factorial allows independent estimates of k main effects, $\binom{k}{2}$ two-factor interactions, ... $\binom{k}{b}$ b-factor interactions, ... , and one k-factor interaction to be obtained. We will show how information on main-effects and low-order interactions may be obtained with fewer than 2^k treatment combinations. It turns out that the effects of major interest are confused with the higher-order effects; however, the latter are assumed small enough to be ignored.

10-2.1   The One-Half Fraction of the 2^k Design

Consider a situation in which three factors, each at two levels, are of interest, but the experimenter cannot afford to run all $2^3 = 8$ treatment combinations. He can, however, afford four runs so this suggests a one-half fraction of a 2^3 design. Because the design contains $2^{3-1} = 4$ treatment combinations, a one-half fraction of the 2^3 design is often called a 2^{3-1} design.

The table of plus and minus signs for the 2^3 design is shown in Table 10-1. Suppose we select the four treatment combinations a, b, c, and abc as our one-half fraction. These treatment combinations are shown in the top half of Table 10-1. We will use both the conventional notation (a, b, c, ...) and the plus and minus notation for the treatment combinations. The equivalence between the two notations is as follows:

Notation 1	Notation 2
a	+ − −
b	− + −
c	− − +
$ab\,c$	+ + +

Notice that the 2^{3-1} design is formed by selecting only those treatment combinations that yield a plus on the ABC effect. Thus ABC is called the *generator* of this particular fraction. Furthermore, the identity element I is also always plus, so we call

$$I = ABC$$

the *defining relation* for our design.

Table 10-1 Plus and Minus Signs for the 2^3 Factorial Design

Treatment Combination	Factorial Effect							
	I	*A*	*B*	*C*	*AB*	*AC*	*BC*	*ABC*
a	+	+	−	−	−	−	+	+
b	+	−	+	−	−	+	−	+
c	+	−	−	+	+	−	−	+
abc	+	+	+	+	+	+	+	+
ab	+	+	+	−	+	−	−	−
ac	+	+	−	+	−	+	−	−
bc	+	−	+	+	−	−	+	−
(1)	+	−	−	−	+	+	+	−

The treatment combinations in the 2^{3-1} design yield three degrees of freedom that we may use to estimate the main effects. Referring to Table 10-1, we obtain the estimates of the main effects as

$$A = \frac{1}{2} [a - b - c + abc]$$
$$B = \frac{1}{2} [-a + b - c + abc]$$
$$C = \frac{1}{2} [a - b + c + abc]$$

It is also easy to verify that the estimates of the two-factor interactions are

$$BC = \frac{1}{2} [a - b - c + abc]$$
$$AC = \frac{1}{2} [-a + b - c + abc]$$
$$AB = \frac{1}{2} [-a - b + c + abc]$$

That is, it is impossible to differentiate between *A* and *BC*, *B* and *AC*, and *C* and *AB*. In fact, we can show that when we estimate *A*, *B*, and *C* we are *really* estimating *A* + *BC*, *B* + *AC*, and *C* + *AB*. Two or more effects that have this property are called *aliases*. In our example *A* and *BC* are aliases, *B* and *AC* are aliases, and *C* and *AB* are aliases.

The alias structure for this design may be easily determined by using the defining relation *I* = *ABC*. Multiplying any effect by the defining relation modulus 2 yields the aliases for that effect. In our example, this yields as the alias of *A*

$$A \cdot I = A \cdot ABC$$

or

$$A = A^2 BC = BC$$

since any effect times the identity I is just the original effect. Similarly, we find the aliases of B and C as

$$B \cdot I = B \cdot ABC$$

$$B = AB^2C = AC$$

and

$$C \cdot I = C \cdot ABC$$

$$C = AB$$

Now suppose that we had chosen the *other* one-half fraction; that is, the treatment combinations in Table 10-1 associated with minus on ABC. This *alternate* one-half fraction consists of the runs

Notation 1	Notation 2
(1)	– – –
ab	+ + –
ac	+ – +
bc	– + +

The defining relation for this design is $I = -ABC$. The aliases are $A = -BC$, $B = -AC$, and $C = -AB$. Thus when we estimate A, B, and C with this particular fraction, we are really estimating $A - BC$, $B - AC$, and $C - AB$. The fraction associated with $I = +ABC$ is called the *principal fraction*.[1]

In general, a one-half fraction of a 2^k design, containing 2^{k-1} runs, is called a 2^{k-1} fractional factorial design. It may be constructed by first partitioning the full 2^k design into two blocks, using the highest-order interaction as the defining contrast. Each block is then a 2^{k-1} fractional factorial design. The defining relation of the 2^{k-1} is $I = \pm ABC \ldots K$, where the sign on the generator depends on the fraction chosen. In practice, we would randomly select which fraction to run.

A 2^{k-1} design may be generated by another method. First, write down the treatment combinations for a *full* 2^{k-1} factorial and then add the kth factor by identifying its plus and minus levels with the plus and minus signs of the highest-order interaction $ABC \ldots (K - 1)$. Therefore, the 2^{3-1} fractional factorial is obtained by writing down the full 2^2 factorial and then equating factor C to the AB interaction. The alternate fraction would be obtained by equating factor C to the $-AB$ interaction. This approach is illustrated in Table 10-2.

It is particularly instructive to note that a 2^{k-1} fractional factorial design may be projected into a *full* factorial design in any $(k - 1)$ of the original k factors. For the 2^{3-1} design, this is evident from examining Table 10-2. Figure 10-1 shows the projec-

---

[1] Some authors call the fraction containing (1) the principal fraction by analogy with the definition of the principal block.

Table 10-2 The Two One-Half Fractions of the 2^3 Design

Full 2^2 Factorial		$2^{3-1}, I = ABC$			$2^{3-1}, I = -ABC$		
A	B	A	B	$C = AB$	A	B	$C = -AB$
$-$	$-$	$-$	$-$	$+$	$-$	$-$	$-$
$+$	$-$	$+$	$-$	$-$	$+$	$-$	$+$
$-$	$+$	$-$	$+$	$-$	$-$	$+$	$+$
$+$	$+$	$+$	$+$	$+$	$+$	$+$	$-$

Figure 10-1. Projection of a 2^{3-1} design into three 2^2 designs.

tion of the 2^{3-1} design with $I = ABC$ into three full 2^2 factorials. Thus if one factor proves unimportant, we may obtain a full factorial design in the remaining two. Furthermore, a 2^{k-1} design may be collapsed into two replicates of a full factorial in any subset of $k - 2$ factors, four replicates of a full factorial in $k - 3$ factors, and so on.

■**EXAMPLE 10-1.** An engineer for the U.S. Postal Service is analyzing the factors that affect the number of pieces of mail per hour fed into a Model 500 edger-feeder machine. He has identified four factors that may be important: the belt angle on the incline conveyor (A), belt speed (B), belt material (C) , and the setting on the metering wheel (D). Each of these factors is set at two levels and a one-half fraction of the 2^4, that is, a 2^{4-1} design is run. The defining relation chosen for the fraction is $I = ABCD$, and the resulting design is shown in Table 10-3.

From inspection of Table 10-3, we see that this 2^{4-1} design was obtained by first writing down the full 2^3 design and equating the fourth factor D to the ABC interaction. Since the generator $ABCD$ is positive, this is the principal fraction. Using the defining relation we note that each main effect is aliased with a three-factor interaction; that is, $A = A^2 BCD = BCD$, $B = AB^2 CD = ACD$, $C = ABC^2 D = ABD$, and $D = ABCD^2 = ABC$. Furthermore, every two-factor interaction is aliased with another

Table 10-3 The 2^{4-1} Design with Defining Relation $I = ABCD$

A	B	C	$D = ABC$	Treatment Combination	Coded Response
−	−	−	−	(1)	74
+	−	−	+	ad	108
−	+	−	+	bd	92
+	+	−	−	ab	130
−	−	+	+	cd	68
+	−	+	−	ac	105
−	+	+	−	bc	95
+	+	+	+	$abcd$	133

two-factor interaction. These alias relationships are $AB = CD$, $AC = BD$, and $BC = AD$. The four main effects plus three two-factor interaction alias pairs account for the seven degrees of freedom for the design.

The analysis of variance for this design is shown in Table 10-4. The sum of squares for any effect is computed in the usual manner; that is, by the formula $SS = [\text{Contrast}]^2 / N$, where $N = 2^{4-1} = 8$. The effect of any factor may be estimated by Effect $= 2[\text{Contrast}]/N$. If three-factor interactions are negligible, we have estimates of the main effects that are clear of any other effects. However, unless we can reasonably assume that two-factor interactions are negligible, there is no estimate of error and no formal analysis of variance can be conducted. It seems clear that the main effects of A (belt angle) and B (belt speed) are large, and further study of these factors should be contemplated.

Table 10-4 Analysis of Variance for the 2^{4-1} Design in Example 10-1

Source of Variation	Sum of Squares	Degrees of Freedom
$A + BCD$	2701.125	1
$B + ACD$	1128.125	1
$C + ABD$	1.125	1
$D + ABC$	1.125	1
$AB + CD$	3.125	1
$AC + BD$	1.125	1
$BC + AD$	28.125	1
Total	3863.875	7

It is also possible to analyze the data as we did in Section 8-2.4, by plotting the estimates of the effects on normal probability paper and declaring significant those effects that do not fall approximately along a straight line. Insignificant effects can then be combined to estimate the error. We may also project the 2^{4-1} fractional factorial design into a single replicate of a full factorial in any subset of three factors, or two replicates of a full factorial in any subset of two factors. Thus, the data in Table 10-3 could be treated as two replicates of a 2^2 design in A and B, the only two factors that seem to have sizable effects. ∎

Yates' Algorithm for Fractional Factorials

We may use Yates' algorithm for the analysis of a 2^{k-1} fractional factorial design by initially considering the data as having been obtained from a full factorial in $k - 1$ variables. The treatment combinations for this full factorial are listed in standard order, and then an additional letter (or letters) is added in parentheses to these treatment combinations to produce the actual treatment combinations run. Yates' algorithm now proceeds as usual. The actual effects estimated are identified by multiplying the effects associated with the treatment combinations in the full 2^{k-1} design by the defining relation of the 2^{k-1} fractional factorial.

The procedure is demonstrated in Table 10-5, using the data from Example 10-1. The data are arranged as a full 2^3 design in the factors A, B, and C. Then the letter d is added in parentheses to yield the actual treatment combinations. The effect estimated by the second row, say, in this table is $A + BCD$, since A and BCD are aliases.

Table 10-5 Yates' Algorithm for the 2^{4-1} Fractional Factorial in Example 10-1

Treatment Combination	Response	(1)	(2)	(3)	Effect	Estimate of Effect $2 \times (3)/N$	Sum of Squares $(3)^2/N$
(1)	74	182	404	805	—	—	—
$a(d)$	108	222	401	147	$A + BCD$	36.75	2701.125
$b(d)$	92	173	72	95	$B + ACD$	23.75	1128.125
ab	130	228	75	5	$AB + CD$	1.25	3.125
$c(d)$	68	34	40	-3	$C + ABD$	-0.75	1.125
ac	105	38	55	3	$AC + BD$	0.75	1.125
bc	95	37	4	15	$BC + AD$	3.75	28.125
$abc(d)$	133	38	1	-3	$ABC + D$	-0.75	1.125

10-2.2 The One-Quarter Fraction of the 2^k Design

For a moderately large number of factors, smaller fractions of the 2^k are frequently useful. Consider a one-quarter fraction of the 2^k design. This design contains 2^{k-2} runs, and is usually called a 2^{k-2} fractional factorial.

The 2^{k-2} design may be constructed by selecting two effects or *generators* and partitioning the full 2^k design into four blocks of 2^{k-2} runs each. Any one of the four blocks is a one-quarter fraction of the 2^k design. If P and Q represent the generators chosen, then $I = P$ and $I = Q$ are called the *generating relations* for the design. The signs of P and Q (either + or −) depend on which one of the four blocks is selected as the fractional design. The fraction for which *both* P and Q are positive is the principal fraction. The *complete* defining relation for the design consists of P, Q, and their *generalized interaction PQ*; that is, the defining relation is $I = P = Q = PQ$. We call the elements P, Q, and PQ in the defining relation *words*. The aliases of any effect are produced upon multiplication of the effect modulus 2 by each word in the defining relation. Clearly, each effect has *three* aliases. The experimenter should be careful in choosing the generators so that potentially important effects are not aliased with each other.

As an example, consider the 2^{6-2} design. Suppose we choose $I = ABCE$ and $I = ACDF$ as the generating relations. Now the generalized interaction of the generators $ABCE$ and $ACDF$ is $BDEF$, and thus the complete defining relation for this design is $I = ABCE = ACDF = BDEF$. To find the aliases of any effect (e.g., A) multiply that effect by each word in the defining relation, producing in this case

$$A = BCE = CDF = ABDEF$$

It is easy to verify that every main effect is aliased by three-factor and five-factor interactions, while two-factor interactions are aliased with each other and with higher-order interactions. Thus, when we estimate A, for example, we are really estimating $A + BCF + CDF + ABDEF$. The complete alias structure of this design is shown in Table 10-6. If three-factor and higher interactions are negligible, this design gives clear estimates of main effects.

Table 10-6 Alias Structure for the 2^{6-2} Design with $I = ABCE = ACDF = BDEF$

Effect	Alias		
A	BCE	CDF	$ABDEF$
B	ACE	DEF	$ABCDF$
C	ABE	ADF	$BCDEF$
D	ACF	BEF	$ABCDE$
E	ABC	BDF	$ACDEF$
F	ACD	BDE	$ABCEF$
AB	CE	$BCDF$	$ADEF$
AC	BE	DF	$ABCDEF$
AD	CF	$BCDE$	$ABEF$
AE	BC	$CDEF$	$ABDF$
AF	CD	$BCEF$	$ABDE$
BD	EF	$ACDE$	$ABCF$
BF	DE	$ABCD$	$ACEF$
ABF	CEF	BCD	ADE
CDE	ABD	AEF	CBF

To actually construct this design, we could derive the four blocks of the 2^6 design with $ABCE$ and $ACDF$ confounded, and then choose the block with treatment combinations that are positive on $ABCE$ and $ACDF$. This would be a 2^{6-2} fractional factorial with generating relations $I = ABCE$ and $I = ACDF$, and since both generators $ABCE$ and $ACDF$ are positive, this is the principal fraction.

A more straightforward approach to constructing the design is to first write down the treatment combinations for a full $2^{6-2} = 2^4$ design in A, B, C, and D. Then the two factors E and F are added by associating their plus and minus levels with the plus and minus signs of the interactions ABC and ACD, respectively. This procedure is shown in Table 10-7.

Table 10-7 Construction of the 2^{6-2} Design with Generators $I = ABCE$ and $I = ACDF$

A	B	C	D	E = ABC	F = ACD	
−	−	−	−	−	−	(1)
+	−	−	−	+	+	aef
−	+	−	−	+	−	be
+	+	−	−	−	+	abf
−	−	+	−	+	+	cef
+	−	+	−	−	−	ac
−	+	+	−	−	+	bcf
+	+	+	−	+	−	abce
−	−	−	+	−	+	df
+	−	−	+	+	−	ade
−	+	−	+	+	+	bdef
+	+	−	+	−	−	abd
−	−	+	+	+	−	cde
+	−	+	+	−	+	acdf
−	+	+	+	−	−	bcd
+	+	+	+	+	+	abcdef

There are, of course, three *alternate* fractions of this particular 2^{6-2} design. They are the fractions with generating relationships $I = ABCE$ and $I = -ACDF$; $I = -ABCE$ and $I = ACDF$; and $I = -ABCE$ and $I = -ACDF$. These fractions may be easily constructed by the method shown in Table 10-7. For example, if we wish to find the fraction for which $I = ABCE$ and $I = -ACDF$, then in the next-to-last column of Table 10-7 we set $F = -ACD$, and the column of levels for factor F becomes

$$+ \quad - \quad + \quad - \quad - \quad + \quad - \quad + \quad - \quad + \quad - \quad + \quad + \quad - \quad + \quad -$$

The complete defining relation for this alternate fraction is $I = ABCE = -ACDF = -BDEF$. Certain signs in the alias structure in Table 10-6 are now changed; for instance the aliases of A are $A = BCE = -CDF = -ABDEF$. Thus the estimate of the A effect actually estimates $A + BCE - CDF - ABDEF$.

Finally, note that the 2^{6-2} fractional factorial will collapse to a single replicate of a 2^4 design in any subset of four factors that is not a word in the defining relation. It will also collapse to a replicated one-half fraction of a 2^4 in any subset of four factors that is a word in the defining relation. Thus, the design in Table 10-7 becomes two replicates of a 2^{4-1} in the factors $ABCE$, $ACDF$, and $BDEF$, since these are the words in the defining relation. There are 12 other combinations of the six factors, such as $ABCD$, $ABCF$, and so on, for which the design projects to a single replicate of the 2^4. This design will also collapse to two replicates of a 2^3 in *any* subset of three of the six factors or four replicates of a 2^2 in any subset of two factors. In general, any 2^{k-2} fractional factorial design can be collapsed into either a full factorial or a fractional factorial in some subset of $r \leqslant k - 2$ of the original factors.

10-2.3   The General 2^{k-p} Fractional Factorial Design

A fractional factorial of the 2^k design containing 2^{k-p} runs is called a $1/2^p$ fraction of the 2^k, or more simply, a 2^{k-p} fractional factorial design. The 2^{k-p} design may be constructed by selecting p independent generators (no chosen generator is a generalized interaction of the others), constructing the 2^p blocks associated with those generators, and then selecting one block as the fractional design. The defining relation for the design consists of the p generators initially chosen and their $2^p - p - 1$ generalized interactions.

The alias structure may be found by multiplying each effect modulus 2 by the defining relation. Care should be exercised in choosing the generators so that effects of potential interest are not aliased with each other. Each effect has $2^p - 1$ aliases. For moderately large values of k, we usually assume higher-order interactions (say third- or fourth-order and higher) to be negligible, and this greatly simplifies the alias structure.

The 2^{k-p} fractional factorial design may be analyzed by Yates' algorithm. This is accomplished by considering the data as from a full 2^r factorial, where $r = k - p$ is a subset of the original k factors. The additional factors are incorporated when identifying the actual effects estimated, as shown in Section 10-2.1. Usually, one may not arbitrarily select the subset of r factors to be initially used. In any event, the design may always be analyzed by first principles; an effect is estimated by

$$\text{Effect} = \frac{2[\text{Contrast}]}{N}$$

and the sum of squares for any effect is

$$SS = \frac{[\text{Contrast}]^2}{N}$$

where $N = 2^{k-p}$ is the total number of observations. The 2^{k-p} design allows only $2^{k-p} - 1$ effects (and their aliases) to be estimated.

The 2^{k-p} design will collapse into either a full factorial or a fractional factorial in any subset of $r \leqslant k - p$ of the original factors. Those subsets of factors providing fractional factorials are subsets appearing as words in the complete defining relation. This is particularly useful in screening experiments, when we suspect at the outset of the experiment that most of the original factors will have small effects. The original 2^{k-p} fractional factorial can then be projected into a full factorial (say) in the most interesting factors. Conclusions drawn from designs of this type should be considered tentative, and subject to further analysis. It is usually possible to find alternative explanations of the data involving higher-order interactions.

A one-eighth fraction of the 2^7 design, or a 2^{7-3} fractional factorial design is shown in Table 10-8. The generating relations for this design are $I = ABCE, I = ABDF$, and $I = ACDG$. Thus the complete defining relation is $I = ABCE = ABDF = ACDG = CDEF = BDEG = BCFG = AEFG$. Neglecting ineractions of order three and higher, this design provides clear estimates of main effects, but two-factor interactions are aliased with each other.

Table 10-8 A 2^{7-3} Fractional Factorial Design

A	B	C	D	E	F	G	
−	−	−	−	−	−	−	(1)
+	−	−	−	+	+	+	aefg
−	+	−	−	+	+	−	bef
+	+	−	−	−	−	+	abg
−	−	+	−	+	−	+	ceg
+	−	+	−	−	+	−	acf
−	+	+	−	−	+	+	bcfg
+	+	−	−	+	−	−	abce
−	−	−	+	−	+	−	df
+	−	−	+	+	−	−	ade
−	+	−	+	+	−	+	bdeg
+	+	−	+	−	+	−	abdf
−	−	+	+	+	+	−	cdef
+	−	+	+	−	−	+	acdg
−	+	+	+	−	−	−	bcd
+	+	+	+	+	+	+	abcdefg

Blocking Fractional Factorials

Occasionally a fractional factorial design requires so many runs that all of them cannot be made under homogeneous conditions. In these situations, fractional factorials may be confounded in blocks.

To illustrate the general procedure, consider the 2^{6-2} fractional factorial design with defining relation $I = ABCE = ACDF = BDEF$, shown in Table 10-7. This fractional design contains 16 treatment combinations, and suppose we wish to run the design in

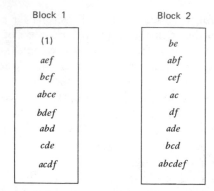

Figure 10-2. The 2^{6-2} design in two blocks with ABF confounded.

two blocks of 8 treatment combinations each. In selecting an interaction to confound with blocks, we note from examining the alias structure in Table 10-6 that there are two alias sets involving only three-factor interactions. Selecting ABF (and its aliases) to be confounded, we obtain the two blocks shown in Figure 10-2. Notice that the principal block contains those treatment combinations that have an even number of letters in common with ABF. These are also the treatment combinations for which $L = x_1 + x_2 + x_6 = 0$ (mod 2).

10-3 SPECIAL TYPES OF 2^{k-p} FRACTIONAL FACTORIAL DESIGNS[1]

Fractional factorial designs can be classified into types. Box and Hunter (1961a,b) classify fractional factorial designs according to their resolution. Their system is as follows:

(i) Resolution III Designs. These are designs in which no main effects is aliased with any other main effect, but main effects are aliased with two-factor interactions and two-factor interactions are aliased with each other. The 2^{3-1} design in Table 10-1 is of resolution III.

(ii) Resolution IV Designs. These are designs in which no main effect is aliased with any other main effect *or* two-factor interaction, but two-factor interactions are aliased with other. The 2^{4-1} design in Table 10-3 is of resolution IV.

(iii) Resolution V Designs. These are designs in which no main effect or two-factor interaction is aliased with any other main effect or two-factor interaction, but two-factor interactions are aliased with three-factor interactions. A 2^{5-1} design is of resolution V.

[1] Some of the material in this section is adapted from "The 2^{k-p} Fractional Factorial Designs," Parts I and II, by G. E. P. Box and J. S. Hunter, *Technometrics*, Vol. 3, pp. 311–352, 449–458, 1961, with permission from the American Statistical Association.

In general, the *resolution* of a design is equal to the smallest number of letters in any word in the defining relation. Consequently the above design types are often called three-letter, four-letter, and five-letter designs, respectively. Roman numeral subscripts are used to identify the resolution of a design. Thus, a 2_{III}^{3-1} design is a 2^{3-1} design of resolution III. For the more highly fractionated designs, more extensive assumptions are required to obtain unique interpretation of the results.

Designs of resolutions III, IV and V have several interesting properties and applications. Some of these results will be presented in this section. For more extensive coverage of this material, refer to Box and Hunter (1961a,b), who first organized and presented these topics in a useful manner.

10-3.1 2^{k-p} Designs of Resolution III

It is possible to construct resolution III designs for investigating up to $k = N - 1$ factors in only N runs, where N is a multiple of 4. These designs are frequently useful in industrial experimentation. Designs in which N is a power of 2 can be constructed by the methods presented earlier in this chapter, and will be presented first. Of particular importance are designs requiring 4 runs for up to 3 factors, 8 runs for up to 7 factors, and 16 runs for up to 15 factors. If $k = N - 1$ the fractional factorial design is said to be *saturated*.

A design for analyzing up to three factors in four runs is the 2_{III}^{3-1} design, presented in Section 10-2.1. Another very useful saturated fractional factorial is a design for studying seven factors in eight runs; that is, the 2_{III}^{7-4} design. This design is a one-sixteenth fraction of the 2^7. It may be constructed by first writing down the plus and minus levels for a full 2^3 in A, B, and C, and then associate the levels of four additional factors with the interactions of the original three as follows: $D = AB$, $E = AC$, $F = BC$, and $G = ABC$. Thus, the generating relations for this design are $I = ABD$, $I = ACE$, $I = BCF$, and $I = ABCG$. The design is shown in Table 10-9.

Table 10-9   The 2_{III}^{7-4} Design with Generators $I = ABD$, $I = ACE$, $I = BCF$, and $I = ABCG$

A	B	C	$D = AB$	$E = AC$	$F = BC$	$G = ABC$	
$-$	$-$	$-$	$+$	$+$	$+$	$-$	*def*
$+$	$-$	$-$	$-$	$-$	$+$	$+$	*afg*
$-$	$+$	$-$	$-$	$+$	$-$	$+$	*beg*
$+$	$+$	$-$	$+$	$-$	$-$	$-$	*abd*
$-$	$-$	$+$	$+$	$-$	$-$	$+$	*cdg*
$+$	$-$	$+$	$-$	$+$	$-$	$-$	*ace*
$-$	$+$	$+$	$-$	$-$	$+$	$-$	*bcf*
$+$	$+$	$+$	$+$	$+$	$+$	$+$	*abcdefg*

The complete defining relation for this design is obtained by multiplying the four generators ABD, ACE, BCF, and $ABCG$ together two at a time, three at a time, and four at a time, yielding

$$I = ABD = ACE = BCF = ABCG = BCDE = ACDF = CDG = ABEF = BEG$$

$$= AFG = DEF = ADEG = CEFG = BDFG = ABCDEFG$$

To find the alias of any effect, simply multiply the effect by each word in the defining relation. Thus, the alias of B is

$$B = AD = ABCE = CF = ACG = CDE = ABCDF = BCDG = AEF = EG$$

$$= ABFG = BDEF = ABDEG = BCEFG = DFG = ACDEFG$$

This design is a one-sixteenth fraction, and since the signs chosen for the generators are positive, this is the principal fraction. It is also of resolution III, since the smallest number of letters in any word of the defining contrast is three. Any one of the 16 different 2_{III}^{7-4} designs could be constructed by using the generators with one of the 16 possible arrangements of signs in $I = \pm ABD, I = \pm ACE, I = \pm BCF, I = \pm ABCG$.

The seven degrees of freedom in this design may be used to estimate the seven main effects. These estimates are obtained as linear combinations of the observations, where the signs in a particular linear combination are given in the associated column of Table 10-9. Thus, to estimate A, use the plus and minus signs in the A column. Each of these effects has 15 aliases; however, if we assume that three-factor and higher interactions are negligible, then considerable simplification in the alias structure results. Making this assumption, each of the linear combinations for the seven main effects in this design estimates

$$l_A = A + BD + CE + FG$$

$$l_B = B + AD + CF + EG$$

$$l_C = C + AE + BF + DG$$

$$l_D = D + AB + CG + EF \qquad (10\text{-}1)$$

$$l_E = E + AC + BG + DF$$

$$l_F = F + BC + AG + DE$$

$$l_G = G + CD + BE + AF$$

where l_i refers to the linear combination of treatment combinations given by column i in Table 10-9.

The saturated 2_{III}^{7-4} design in Table 10-9 can be used to obtain resolution III designs for studying fewer than seven factors in eight runs. For example, to generate a design for six factors in eight runs, simply drop any one column in Table 10-9, for example, column G. This produces the design shown in Table 10-10.

Table 10-10 A 2_{III}^{6-3} Design with Generators $I = ABD$, $I = ACE$, and $I = BCF$

A	B	C	$D = AB$	$E = AC$	$F = BC$	
−	−	−	+	+	+	def
+	−	−	−	−	+	af
−	+	−	−	+	−	be
+	+	−	+	−	−	abd
−	−	+	+	−	−	cd
+	−	+	−	+	−	ace
−	+	+	−	−	+	bcf
+	+	+	+	+	+	abcdef

It is easy to verify that this design is also of resolution III, in fact, it is a 2_{III}^{6-3} or a one-eighth fraction of the 2^6. The defining relation for the 2_{III}^{6-3} design is equal to the defining relation for the original 2_{III}^{7-4} design with any words containing the letter G deleted. Thus, the defining relation for our new design is

$$I = ABD = ACE = BCF = BCDE = ACDF = ABEF = DEF$$

In general, when d factors are dropped to produce a new design, the new defining relation is obtained as those words in the original defining relation that do not contain any dropped letters. When constructing designs by this method, care should be exercised to obtain the best arrangement possible. If we drop columns B, D, F, and G from Table 10-9, we obtain a design for three factors in eight runs, yet the treatment combinations correspond to two replicates of a 2^2. The experimenter would probably prefer to run a full 2^3 design in A, C, and E.

It is also possible to obtain a resolution III design for studying up to 15 factors in 16 runs. This saturated 2_{III}^{15-11} design can be generated by first writing down the 16 treatment combinations associated with a 2^4 in A, B, C, and D, and then equating 11 new factors with the 2, 3, and 4-factor interactions of the original 4. A similar procedure can be used for the 2_{III}^{31-26} design, which allows up to 31 factors to be studied in 32 runs.

Combining Fractions to Separate Effects

By combining fractional factorial designs in which certain signs are switched, we can systematically isolate effects of potential interest. The alias structure for any fraction with the signs for one or more factors reversed is obtained by making changes of sign on the appropriate factors in the alias structure of the original fraction.

Consider the 2_{III}^{7-4} design in Table 10-9. Suppose that along with this principal fraction a second fractional design with the signs reversed in the column for factor D is also run. That is, the column for D in the second fraction is

$$- \quad + \quad + \quad - \quad - \quad + \quad + \quad -$$

The effects that may be estimated from the first fraction are shown in Equation 10-1, and from the second fraction we obtain.

$$l_A^* = A - BD + CE + FG$$

$$l_B^* = B - AD + CF + EG$$

$$l_C^* = C + AE + BF - DG$$

$$l_D^* = -D + AB + CG + EF \qquad (10\text{-}2)$$

$$l_E^* = E + AC + BG - DF$$

$$l_F^* = F + BC + AG - DE$$

$$l_G^* = G - CD + BE + AF$$

assuming that three-factor and higher interactions are insignificant. Now from the two linear combinations of effects $\frac{1}{2}(l_i + l_i^*)$ and $\frac{1}{2}(l_i - l_i^*)$ we obtain

i	From $\frac{1}{2}(l_i + l_i^*)$	From $\frac{1}{2}(l_i - l_i^*)$
A	$A + CE + FG$	BD
B	$B + CF + EG$	AD
C	$C + AE + BF$	DG
D	$AB + CG + EF$	D
E	$E + AC + BG$	DF
F	$F + BC + AG$	DE
G	$G + BE + AF$	CD

Thus we have isolated the main effect of D and all of its two-factor interactions. In general, if we add to a fractional factorial design of resolution III or higher a further fraction with the signs of a *single factor* reversed, then the combined design will provide estimates of the main effect of that factor and its two-factor interactions.

Now suppose we add to any fractional factorial design a second fraction in which the signs for *all factors* are reversed. This procedure breaks the alias links between main effects and two-factor interactions. That is, we may use the *combined* design to estimate all main effects clear of any two-factor interactions. The following example illustrates the technique.

■**EXAMPLE 10-2.** A human performance analyst is conducting an experiment to study eye focus time, and has built an apparatus in which several factors can be controlled during the test. The factors he initially regards as important are acuity or sharpness of vision (A), distance from target to eye (B), target shape (C), illumination level (D), target size (E), target density (F), and subject (G). Two levels of each factor are considered. He suspects that only a few of these seven factors are of major importance, and that high-order interactions between the factors can be neglected. On the basis of this assumption, the analyst decides to run a screening experiment to identify the most important factors, and then concentrate further study on those. To screen these seven factors, he runs the treatment combinations from the 2_{III}^{7-4} design in Table 10-9 in random order, obtaining the focus times in milliseconds shown in Table 10-11.

Table 10-11 2_{III}^{7-4} Design for the Eye Focus Time Experiment

A	B	C	D = AB	E = AC	F = BC	G = ABC		Response
−	−	−	+	+	+	−	def	85.5
+	−	−	−	−	+	+	afg	75.1
−	+	−	−	+	−	+	beg	93.2
+	+	−		−	−	−	abd	145.4
−	−	+	+	−	−	+	cdg	83.7
+	−	+	−	+	−	−	ace	77.6
−	+	+	−	−	+	−	bcf	95.0
+	+	+	+	+	+	+	abcdefg	141.8

Seven main effects and their aliases may be estimated from this data. From Equation 10-1, we see that the effects are

$$l_A = A + BD + CE + FG = 20.63$$

$$l_B = B + AD + CF + EG = 38.38$$

$$l_C = C + AE + BF + DG = -0.28$$

$$l_D = D + AB + CG + EF = 28.88$$

$$l_E = E + AC + BG + DF = -0.28$$

$$l_F = F + BC + AG + DE = -0.63$$

$$l_G = G + CD + BE + AF = -2.43.$$

Table 10-12 Second 2_{III}^{7-4} Design for the Eye Focus Time Experiment

A	B	C	D = AB	E = AC	F = BC	G = ABC		Response
+	+	+	−	−	−	+	abcg	91.3
−	+	+	+	+	−	−	bcde	136.7
+	−	+	+	−	+	−	acdf	82.4
−	−	+	−	+	+	+	cefg	73.4
+	+	−	−	+	+	−	abef	94.1
−	+	−	+	−	+	+	bdfg	143.8
+	−	−	+	+	−	+	adeg	87.3
−	−	−	−	−	−	−	(1)	71.9

The three largest effects are l_A, l_B, and l_D. The simplest interpretation of the data is that the main effects of $A, B,$ and D are all significant. However, this interpretation is not unique, as one could also logically conclude that $A, B,$ and the AB interaction, or perhaps $B, D,$ and the BD interaction, or perhaps $A, D,$ and the AD interaction are the true effects.

To separate main effects and two-factor interactions, a second fraction is run with all signs reversed. This design is shown in Table 10-12 along with the observed responses. The effects estimated by this fraction are

$$l_A^* = -A + BD + CE + FG = 17.68$$

$$l_B^* = -B + AD + CF + EG = -37.73$$

$$l_C^* = -C + AE + BF + DG = 3.33$$

$$l_D^* = -D + AB + CG + EF = -29.88$$

$$l_E^* = -E + AC + BG + DF = -0.53$$

$$l_F^* = -F + BC + AG + DE = -1.63$$

$$l_G^* = -G + CD + BE + AF = -2.68$$

Now, on combining the two fractions and forming the linear combinations $\frac{1}{2}(l_i + l_i^*)$ and $\frac{1}{2}(l_i - l_i^*)$, we obtain

i	From $\frac{1}{2}(l_i + l_i^*)$	From $\frac{1}{2}(l_i - l_i^*)$
A	**BD** + CE + FG = 19.15	A = 1.48
B	AD + CF + EG = 0.33	**B** = 38.05
C	AE + BF + DG = 1.53	C = -1.80
D	AB + CG + EF = -0.50	**D** = 29.38
E	AC + BG + DF = -0.40	E = 0.13
F	BC + AG + DE = -1.53	F = 0.50
G	CD + BE + AF = -2.55	G = 0.13

The two largest effects are B and D. Furthermore, the third largest effect is $BD + CE + FG$, so it seems reasonable to attribute this to the BD interaction. The analyst used the two factors distance (B) and illumination level (D) in subsequent experiments with the other factors A, C, E and F at standard settings and verified the results obtained here. He decided to use subjects as blocks in these new experiments, rather than ignore a potential subject effect, since several different subjects had to be used to complete the experiment. ∎

Plackett-Burman Designs
These designs, due to Plackett and Burman (1946), are two-level fractional factorial designs for studying $k = N - 1$ variables in N runs, where N is a multiple of 4. If N is a

Table 10-13 Plus and Minus Signs for the Plackett-Burman Designs

```
k = 11   N = 12   + + - + + + - - - + -
k = 19   N = 20   + + - - + + - + - + - + - - - - + + -
k = 23   N = 24   + + + + + - + - + + - - + + - - + - + - - - - -
k = 35   N = 36   - + - + + + - - - + + + + + - + + + - - + - - - - + - + - + + - - + -
```

$$k = 27, \quad N = 28$$

```
+ - + + + + - - -    - + - - - + - - -    - - - - + + +
+ + - + + + - - -    - - + + - - + - -    - + + + + - + + -
- + + + + + - - -    + - - - + - - + -    + - + - + + - + +
- - - + - + + + +    - - + - + - - - +    + - - + + - + +
- - - + + - + + +    + - - - - + + - -    + + - - - + + + -
- - - - + + + + +    - + - + - - - + -    - + + + - + - + +
+ + + - - - + - +    - - + - - + - + -    + - + + - + + + -
+ + + - - - + + -    + - - + - - - - +    + + - + + - - + +
+ + + - - - - + +    - + - - + - + - -    - + + - + + + - -
```

power of 2, these designs are identical to those presented earlier in this section. However, for $N = 12, 20, 24, 28$, and 36 the Plackett-Burman designs are frequently useful.

The upper half of Table 10-13 presents rows of plus and minus signs used to construct the Plackett-Burman designs for $N = 12, 20, 24$, and 36, while the lower half of the table presents blocks of plus and minus signs for constructing the design for $N = 28$. The designs for $N = 12, 20, 24$, and 36 are obtained by writing the appropriate row in Table 10-13 as a column. A second column is then generated from this first one by moving the elements of the column down one position and placing the last element in

Table 10-14 Plackett-Burman Design for $N = 12, k = 11$

A	B	C	D	E	F	G	H	I	J	K
+	-	+	-	-	-	+	+	+	-	+
+	+	-	+	-	-	-	+	+	+	-
-	+	+	-	+	-	-	-	+	+	+
+	-	+	+	-	+	-	-	-	+	+
+	+	-	+	+	-	+	-	-	-	+
+	+	+	-	+	+	-	+	-	-	-
-	+	+	+	-	+	+	-	+	-	-
-	-	+	+	+	-	+	+	-	+	-
-	-	-	+	+	+	-	+	+	-	+
+	-	-	-	+	+	+	-	+	+	-
-	+	-	-	-	+	+	+	-	+	+
-	-	-	-	-	-	-	-	-	-	-

the first position. A third column is produced from the second similarly, and the process continued until column k is generated. A row of minus signs is then added, completing the design. For $N = 28$, the three blocks X, Y, and Z are written down in the order

$$X \quad Y \quad Z$$

$$Z \quad X \quad Y$$

$$Y \quad Z \quad X$$

and a row of minus signs added to these 27 rows. The design for $N = 12$ runs and $k = 11$ factors is shown in Table 10-14.

10-3.2 Resolution IV and V Designs

A 2^{k-p} fractional factorial design is of resolution IV if main effects are clear of two-factor interactions and some two-factor interactions are aliased with each other. Thus, if three-factor and higher interactions are suppressed, main effects may be estimated directly in a 2_{IV}^{k-p} design. The 2^{6-2} design in Table 10-7 is of resolution IV. Furthermore, the two combined fractions of the 2_{III}^{7-4} design in Example 10-2 yields a 2_{IV}^{7-3} design.

Any 2_{IV}^{k-p} design must contain at least $2k$ runs. Resolution IV designs that contain exactly $2k$ runs are called *minimal* designs. Resolution IV designs may be obtained from resolution III designs by the process of *fold over*. To fold over a 2_{III}^{k-p} design simply add to the original fraction a second fraction with all signs reversed. Then the plus signs in the identity column I in the first fraction are switched in the second fraction, and a $(k + 1)$st factor associated with this column. The result is a 2_{IV}^{k+1-p} fractional factorial design. The process is demonstrated in Table 10-15 for the 2_{III}^{3-1} design. It is easy to verify that the resulting design is a 2_{IV}^{4-1} design with generating relation $I = ABCD$.

Table 10-15 A 2_{IV}^{4-1} Design Obtained by Fold Over

	D			
	I	A	B	C
Original 2_{III}^{3-1} $I = ABC$	+	−	−	+
	+	+	−	−
	+	−	+	−
	+	+	+	+
Second 2_{III}^{3-1} with signs switched	−	+	+	−
	−	−	+	+
	−	+	−	+
	−	−	−	−

Resolution V designs are fractional factorials in which main effects and two-factor interactions do not have other main effects and two-factor interactions as their aliases. The smallest word in the defining relation of such a design must have five letters. The 2^{5-1} with generating relation $I = ABCDE$ is of resolution V. Another example is the 2_V^{8-2} design with generating relations $I = ABCDG$ and $I = ABEFH$. Further examples of these designs are given by Box and Hunter (1961b).

10-4 FRACTIONAL REPLICATION OF THE 3^k FACTORIAL DESIGN

The concept of fractional replication can be extended to the 3^k factorial designs. As a complete replicate of the 3^k can require a rather large number of runs even for moderate values of k, fractional replication of these designs is particularly attractive.

10-4.1 The One-Third Fraction of the 3^k Design

The largest useful fraction of the 3^k design is a one-third fraction containing 3^{k-1} runs. Consequently, we refer to this as a 3^{k-1} fractional factorial design. To construct a 3^{k-1} fractional factorial design select a two-degree-of-freedom component of interaction (generally, the highest-order interaction) and partition the *full* 3^k design into three blocks. Each of the three resulting blocks is a 3^{k-1} fractional factorial design, and any one of the blocks may be selected for use. If $AB^{\alpha_2}C^{\alpha_3} \cdots K^{\alpha_k}$ is the component of interaction used to define the blocks, then $I = AB^{\alpha_2}C^{\alpha_3} \cdots K^{\alpha_k}$ is called the *defining relation* of the fractional factorial design. Each main effect or component of interaction estimated from the 3^{k-1} design has two aliases, which may be found by multiplying the effect by *both* I and I^2 modulus 3.

As an example, consider a one-third fraction of the 3^3 design. We may select any component of the ABC interaction to construct the design; that is, either ABC, AB^2C, ABC^2, or AB^2C^2. Thus, there are actually 12 *different* one-third fractions of the 3^3 design defined by

$$x_1 + \alpha_2 x_2 + \alpha_3 x_3 = u \,(\text{mod } 3)$$

where $\alpha_i = 1$ or 2 and $u = 0$, 1, or 2. Suppose we select the component AB^2C^2. Each fraction of the resulting 3^{3-1} design will contain exactly $3^2 = 9$ treatment combinations which must satisfy

$$x_1 + 2x_2 + 2x_3 = u \,(\text{mod } 3)$$

where $u = 0$, 1, or 2. It is easy to verify that the three one-third fractions are as shown in Figure 10-3.

Design 1	Design 2	Design 3
000	200	100
012	212	112
101	001	201
202	102	002
021	221	121
110	010	210
122	022	222
211	111	011
220	120	020

Figure 10-3. The three one-third fractions of the 3^3 design with defining relation $I = AB^2C^2$.

If any one of the 3^{3-1} designs in Figure 10-3 is run, the resulting alias structure is

$$A = A(AB^2C^2) = A^2B^2C^2 = ABC$$

$$A = A(AB^2C^2)^2 = A^3B^4C^4 = BC$$

$$B = B(AB^2C^2) = AB^3C^2 = AC^2$$

$$B = B(AB^2C^2)^2 = A^2B^5C^4 = ABC^2$$

$$C = C(AB^2C^2) = AB^2C^3 = AB^2$$

$$C = C(AB^2C^2)^2 = A^2B^4C^5 = AB^2C$$

$$AB = AB(AB^2C^2) = A^2B^3C^2 = AC$$

$$AB = AB(AB^2C^2)^2 = A^3B^5C^4 = BC^2$$

Consequently, the four effects that are actually estimated from the eight degrees of freedom in the design are $A + BC + ABC$, $B + AC^2 + ABC^2$, $C + AB^2 + AB^2C$, and $AB + AC + BC^2$. This design would only be of practical value if all interactions were small relative to the main effects.

Before leaving the 3^{3-1} design, note that if we let A denote the row and B denote the column, then the design can be written as

000	012	021
101	110	122
202	211	220

which is a 3×3 Latin square. The assumption of negligible interactions required for unique interpretations of the 3^{3-1} design is paralleled in the Latin square design. However, the two designs arise from different motives, one as a consequence of fractional replication and the other from randomization restrictions. From Table 4-15 we observe that there are only twelve 3 X 3 Latin squares, and each one corresponds to one of the twelve different 3^{3-1} fractional factorial designs.

The treatment combinations in a 3^{k-1} design with defining relation $I = AB^{\alpha_2}C^{\alpha_3} \cdots K^{\alpha_k}$ can be generated by another method. First, write down the 3^{k-1} runs for a *full* 3^{k-1} design in the usual 0, 1, 2 notation. Then introduce the kth factor by equating its levels x_k to the appropriate component of the highest-order interaction, say $AB^{\alpha_2}C^{\alpha_3} \cdots (K-1)^{\alpha_{k-1}}$, through the relationship

$$x_k = \beta_1 x_1 + \beta_2 x_2 + \cdots + \beta_{k-1}x_{k-1} \tag{10-3}$$

where $\beta_i = (3 - \alpha_k)\alpha_i \pmod 3$ for $1 \leqslant i \leqslant k - 1$.

As an illustration, we will use this method to generate the 3^{4-1} design with defining relation $I = AB^2CD$ shown in Table 10-16. It is easy to verify that the first three digits of each treatment combination in this table are the 27 runs of a full 3^3 design. For AB^2CD we have $\alpha_1 = \alpha_3 = \alpha_4 = 1$ and $\alpha_2 = 2$. This implies that $\beta_1 = (3-1)\alpha_1 \pmod 3 = (3-1)(1) = 2, \beta_2 = (3-1)\alpha_2 \pmod 3 = (3-1)(2) = 4 = 1 \pmod 3$, and $\beta_3 = (3-1)\alpha_3 \pmod 3 = (3-1)(1) = 2$. Thus, (10-3) becomes

$$x_4 = 2x_1 + x_2 + 2x_3 \tag{10-4}$$

The levels of the fourth factor satisfy (10-4). For example, we have 2(0) + 1(0) + 2(0) = 0, 2(0) + 1(1) + 2(0) = 1, 2(1) + 1(1) + 2(0) = 3 = 0, and so on.

Table 10-16   A 3^{4-1} Design with $I = AB^2CD$

0000	0012	2221
0101	0110	0021
1100	0211	0122
1002	1011	0220
0202	1112	1020
1201	1210	1121
2001	2010	1222
2102	2111	2022
2200	2212	2121

The resulting 3^{4-1} design has 26 degrees of freedom that may be used to compute sums of squares for 13 main effects and components of interactions (and their aliases). The aliases of any effect are found in the usual manner, for example, the aliases of A are $A(AB^2CD) = ABC^2D^2$ and $A(AB^2CD)^2 = BC^2D^2$. One may verify that the four main effects are clear of any two-factor interactions, but that some two-factor interactions are aliased with each other.

The statistical analysis of a 3^{k-1} design is accomplished by the usual analysis of variance procedures for factorial experiments. Sums of squares for components of interaction may be computed as in Section 8-3. Remember in interpreting results that components of interactions have no practical interpretation.

10-4.2 Other 3^{k-p} Fractional Factorial Design

For moderate to large values of k, even further fractionation of the 3^k design is often desirable. In general, we may construct a $(\frac{1}{3})^p$ fraction of the 3^k design for $p < k$, where the fraction contains 3^{k-p} runs. Such a design is called a 3^{k-p} fractional factorial design. Thus, a 3^{k-2} design is a one-ninth fraction, a 3^{k-3} design is a $(\frac{1}{27})$th fraction, and so on.

The procedure for constructing a 3^{k-p} fractional factorial design is to select p components of interaction and use these effects to partition the 3^k treatment combinations into 3^p blocks. Each block is now a 3^{k-p} fractional factorial design. The defining relation I of any fraction consists of the p effects initially chosen and their $(3^p - 2p - 1)/2$ generalized interactions. The alias of any main effect or component of interaction is produced on multiplication modulus 3 of the effect by I and I^2.

We may also generate the runs defining a 3^{k-p} fractional factorial design by first writing down the treatment combinations of a *full* 3^{k-p} factorial design and then introducing the additional p factors by equating them to components of interaction, as we did in Section 10-4.1.

We will illustrate the procedure by constructing a 3^{4-2} design; that is, a one-ninth fraction of the 3^4. Let AB^2C and BCD be the two components of interaction chosen to construct the design. Their generalized interactions are $(AB^2C)(BCD) = AC^2D$ and $(AB^2C)(BCD)^2 = ABD^2$. Thus, the defining relation for this design is $I = AB^2C = BCD = AC^2D = ABD^2$. The nine treatment combinations in the design are found by writing down a 3^2 design in the factors A and B, and then adding two new factors by setting

$$x_3 = 2x_1 + x_2$$

$$x_4 = 2x_2 + 2x_3$$

This is equivalent to using AB^2C and BCD to partition the full 3^4 design into nine blocks and then selecting one of these blocks as the desired fraction. The complete design is shown in Table 10-17.

Table 10-17 A 3^{4-2} Design with $I = AB^2C$ and $I = BCD$

0000	0111	0222
1021	1102	1210
2012	2120	2201

This design has eight degrees of freedom that may be used to estimate four main effects and their aliases. The aliases of any effect may be found by multiplying the effect modulus 3 by AB^2C, BCD, AC^2D, ABD^2, and their squares. The complete alias structure for the design is given in Table 10-18.

Table 10-18 Alias Structure for the 3^{4-2} Design in Table 10-17

Effect	Aliases							
	I				I^2			
A	ABC^2	$ABCD$	ACD^2	AB^2D	BC^2	$AB^2C^2D^2$	CD^2	BD^2
B	AC	BC^2D^2	ABC^2D	AB^2D^2	ABC	CD	AB^2C^2D	AD^2
C	AB^2C^2	BC^2D	AD	$ABCD^2$	AB^2	BD	ACD	ABC^2D^2
D	AB^2CD	BCD^2	AC^2D^2	AB	AB^2CD^2	BC	AC^2	ABD

From the alias structure, we see that this design is useful only in the absence of inter-action. Furthermore, if A denotes the rows and B denotes the columns, then from examining Table 10-17, we see that the 3^{4-2} design is also a Graeco-Latin square.

The publication by Connor and Zelen (1959) contains an extensive selection of designs for $4 \leqslant k \leqslant 10$. This pamphlet was prepared for the National Bureau of Standards, and is the most complete table of fractional 3^{k-p} plans available.

10-5 PROBLEMS

10-1. Suppose that in Problem 8-2 it was only possible to run a one-half fraction of the 2^4 design. Construct the design and perform the statistical analysis, using the data from replicate I.

10-2. Suppose that in Problem 8-6 only a one-quarter fraction of the 2^5 design could be run. Construct the design and perform the analysis, using the data from replicate I.

10-3. Consider the data in Problem 8-7. Suppose that only a one-quarter fraction of the 2^5 design could be run. Construct the design and analyze the data.

10-4. Construct a 2^{7-2} design by choosing two four-factor interactions as the inde-pendent generators. Write down the complete alias structure for this design. Outline the analysis of variance table.

10-5. Consider the 2^5 design in Problem 8-7. Suppose that only a one-half fraction could be run. Furthermore, two days were required to take the 16 observa-tions, and it was necessary to confound the 2^{5-1} design in two blocks. Con-struct the design and analyze the data.

10-6. Analyze the data in Problem 8-9 as if it came from a 2^{4-1} with $I = ABCD$. Project the design into a full factorial in the subset of the original four factors that appear significant.

10-7. Repeat Problem 10-6 using $I = -ABCD$. Does use of the alternate fraction change your interpretation of the data?

10-8. Project the 2^{4-1} design in Example 10-1 into two replicates of a 2^2 in the factors A and B. Analyze the data and draw conclusions.

10-9. Construct a 2_{III}^{6-3} design. Determine the effects that may be estimated if a second fraction of this design is run with all signs reversed.

10-10. Consider the 2_{III}^{6-3} design in Problem 10-9. Determine the effects that may be estimated if a second fraction of this design is run with the signs for factor A reversed.

10-11. Fold over the 2_{III}^{7-4} design in Table 10-11. Verify that the resulting design is a 2_{IV}^{8-4} design. Is this a minimal design?

10-12. Fold over a 2_{III}^{5-2} design. Verify that the resulting design is a 2_{IV}^{6-2} design. Compare this design to the 2_{IV}^{6-2} design in Table 10-7.

10-13. An industrial engineer is conducting an experiment using a Monte Carlo simulation model of an inventory system. The independent variables in his model are the order quantity (A), the reorder point (B), the setup cost (C), the backorder cost (D), and the carrying cost rate (E). The response variable is average annual cost. To conserve computer time, he decides to investigate these factors using a 2_{III}^{5-2} design, with $I = ABD$ and $I = BCE$. The results he obtains are $de = 95$, $ae = 134$, $b = 158$, $abd = 190$, $cd = 92$, $ac = 187$, $bce = 155$, and $abcde = 185$.

(a) Verify that the treatment combinations given are correct. Estimate the effects, assuming three-factor and higher interactions are negligible.

(b) Suppose that a second fraction is added to the first, for example, $ade = 136$, $e = 93$, $ab = 187$, $bd = 153$, $acd = 139$, $c = 99$, $abce = 191$, and $bcde = 150$. How was this second fraction obtained? Add this data to the original fraction, and estimate the effects.

(c) Suppose that the fraction $abc = 189$, $ce = 96$, $bcd = 154$, $acde = 135$, $abe = 193$, $bde = 152$, $ad = 137$, and $(1) = 98$ was run. How was this fraction obtained? Add this data to the original fraction and estimate the effects.

10-14. *Irregular Fractions of the 2^k.* [John (1971)]. Consider a 2^4 design. We must estimate the four main effects and the six two-factor interactions, but the full 2^4 factorial cannot be run. The largest possible block size contains 12 runs. These 12 runs can be obtained from the four $(\frac{1}{4})$th replicates defined by $I = \pm AB = \pm ACD = \pm BCD$ by omitting the principal fraction. Show how the remaining three 2^{4-2} fractions can be combined to estimate the required effects, assuming three-factor and higher interactions are negligible. This design could be thought of as a three-quarter fraction.

10-15. Consider the data from replicate I of Problem 8-18. Suppose that only a one-third fraction of this design with $I = AB^2C$ is run. Construct the design, determine the alias structure, and analyze the data.

10-16. Construct a 3^{4-1} design with defining relation $I = ABCD^2$. Write out the alias structure for this design.

10-17. Verify that the design in Problem 10-16 is a resolution IV design.

10-18. Consider the data from replicate I of Problem 8-19. Suppose that only a one-third fraction of this design with $I = ABC$ is run. Construct the design, determine the alias structure, and analyze the data.

10-19. Construct a 3^{4-1} design with $I = ABCD$. Write out the alias structure for this design.

10-20. Verify that the design in Problem 10-19 is a resolution IV design.

10-21. Construct a 3^{5-2} design with $I = ABC$ and $I = CDE$. Write out the alias structure for this design.

10-22. Construct a 3^{9-6} design, and verify that it is a resolution III design.

10-23. Construct a 5^{3-1} design with $I = AB^2C^3DE$ as the defining relation. Write out the alias structure for this design.

10-24. Discuss fractional replication in the 4^k series of designs.

Chapter Eleven
Nested or Hierarchial Designs

11-1 INTRODUCTION

In certain multifactor experiments the levels of one factor (e.g., factor B) are similar but not identical for different levels of another factor (e.g., A). Such an arrangement is called a *nested* or *hierarchial* design, with the levels of factor B nested under the levels of factor A. For example, consider a company that purchases its raw material from three different suppliers. The company wishes to determine if the purity of the raw material is the same from each supplier. There are four batches of raw material available from each supplier, and three determinations of purity are to be taken from each batch. The physical situation is depicted in Figure 11-1.

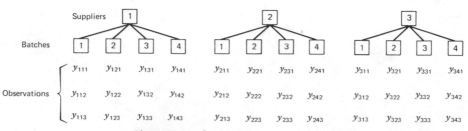

Figure 11-1. A two-stage nested design.

266

This is a *two-stage nested* or *hierarchial design*, with batches nested within suppliers. At first glance, you may ask why the two factors suppliers and batches are not crossed. If the factors were crossed, then batch 1 would always refer to the same batch, batch 2 would always refer to the same batch, and so on. This is clearly not the case, as the batches from each supplier are unique for that particular supplier. That is, batch 1 from supplier 1 has no connection with batch 1 from any other supplier, batch 2 from supplier 1 has no connection with batch 2 from any other supplier, etc. To emphasize the fact that batches from each supplier are different batches, we may renumber the batches as 1, 2, 3, and 4 from supplier 1; 5, 6, 7, and 8 from supplier 2; and 9, 10, 11, and 12 from supplier 3, as shown in Figure 11-2.

Figure 11-2. Alternate layout for the two-stage nested design.

Sometimes we may be uncertain as to whether a factor is crossed or nested. If the levels of the factor can be renumbered arbitrarily as in Figure 11-2, then the factor is nested.

11-2 ANALYSIS OF NESTED DESIGNS

The linear statistical model for the two-stage nested design is

$$y_{ijk} = \mu + \tau_i + \beta_{j(i)} + \epsilon_{(ij)k} \begin{cases} i = 1, 2, \ldots, a \quad suppliers \\ j = 1, 2, \ldots, b \quad batches \\ k = 1, 2, \ldots, n \end{cases} \quad (11\text{-}1)$$

That is, there are a levels of factor A, b levels of factor B nested under each level of A, and n replicates. The subscript $j(i)$ indicates that the jth level of factor B is nested under the ith level of factor A. It is convenient to think of the replicates as nested within the combination of levels of A and B; thus the subscript $(ij)k$ is used for the error term. This is a *balanced* nested design, since there are an equal number of levels of B within each level of A, and an equal number of replicates. ~~Since every level of factor B does not appear with every level of factor A~~ there can be no interaction between A and B.

We may write the total corrected sum of squares as

$$\sum_{i=1}^{a} \sum_{j=1}^{b} \sum_{k=1}^{n} (y_{ijk} - \bar{y}_{...})^2 = \sum_{i=1}^{a} \sum_{j=1}^{b} \sum_{k=1}^{n} [(\bar{y}_{i..} - \bar{y}_{...}) + (\bar{y}_{ij.} - \bar{y}_{i..}) + (y_{ijk} - \bar{y}_{ij.})]^2$$

$$(11\text{-}2)$$

since the levels of B appear with only one level of A ... (handwritten annotation)

Expanding the right-hand side of (11-2) yields

$$\sum_{i=1}^{a}\sum_{j=1}^{b}\sum_{k=1}^{n}(y_{ijk}-\bar{y}_{...})^2 = bn\sum_{i=1}^{a}(\bar{y}_{i..}-\bar{y}_{...})^2 + n\sum_{i=1}^{a}\sum_{j=1}^{b}(\bar{y}_{ij.}-\bar{y}_{i..})^2$$

$$+\sum_{i=1}^{a}\sum_{j=1}^{b}\sum_{k=1}^{n}(y_{ijk}-\bar{y}_{ij.})^2 \quad (11\text{-}3)$$

since the three cross-product terms are zero. Equation 11-3 indicates that the total sum of squares can be partitioned into a sum of squares due to factor A, a sum of squares due to factor B under the levels of A, and a sum of squares due to error. Symbolically, we may write Equation 11-3 as

$$SS_T = SS_A + SS_{B(A)} + SS_E \quad (11\text{-}4)$$

There are $abn - 1$ degrees of freedom for SS_T, $a - 1$ degrees of freedom for SS_A, $a(b - 1)$ degrees of freedom for $SS_{B(A)}$, and $ab(n - 1)$ degrees of freedom for error. Note that $abn - 1 = (a - 1) + a(b - 1) + ab(n - 1)$. If the errors are NID$(0, \sigma^2)$ we may divide each sum of squares on the right of (11-4) by its degrees of freedom to yield independently distributed mean squares such that the ratio of any two mean squares is distributed as F.

The appropriate statistics for testing the effects of factors A and B depend on whether the levels of A and B are fixed or random. If factors A and B are fixed, we assume that $\sum_{i=1}^{a}\tau_i = 0$ and $\sum_{j=1}^{b}\beta_{j(i)} = 0$ $(i = 1, 2, \ldots, a)$. That is, the A treatment effects sum to zero, and the B treatment effects sum to zero within each level of A. Alternatively, if A and B are random, then we assume that τ_i is NID$(0, \sigma_\tau^2)$ and $\beta_{j(i)}$ is NID$(0, \sigma_\beta^2)$. Mixed models with A fixed and B random are also widely encountered. The expected mean squares can be determined by a straightforward application of the rules in Chapter 7. Table 11-1 gives the expected mean squares for these situations.

Table 11-1 Expected Mean Squares in the Two-Stage Nested Design

$E(MS)$	A Fixed B Fixed	A Fixed B Random	A Random B Random
$E(MS_A)$	$\sigma^2 + \dfrac{bn\sum\tau_i^2}{a-1}$	$\sigma^2 + n\sigma_\beta^2 + \dfrac{bn\sum\tau_i^2}{a-1}$	$\sigma^2 + n\sigma_\beta^2 + bn\sigma_\tau^2$
$E(MS_{B(A)})$	$\sigma^2 + \dfrac{n\sum\sum\beta_{j(i)}^2}{a(b-1)}$	$\sigma^2 + n\sigma_\beta^2$	$\sigma^2 + n\sigma_\beta^2$
$E(MS_E)$	σ^2	σ^2	σ^2

Table 11-1 indicates that if the levels of A and B are fixed, then H_0: $\tau_i = 0$ is tested by MS_A/MS_E and H_0: $\beta_{j(i)} = 0$ is tested by $MS_{B(A)}/MS_E$. If A is a fixed factor and B is random, then H_0: $\tau_i = 0$ is tested by $MS_A/MS_{B(A)}$ and H_0: $\sigma_\beta^2 = 0$ is tested by

$MS_{B(A)}/MS_E$. Finally, if both A and B are random factors, we test $H_0: \sigma_\tau^2 = 0$ by $MS_A/MS_{B(A)}$ and $H_0: \sigma_\beta^2 = 0$ by $MS_{B(A)}/MS_E$. The test procedure is summarized in an analysis of variance table as shown in Table 11-2. Computing formulas for the sums of squares may be obtained by expanding the quantities in Equation 11-3 and simplifying. They are

$$SS_A = \sum_{i=1}^{a} \frac{y_{i..}^2}{bn} - \frac{y_{...}^2}{abn} \tag{11-5}$$

$$SS_{B(A)} = \sum_{i=1}^{a} \sum_{j=1}^{b} \frac{y_{ij.}^2}{n} - \sum_{i=1}^{a} \frac{y_{i..}^2}{bn} \tag{11-6}$$

$$SS_E = \sum_{i=1}^{a} \sum_{j=1}^{b} \sum_{k=1}^{n} y_{ijk}^2 - \sum_{i=1}^{a} \sum_{j=1}^{b} \frac{y_{ij.}^2}{n} \tag{11-7}$$

$$SS_T = \sum_{i=1}^{a} \sum_{j=1}^{b} \sum_{k=1}^{n} y_{ijk}^2 - \frac{y_{...}^2}{abn} \tag{11-8}$$

We see that Equation 11-6 for $SS_{B(A)}$ can be written as

$$SS_{B(A)} = \sum_{i=1}^{a} \left[\sum_{j=1}^{b} \frac{y_{ij.}^2}{n} - \frac{y_{i..}^2}{bn} \right]$$

This expresses the idea that $SS_{B(A)}$ is the sum of squares between levels of B for each level of A, summed over all levels of A.

Table 11-2 Analysis of Variance Table for the Two-Stage Nested Design

Source of Variation	Sum of Squares	Degrees of Freedom	Mean Square
A	$\sum_i \dfrac{y_{i..}^2}{bn} - \dfrac{y_{...}^2}{abn}$	$a - 1$	MS_A
B within A	$\sum_i \sum_j \dfrac{y_{ij.}^2}{n} - \sum_i \dfrac{y_{i..}^2}{bn}$	$a(b - 1)$	$MS_{B(A)}$
Error	$\sum_i \sum_j \sum_k y_{ijk}^2 - \sum_i \sum_j \dfrac{y_{ij.}^2}{n}$	$ab(n - 1)$	MS_E
Total	$\sum_i \sum_j \sum_k y_{ijk}^2 - \dfrac{y_{...}^2}{abn}$	$abn - 1$	

■**EXAMPLE 11-1.** Consider the company described in Section 11-1, which buys raw material in batches from three different suppliers. We wish to determine if all suppliers provide material of the same purity. Four batches of raw material are selected at random from each supplier and three determinations of purity made on each batch. This is, of course, a two-stage nested design. The data, after coding by subtracting 93, are shown in Table 11-3.

Table 11-3 Coded Purity Data for Example 11-1 (code: y_{ijk} = purity − 93)

Suppliers		1				2				3			
Batches		1	2	3	4	1	2	3	4	1	2	3	4
		1	−2	−2	1	1	0	−1	0	2	−2	1	3
		−1	−3	0	4	−2	4	0	3	4	0	−1	2
		0	−4	1	0	−3	2	−2	2	0	2	2	1
Batch totals	$y_{ij.}$	0	−9	−1	5	−4	6	−3	5	6	0	2	6
Supplier totals	$y_{i..}$		−5				4				14		

The sums of squares are computed as follows:

$$SS_T = \sum_{i=1}^{a} \sum_{j=1}^{b} \sum_{k=1}^{n} y_{ijk}^2 - \frac{y_{...}^2}{abn}$$

$$= 153.00 - \frac{(13)^2}{36} = 148.31,$$

$$SS_A = \sum_{i=1}^{a} \frac{y_{i..}^2}{bn} - \frac{y_{...}^2}{abn}$$

$$= \frac{(-5)^2 + (4)^2 + (14)^2}{(4)(3)} - \frac{(13)^2}{36}$$

$$= 19.75 - 4.69 = 15.06,$$

$$SS_{B(A)} = \sum_{i=1}^{a} \sum_{j=1}^{b} \frac{y_{ij.}^2}{n} - \sum_{i=1}^{a} \frac{y_{i..}^2}{bn}$$

$$= \frac{(0)^2 + (-9)^2 + (-1)^2 + \cdots + (2)^2 + (6)^2}{3} - 19.75$$

$$= 89.67 - 19.75 = 69.92$$

and

$$SS_E = \sum_{i=1}^{a} \sum_{j=1}^{b} \sum_{k=1}^{n} y_{ijk}^2 - \sum_{i=1}^{a} \sum_{j=1}^{b} \frac{y_{ij.}^2}{n}$$

$$= 153.00 - 89.67 = 63.33$$

The analysis of variance is summarized in Table 11-4. Suppliers are fixed and batches are random, so the expected mean squares are obtained from the middle column of Table 11-1. They are repeated for convenience in Table 11-4. The analysis of variance indicates that at the five percent level, there is no significant effect due to suppliers on purity, but that the purity of batches of raw material from the same supplier does differ significantly.

Table 11-4 Analysis of Variance for the Data in Example 11-1

Source of Variation	Sum of Squares	Degrees of Freedom	Mean Square	Expected Mean Square	F_0
Suppliers	15.06	2	7.53	$\sigma^2 + 3\sigma_\beta^2 + 6 \sum \tau_i^2$	0.97
Batches (within suppliers)	69.92	9	7.77	$\sigma^2 + 3\sigma_\beta^2$	2.94[a]
Error	63.33	24	2.64	σ^2	
Total	148.31	35			

[a]Significant at 5 percent.

Notice what would have happened if we had incorrectly analyzed this design as a two-factor factorial experiment. If batches are considered crossed with suppliers, we obtain batch totals of 2, -3, -2, and 16, with each batch $\times$ suppliers cell containing three replicates. Thus, a sum of squares due to batches and an interaction sum of squares can be computed. The complete factorial analysis of variance is shown in Table 11-5.

Table 11-5 Incorrect Analysis of the Two-Stage Nested Design in Example 11-1 as a Factorial (Suppliers fixed, Batches random).

Source of Variation	Sum of Squares	Degrees of Freedom	Mean Square	F_0
Suppliers (S)	15.06	2	7.53	1.02
Batches (B)	25.64	3	8.55	3.24[a]
S $\times$ B interaction	44.28	6	7.38	2.80[a]
Error	63.33	24	2.64	
Total	148.31	35		

[a]Significant at 5 percent.

This analysis indicates that batches differ significantly at the 5 percent level, and that there is a significant interaction between batches and suppliers. However, it is impossible to give a meaningful practical interpretation of the batches $\times$ suppliers interaction. Also, notice that from a comparison of Tables 11-4 and 11-5,

$$SS_B + SS_{S \times B} = 25.64 + 44.28 = 69.92 \equiv SS_{B(A)}$$

That is, the sum of squares for batches within suppliers consists of the sum of the batch and batches × suppliers interaction sums of squares. The degrees of freedom have a similar property; that is,

Batches		Batches × suppliers		Batches within suppliers
3	+	6	=	9

Therefore, we see that a computer program for analyzing factorial designs could also be used for the analysis of nested designs by pooling the "main effect" of the nested factor and interactions of that factor with the factor under which it is nested. This is useful information, since a computer program specifically written for the analysis of nested designs may not be readily available while programs for analyzing factorial designs are relatively accessable.   ■

If both factors A and B are fixed, we may estimate the parameters in the balanced two-stage nested design model

$$y_{ijk} = \mu + \tau_i + \beta_{j(i)} + \epsilon_{(ij)k}$$

The least squares normal equations for this model are shown in Equation 11-9. For clarity in reading (11-9), we have shown the parameter associated with each normal equation immediately to the left of that equation. For example, the first normal equation refers to μ, the second to τ_1, the third to τ_2, and so on.

The normal equations appear rather formidable; however, they may be easily derived using the rules of Section 3-10. As an illustration consider the normal equation for τ_1, which is the second normal equation in (11-9). The right-hand side of this equation is $y_{1..}$, since $y_{1..}$ is the total of all observations containing τ_1. Furthermore, in $y_{1..}$ the parameter μ appears bn times, τ_1 appears bn times, and each $\beta_{j(1)}$ $(j = 1, 2, \ldots, b)$ appears n times. Hence the normal equation for τ_1 should be $bn\hat{\mu} + bn\hat{\tau}_1 + n\sum_{j=1}^{b} \hat{\beta}_{j(1)} = y_{1..}$, as shown in (11-9).

There are $1 + a + ab$ normal equations in $1 + a + ab$ unknowns ($\hat{\mu}$, $\hat{\tau}_i$, and $\hat{\beta}_{j(i)}$ for $i = 1, 2, \ldots, a$ and $j = 1, 2, \ldots, b$). However, adding the equations associated with the τ_i yields the first normal equation, and adding the β-equations in each level of τ_i yields the normal equation for that τ_i. Therefore, there are $1 + a$ linear dependencies in the normal equations. The usual restrictions imposed on the estimates of the parameters are

$$\sum_{i=1}^{a} \hat{\tau}_i = 0$$

and

$$\sum_{j=1}^{b} \hat{\beta}_{j(i)} = 0 \qquad i = 1, 2, \ldots, a$$

These constraints greatly simplify the normal equations, resulting in the solution $\hat{\mu} = \bar{y}_{...}$ and

$$\hat{\tau}_i = \bar{y}_{i..} - \bar{y}_{...} \qquad i = 1, 2, \ldots, a \tag{11-10}$$

$$\hat{\beta}_{j(i)} = \bar{y}_{ij.} - \bar{y}_{i..} \qquad i = 1, 2, \ldots, a \qquad j = 1, 2, \ldots, b \tag{11-11}$$

$$\mu:\ abn\hat{\mu} + bn\hat{\tau}_1 + bn\hat{\tau}_2 + \cdots + bn\hat{\tau}_a + n\hat{\beta}_{1(1)} + n\hat{\beta}_{2(1)} + \cdots + n\hat{\beta}_{b(1)} + n\hat{\beta}_{1(2)} + n\hat{\beta}_{2(2)} + \cdots + n\hat{\beta}_{b(2)} + \cdots + n\hat{\beta}_{1(a)} + n\hat{\beta}_{2(a)} + \cdots + n\hat{\beta}_{b(a)} = y_{\ldots}$$

$$\tau_1:\ bn\hat{\mu} + bn\hat{\tau}_1 + n\hat{\beta}_{1(1)} + n\hat{\beta}_{2(1)} + \cdots + n\hat{\beta}_{b(1)} = y_{1\ldots}$$

$$\tau_2:\ bn\hat{\mu} + n\hat{\tau}_2 + n\hat{\beta}_{1(2)} + n\hat{\beta}_{2(2)} + \cdots + n\hat{\beta}_{b(2)} = y_{2\ldots}$$

$$\ \cdot\ \cdot\ \cdot\ \cdot$$

$$\tau_a:\ bn\hat{\mu} + bn\hat{\tau}_a + n\hat{\beta}_{1(a)} + n\hat{\beta}_{2(a)} + \cdots + n\hat{\beta}_{b(a)} = y_{a\ldots}$$

$$\beta_{1(1)}:\ n\hat{\mu} + n\hat{\tau}_1 + n\hat{\beta}_{1(1)} = y_{11\cdot}$$

$$\beta_{2(1)}:\ n\hat{\mu} + n\hat{\tau}_1 + n\hat{\beta}_{2(1)} = y_{21\cdot}$$

$$\ \cdot\ \cdot\ \cdot\ \cdot$$

$$\beta_{b(1)}:\ n\hat{\mu} + n\hat{\tau}_1 + n\hat{\beta}_{b(1)} = y_{b1\cdot}$$

$$\beta_{1(2)}:\ n\hat{\mu} + n\hat{\tau}_2 + n\hat{\beta}_{1(2)} = y_{12\cdot} \qquad (11\text{-}9)$$

$$\beta_{2(2)}:\ n\hat{\mu} + n\hat{\tau}_2 + n\hat{\beta}_{2(2)} = y_{22\cdot}$$

$$\ \cdot\ \cdot\ \cdot\ \cdot$$

$$\beta_{2(2)}:\ n\hat{\mu} + n\hat{\tau}_2 + n\hat{\beta}_{b(2)} = y_{b2\cdot}$$

$$\beta_{1(a)}:\ n\hat{\mu} + n\hat{\tau}_a + n\hat{\beta}_{1(a)} = y_{1a\cdot}$$

$$\beta_{2(a)}:\ n\hat{\mu} + n\hat{\tau}_a + n\hat{\beta}_{2(a)} = y_{2a\cdot}$$

$$\ \cdot\ \cdot\ \cdot\ \cdot$$

$$\beta_{b(a)}:\ n\hat{\mu} + n\hat{\tau}_a + n\hat{\beta}_{b(a)} = y_{ba\cdot}$$

Notice that this solution to the normal equations has considerable intuitive appeal; the A treatment effects are estimated by the average of all observations under each level of A minus the grand average, and the B treatment effects in each level of A are estimated by the corresponding cell average minus the average under that level of A. Using this solution, the general regression significance test can be applied to develop the analysis of variance for the balanced two-stage nested design (see Problem 11-5).

For the random effects case, the analysis of variance method can be used to estimate the variance components σ^2, σ_β^2, and σ_τ^2. From the expected mean squares in the last column of Table 11-1, we obtain

$$\hat{\sigma}^2 = MS_E \tag{11-12}$$

$$\hat{\sigma}_\beta^2 = \frac{MS_{B(A)} - MS_E}{n} \tag{11-13}$$

and

$$\hat{\sigma}_\tau^2 = \frac{MS_A - MS_{B(A)}}{bn} \tag{11-14}$$

Many applications of nested designs involve a mixed model, with the main factor (A) fixed and the nested factor (B) random. This is the case for the problem described in Example 11-1; suppliers (factor A) are fixed, while batches of raw material (factor B) are random. The effects of the suppliers may be estimated by

$$\hat{\tau}_1 = \bar{y}_{1..} - \bar{y}_{...} = \frac{-5}{12} - \frac{13}{36} = \frac{-28}{36}$$

$$\hat{\tau}_2 = \bar{y}_{2..} - \bar{y}_{...} = \frac{4}{12} - \frac{13}{36} = \frac{-1}{36}$$

$$\hat{\tau}_3 = \bar{y}_{3..} - \bar{y}_{...} = \frac{14}{12} - \frac{13}{36} = \frac{29}{36}$$

To estimate the variance components σ^2 and σ_β^2 we eliminate the line in the analysis of variance table pertaining to suppliers, and apply the analysis of variance method to the next two lines. This yields

$$\hat{\sigma}^2 = MS_E = 2.64$$

and

$$\hat{\sigma}_\beta^2 = \frac{MS_{B(A)} - MS_E}{n} = \frac{7.77 - 2.64}{3} = 1.71$$

11-3 THE GENERAL *m*-STAGE NESTED DESIGN

The results of Section 11-2 can be easily extended to the case of *m* completely nested factors. Such a design would be called an *m*-stage nested design. As an example, suppose a foundry wishes to investigate the hardness of two metal alloys. Three heats of each alloy are prepared, and two ingots selected at random from each heat for testing with two hardness measurements made on each ingot. The situation is illustrated in Figure 11-3.

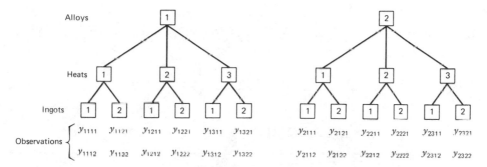

Figure 11-3. A three-stage nested design.

In this experiment, heats are nested under the levels of the factor alloys, and ingots are nested under the levels of the factor heats. Thus, this is a three-stage nested design with two replicates. The levels of alloys and heats are fixed, while the ingots are chosen at random.

The model for the general three-stage nested design is

$$y_{ijkl} = \mu + \tau_i + \beta_{j(i)} + \gamma_{k(ij)} + \epsilon_{(ijk)l} \begin{cases} i = 1, 2, \ldots, a \\ j = 1, 2, \ldots, b \\ k = 1, 2, \ldots, c \\ l = 1, 2, \ldots, n \end{cases} \quad (11\text{-}15)$$

For our example, τ_i is the effect of the *i*th alloy, $\beta_{j(i)}$ is the effect of the *j*th heat within the *i*th alloy, $\gamma_{k(ij)}$ is the effect of the *k*th ingot within the *j*th heat and *i*th alloy, and $\epsilon_{(ijk)l}$ is the usual NID$(0, \sigma^2)$ error term. Extension of this model to *m* factors is straightforward.

The calculation of sums of squares and the analysis of variance for the *m*-stage nested design is similar to the analysis presented in Section 11-2. For example, the analysis of variance for the three-stage nested design is summarized in Table 11-6. Computational formulas for the sums of squares are also shown in this table. Notice that they are a simple extension of the computational formulas for the two-stage nested design.

Table 11-6 Analysis of Variance for the Three-Stage Nested Design

Source of Variation	Sum of Squares	Degrees of Freedom	Mean Square
A	$\displaystyle\sum_i \frac{y_{i...}^2}{bcn} - \frac{y_{....}^2}{abcn}$	$a - 1$	MS_A
B (within A)	$\displaystyle\sum_i \sum_j \frac{y_{ij..}^2}{cn} - \sum_i \frac{y_{i...}^2}{bcn}$	$a(b-1)$	$MS_{B(A)}$
C (within B)	$\displaystyle\sum_i \sum_j \sum_k \frac{y_{ijk.}^2}{n} - \sum_i \sum_j \frac{y_{ij..}^2}{cn}$	$ab(c-1)$	$MS_{C(B)}$
Error	$\displaystyle\sum_i \sum_j \sum_k \sum_l y_{ijkl}^2 - \sum_i \sum_j \sum_k \frac{y_{ijk.}^2}{n}$	$abc(n-1)$	MS_E
Total	$\displaystyle\sum_i \sum_j \sum_k \sum_l y_{ijkl}^2 - \frac{y_{....}^2}{abcn}$	$abcn - 1$	

To determine the proper test statistics we must find the expected mean squares using the methods of Chapter 7. For example, if factors A and B are fixed and factor C is random, then we may derive the expected mean squares as shown in Table 11-7. This table shows the proper test statistics for the example at the start of this section, where alloys (A) and heats (B) were fixed, and ingots (C) were random.

Table 11-7 Expected Mean Square Derivation for a Three-Stage Nested Design, A and B fixed and C Random

Factor	F a i	F b j	R c k	R n l	Expected Mean Square
τ_i	0	b	c	n	$\sigma^2 + n\sigma_\gamma^2 + \dfrac{bcn \sum \tau_i^2}{a-1}$
$\beta_{j(i)}$	1	0	c	n	$\sigma^2 + n\sigma_\gamma^2 + \dfrac{cn \sum \sum \beta_{j(i)}^2}{a(b-1)}$
$\gamma_{k(ij)}$	1	1	1	n	$\sigma^2 + n\sigma_\gamma^2$
$\epsilon_{l(ijk)}$	1	1	1	1	σ^2

11-4 DESIGNS WITH NESTED AND CROSSED FACTORS

Occasionally in a multifactor experiment, some factors will be crossed and other factors will be nested. Anderson and McLean (1974) and Hicks (1973) call such designs nested-factorial designs. The statistical analysis of one such design with three factors is illustrated in the following example.

■**EXAMPLE 11-2.** An industrial engineer is studying the assembly of an electronic component to improve the speed of the assembly operation. He has designed three assembly fixtures and two workplace layouts that seem promising. Operators are required to perform the assembly, and it is decided to randomly select four operators for each fixture-layout combination. However, because the workplaces are in different locations within the plant it is difficult to use the *same* four operators for each layout. Therefore, the four operators chosen for layout 1 are different individuals from the four operators chosen for layout 2. The treatment combinations in this design are run in random order and two replicates are obtained. The assembly times are measured in seconds, and are shown in Table 11-8.

Table 11-8 Assembly Time Data for Example 11-2

Operator	Layout 1				Layout 2				$y_{i...}$
	1	2	3	4	1	2	3	4	
Fixture 1	22	23	28	25	26	27	28	24	404
	24	24	29	23	28	25	25	23	
Fixture 2	30	29	30	27	29	30	24	28	447
	27	28	32	25	28	27	23	30	
Fixture 3	25	24	27	26	27	26	24	28	401
	21	22	25	23	25	24	27	27	
Operator totals, $y_{.jk.}$	149	150	171	149	163	159	151	160	
Layout totals, $y_{.j..}$		619				633			$1252 = y_{...}$

We see that operators are nested within the levels of layouts, while fixtures and layouts are crossed. Thus, this design has both nested and crossed factors. The linear statistical model for this design is

$$y_{ijkl} = \mu + \tau_i + \beta_j + \gamma_{k(j)} + (\tau\beta)_{ij} + (\tau\gamma)_{ik(j)} + \epsilon_{(ijk)l}, \quad \begin{cases} i = 1, 2, 3 \\ j = 1, 2 \\ k = 1, 2, 3, 4 \\ l = 1, 2 \end{cases}$$

where τ_i is the effect of the ith fixture, β_j is the effect of the jth layout, $\gamma_{k(j)}$ is the effect of the kth operator within the jth level of layout, $(\tau\beta)_{ij}$ is the fixture × layout interaction, $(\tau\gamma)_{ik(j)}$ is the fixture × operators within layouts interaction, and $\epsilon_{(ijk)l}$ is the usual error term. Notice that no layout × operator interaction can exist, because all operators do not use all layouts. Similarly, there can be no three-way fixture × layout × operator interaction. Furthermore, fixtures and layouts are fixed factors, while operators is a random factor. The expected mean squares are derived in Table 11-9. The proper test statistic for any effect or interaction can be found from inspection of this table.

Table 11-9   Expected Mean Square Derivation for Example 11-2

Factor	F 3 i	F 2 j	R 4 k	R 2 l	Expected Mean Square
τ_i	0	2	4	2	$\sigma^2 + 2\sigma_{\tau\gamma}^2 + 8\sum \tau_i^2$
β_j	3	0	4	2	$\sigma^2 + 6\sigma_\gamma^2 + 24\sum \beta_j^2$
$\gamma_{k(j)}$	3	1	1	2	$\sigma^2 + 6\sigma_\gamma^2$
$(\tau\beta)_{ij}$	0	0	4	2	$\sigma^2 + 2\sigma_{\tau\gamma}^2 + 4\sum\sum (\tau\beta)_{ij}^2$
$(\tau\gamma)_{ik(j)}$	0	1	1	2	$\sigma^2 + 2\sigma_{\tau\gamma}^2$
$\epsilon_{(ijk)l}$	1	1	1	1	σ^2

The total sum of squares is

$$SS_T = \sum_{i=1}^{a} \sum_{j=1}^{b} \sum_{k=1}^{c} \sum_{l=1}^{n} y_{ijkl}^2 - \frac{y_{....}^2}{abcn}$$

$$= 32,956.00 - \frac{(1252)^2}{48} = 299.67$$

The sums of squares for fixtures and layouts are computed in the usual manner, that is,

$$SS_{\text{Fixtures}} = \sum_{i=1}^{a} \frac{y_{i...}^2}{bcn} - \frac{y_{....}^2}{abcn}$$

$$= 32,739.13 - \frac{(1252)^2}{48} = 82.80$$

and

$$SS_{\text{Layouts}} = \sum_{j=1}^{b} \frac{y_{.j..}^2}{acn} - \frac{y_{....}^2}{abcn}$$

$$= 32,660.42 - \frac{(1252)^2}{48} = 4.08$$

The fixtures X layouts or FL interaction is computed from the fixtures X layouts cell totals (which are 198, 206, 228, 219, 193, and 208) as follows:

$$SS_{FL} = \sum_{i=1}^{a} \sum_{j=1}^{b} \frac{y_{ij..}^2}{cn} - \frac{y_{....}^2}{abcn} - SS_{\text{Fixtures}} - SS_{\text{Layouts}}$$

$$= 32,762.25 - \frac{(1252)^2}{48} - 82.80 - 4.08 = 19.04$$

To find the sum of squares for operators within layouts we use the operator totals. The sum of squares is

$$SS_{O(L)} = \sum_{j=1}^{a} \sum_{k=1}^{b} \frac{y_{.jk.}^2}{an} - \sum_{j=1}^{b} \frac{y_{.j..}^2}{acn}$$

$$= 32,732.33 - 32,660.42 = 71.91$$

Notice that apart from the additional factor fixtures this is the same computational formula used to obtain the sum of squares for any nested factor. $SS_{O(L)}$ has $b(c-1) = 2(4-1) = 6$ degrees of freedom.

We do not have available the computational formula or the number of degrees of freedom for the fixtures X operators within layouts interaction. To determine the number of degrees of freedom and a computational formula for the sum of squares due to this effect, we shall use the rules of Chapter 7. The term in the model for this effect is $(\tau\gamma)_{ik(j)}$, and using rule 4 from Chapter 7, we obtain $b(a-1)(c-1) = 2(3-1)(4-1) = 12$ degrees of freedom. The sum of squares is obtained by using rule 5 in Chapter 7, starting with the symbolic form $b(a-1)(c-1) = abc - bc - ab + b$. Then

$$SS_{FO(L)} = \overset{abc}{\sum_{i=1}^{a} \sum_{j=1}^{b} \sum_{k=1}^{c} \frac{y_{ijk.}^2}{n}} - \overset{-bc}{\sum_{j=1}^{b} \sum_{k=1}^{c} \frac{y_{.jk.}^2}{an}} - \overset{-ab}{\sum_{i=1}^{a} \sum_{j=1}^{b} \frac{y_{ij..}^2}{cn}} + \overset{+b}{\sum_{j=1}^{b} \frac{y_{.j..}^2}{acn}}$$

$$= 32,900.00 - 32,732.33 - 32,762.25 + 32,660.42$$

$$= 65.84$$

Finally, the error sum of squares is obtained by subtraction as

$$SS_E = SS_T - SS_{\text{Fixtures}} - SS_{\text{Layouts}} - SS_{FL} - SS_{O(L)} - SS_{FO(L)}$$

$$= 299.67 - 82.80 - 4.08 - 19.04 - 71.91 - 65.84$$

$$= 56.00$$

The complete analysis of variance is shown in Table 11-10. We see that assembly fixtures are significant, and that operators within layouts also differ significantly. There is also a significant interaction between fixtures and operators within layouts, indicating that the effect of the different fixtures is not the same for all operators. The workplace layouts seem to have little effect on the assembly time. ∎

Table 11-10   Analysis of Variance for Example 11-2

Source of Variation	Sum of Squares	Degrees of Freedom	Mean Square	F_0
Fixtures (F)	82.80	2	41.40	7.54[b]
Layouts (L)	4.08	1	4.09	0.34
Operators (within layouts), $O(L)$	71.91	6	11.99	5.15[b]
FL	19.04	2	9.52	1.73
FO(L)	65.84	12	5.49	2.36[a]
Error	56.00	24	2.33	
Total	299.67	47		

[a]Significant at 5 percent.
[b]Significant at 1 percent.

Computer programs for the analysis of factorial experiments may also be used to analyze designs with nested and crossed factors. For instance, the experiment in Example 11-2 could be considered as a three-factor factorial in the factors fixtures (F), operators (O), and layouts (L). Then certain sums of squares and degrees of freedom from the factorial analysis would be pooled to form the appropriate quantities required for the design with nested and crossed factors as indicated below:

Factorial Analysis		Nested-Factorial Analysis	
Sum of Squares	Degrees of Freedom	Sum of Squares	Degrees of Freedom
SS_F	2	SS_F	2
SS_L	1	SS_L	1
SS_{FL}	2	SS_{FL}	2
SS_O	3	$SS_{O(L)} = SS_O + SS_{LO}$	6
SS_{LO}	3		
SS_{FO}	6	$SS_{FO(L)} = SS_{FO} + SS_{FOL}$	12
SS_{FOL}	6		
SS_E	24	SS_E	24
SS_T	47	SS_T	47

11-5   PROBLEMS

11-1.   An ammunition manufacturer is studying the burning rate of powder from three production processes. Four batches of powder are randomly selected from the output of each process and three determinations of burning rate were made on each batch. The results are shown below. Analyze the data and draw conclusions.

Process	1				2				3			
Batch	1	2	3	4	1	2	3	4	1	2	3	4
	25	19	15	15	19	23	18	35	14	35	38	25
	30	28	17	16	17	24	21	27	15	21	54	29
	26	20	14	13	14	21	17	25	20	24	50	33

11-2. The surface finish of metal parts made on four machines is being studied. An experiment in which each machine is run by three different operators and two specimens from each operator are collected and tested. Because of the location of the machines, different operators are used on each machine, and operators are chosen at random. The data are shown below. Analyze the data and draw conclusions.

Machine	1			2			3			4		
Operator	1	2	3	1	2	3	1	2	3	1	2	3
	79	94	46	92	85	76	88	53	46	36	40	62
	62	74	57	99	79	68	75	56	57	53	56	47

11-3. A manufacturing engineer is studying the dimensional variability of a particular component that is produced on three machines. Each machine has two spindles, and four components are randomly selected from each spindle. The results are shown below. Analyze these data, assuming that machines and spindles are fixed factors.

Machine	1		2		3	
Spindle	1	2	1	2	1	2
	12	8	14	12	14	16
	9	9	15	10	10	15
	11	10	13	11	12	15
	12	8	14	13	11	14

11-4. In order to simplify production scheduling, an industrial engineer is studying the possibility of assigning one time standard to a particular class of jobs, believing that differences between jobs are negligible. To see if this simplification is possible, six jobs are randomly selected. Each job is given to a different group of three operators. Each operator completes the job twice, but at different times during the week, and the following results were obtained. What are your conclusions about the use of a common time standard for all jobs in this class? What value would you use for the standard?

Job	Operator 1		Operator 2		Operator 3	
1	158.3	159.4	159.2	159.6	158.9	157.8
2	154.6	154.9	157.7	156.8	154.8	156.3
3	162.5	162.6	161.0	158.9	160.5	159.5
4	160.0	158.7	157.5	158.9	161.1	158.5
5	156.3	158.1	158.3	156.9	157.7	156.9
6	163.7	161.0	162.3	160.3	162.6	161.8

11-5. Use the general regression significance test to develop the analysis of variance for the balanced two-stage nested design.

11-6. Consider the three-stage nested design shown in Figure 11-3. Using the data below, analyze the design, assuming that alloys and heats are fixed while ingots are random.

Alloys	1						2					
Heats	1		2		3		1		2		3	
Ingots	1	2	1	2	1	2	1	2	1	2	1	2
	40	27	95	69	65	78	22	23	83	75	61	35
	63	30	67	47	54	45	10	39	62	64	77	42

11-7. Derive the expected mean squares for a balanced three-stage nested design assuming that A is fixed and that B and C are random. Obtain formulas for estimating the variance components.

11-8. Derive the expected mean squares for a balanced three-stage nested design if all three factors are random. Obtain formulas for estimating the variance components.

11-9. Verify the expected mean squares given in Table 11-1.

11-10. *Unbalanced Nested Designs.* Consider an unbalanced two-stage nested design with b_i levels of B under the ith level of A, and n_{ij} replicates in the (ij)th cell.
  (a) Write down the least squares normal equations for this situation. Solve the normal equations.
  (b) Construct the analysis of variance table for the unbalanced two-stage nested design.
  (c) Analyze the following data, using the results in part (b).

Factor A	1		2		
Factor B	1	2	1	2	3
	6	-3	5	2	1
	4	1	7	4	0
	8		9	3	-3
			6		

11-11. *Variance Components in the Unbalanced Two-Stage Nested Design.*
Consider the model

$$y_{ijk} = \mu + \tau_i + \beta_{j(i)} + \epsilon_{k(ij)} \begin{cases} i = 1, 2, \ldots, a \\ j = 1, 2, \ldots, b_i \\ k = 1, 2, \ldots, n_{ij} \end{cases}$$

where A and B are random factors. Show that

$$E(MS_A) = \sigma^2 + c_1 \sigma_\beta^2 + c_2 \sigma_\tau^2$$

$$E(MS_{B(A)}) = \sigma^2 + c_0 \sigma_\beta^2$$

$$E(MS_E) = \sigma^2$$

where

$$c_0 = \frac{N - \sum\limits_{i=1}^{a} \left(\sum\limits_{j=1}^{b_i} n_{ij}^2/n_{i.} \right)}{b_. - a}$$

$$c_1 = \frac{\sum\limits_{i=1}^{a} \left(\sum\limits_{j=1}^{b_i} n_{ij}^2/n_{i.} \right) - \sum\limits_{i=1}^{a} \sum\limits_{j=1}^{b_i} n_{ij}^2/N}{a - 1}$$

$$c_2 = \frac{N - \dfrac{\sum\limits_{i=1}^{a} n_{i.}^2}{N}}{a - 1}$$

11-12. A mechanical engineer is testing the performance of three machines. Each machine can be operated at two power settings. Furthermore, a machine has three stations on which the product is formed. An experiment is conducted where each machine is tested at both power settings, and three observations are taken from each station. The runs are made in random order, and the results are shown below. Analyze this experiment, assuming that all three factors are fixed.

Machine	1			2			3		
Station	1	2	3	1	2	3	1	2	3
Power setting 1	34.1	33.7	36.2	32.1	33.1	32.8	32.9	33.8	33.6
	30.3	34.9	36.8	33.5	34.7	35.1	33.0	33.4	32.8
	31.6	35.0	37.1	34.0	33.9	34.3	33.1	32.8	31.7
Power setting 2	24.3	28.1	25.7	24.1	24.1	26.0	24.2	23.2	24.7
	26.3	29.3	26.1	25.0	25.1	27.1	26.1	27.4	22.0
	27.1	28.6	24.9	26.3	27.9	23.9	25.3	28.0	24.8

11-13. Suppose in Problem 11-12 that a large number of power settings could have been used, and the two selected for the experiment were chosen randomly. Obtain the expected mean squares for this situation and modify the previous analysis appropriately.

11-14. A structural engineer is studying the strength of aluminum alloy purchased from three vendors. Each vendor submits the alloy in standard-sized bars, either 1.0, 1.50, or 2.0 inches. Processing of different size bar stock from a common ingot represents different forging techniques, and so this factor may be important. Furthermore, the bar stock is forged from ingots made in different heats. Each vendor submits two test specimens of each size bar stock from three heats. The resulting strength data is shown below. Analyze the data assuming that vendors and bar size are fixed and heats are random.

Vendor	1			2			3		
Heat	1	2	3	1	2	3	1	2	3
Bar size: 1″	1.230	1.346	1.235	1.301	1.346	1.315	1.247	1.275	1.324
	1.259	1.400	1.206	1.263	1.392	1.320	1.296	1.268	1.315
$1\frac{1}{2}″$	1.316	1.329	1.250	1.274	1.384	1.346	1.273	1.260	1.392
	1.300	1.362	1.239	1.268	1.375	1.357	1.264	1.265	1.364
2″	1.287	1.346	1.273	1.247	1.362	1.336	1.301	1.280	1.319
	1.292	1.382	1.215	1.215	1.328	1.342	1.262	1.271	1.323

11-15. Suppose, in Problem 11-14, that bar stock may be purchased in many sizes and the three sizes actually used in the experiment were selected randomly. Obtain the expected mean squares for this situation and modify the previous analysis appropriately.

Chapter Twelve
Multifactor Experiments With Randomization Restrictions

In Chapters 4 and 5 we presented designs for experiments with *randomization restrictions;* that is, experiments in which we cannot assign the treatment combinations to the experimental units in a completely random manner. The randomized block and Latin square designs are effective design techniques for dealing with one and two randomization restrictions, respectively. These designs were discussed in Chapters 4 and 5 relative to a single factor. However, there are many *multifactor* experimental situations in which it is necessary to restrict complete randomization. These design problems are the subject of this chapter. For further reading on these designs, see Ostle (1963), John (1971), Hicks (1973), and Anderson and McLean (1974).

12-1 RANDOMIZED BLOCKS AND LATIN SQUARES AS MULTIFACTOR DESIGNS

Consider a factorial experiment with two factors, say A and B. The linear statistical model for this design is

$$y_{ijk} = \mu + \tau_i + \beta_j + (\tau\beta)_{ij} + \epsilon_{ijk} \begin{cases} i = 1, 2, \ldots, a \\ j = 1, 2, \ldots, b \\ k = 1, 2, \ldots, n \end{cases} \qquad (12\text{-}1)$$

where τ_i, β_j, and $(\tau\beta)_{ij}$ represent the effects of factors A, B, and the AB interaction, respectively. The statistical analysis for this design is given in Chapter 6. Now suppose that in order to run this experiment a particular raw material is required. This raw material is available in batches that are not large enough to allow *all abn* treatment combinations to be run from the *same* batch. However, if a batch contains enough material for *ab* observations, then an alternative design is to run each of the n replicates using a separate batch of raw material. Consequently, the batches of raw material represent a randomization restriction or a *block*, and a single replicate of a complete factorial experiment has been run within each block. The model for this new design is

$$y_{ijk} = \mu + \tau_i + \beta_j + (\tau\beta)_{ij} + \delta_k + \epsilon_{ijk} \begin{cases} i = 1, 2, \ldots, a \\ j = 1, 2, \ldots, b \\ k = 1, 2, \ldots, n \end{cases} \quad (12\text{-}2)$$

where δ_k is the effect of the kth block. Of course, within a block the order of running the treatment combinations is completely randomized.

The model (12-2) assumes that interaction between blocks and treatments is negligible. This was assumed previously in the analysis of randomized block designs. If these interactions do exist, they cannot be separated from the error component. In fact, the error term in this model really consists of the $(\tau\delta)_{ik}$, $(\beta\delta)_{jk}$, and $(\tau\beta\delta)_{ijk}$ interactions. The analysis of variance is shown in Table 12-1. The layout closely re-

Table 12-1 Analysis of Variance for a Two-Factor Factorial in a Randomized Complete Block

Source of Variation	Sum of Squares	Degrees of Freedom
Blocks	$\sum_k \dfrac{y_{..k}^2}{ab} - \dfrac{y_{...}^2}{abn}$	$n - 1$
A	$\sum_i \dfrac{y_{i..}^2}{bn} - \dfrac{y_{...}^2}{abn}$	$a - 1$
B	$\sum_j \dfrac{y_{.j.}^2}{an} - \dfrac{y_{...}^2}{abn}$	$b - 1$
AB	$\sum_i \sum_j \dfrac{y_{ij.}^2}{n} - \dfrac{y_{...}^2}{abn} - SS_A - SS_B$	$(a - 1)(b - 1)$
Error	Subtraction	$(ab - 1)(n - 1)$
Total	$\sum_i \sum_j \sum_k y_{ijk}^2 - \dfrac{y_{...}^2}{abn}$	$abn - 1$

Table 12-2 Expected Mean Square Derivation, Factorial Experiment in a Randomized Block

Expanded Model

Factor	Degrees of Freedom	a F i	b R j	n R k	1 R b	Expected Mean Square
τ_i	$a-1$	0	b	n	1	$\sigma^2 + \sigma^2_{\tau\beta\delta} - b\sigma^2_{\tau\delta} + n\sigma^2_{\tau\beta} + \dfrac{bn\sum\tau_i^2}{a-1}$
β_j	$b-1$	a	1	n	1	$\sigma^2 + a\sigma^2_{\beta\delta} + an\sigma^2_\beta$
$(\tau\beta)_{ij}$	$(a-1)(b-1)$	0	1	n	1	$\sigma^2 + \sigma^2_{\tau\beta\delta} + n\sigma^2_{\tau\beta}$
δ_k	$(n-1)$	a	b	1	1	$\sigma^2 + a\sigma^2_{\tau\delta} + ab\sigma^2_\delta$
$(\tau\delta)_{ik}$	$(a-1)(n-1)$	0	b	1	1	$\sigma^2 + \sigma^2_{\tau\beta\delta} + b\sigma^2_{\tau\delta}$
$(\beta\delta)_{jk}$	$(b-1)(n-1)$	a	1	1	1	$\sigma^2 + a\sigma^2_{\beta\delta}$
$(\tau\beta\delta)_{ijk}$	$(a-1)(b-1)(n-1)$	0	1	1	1	$\sigma^2 + \sigma^2_{\tau\beta\delta}$
$\epsilon_{(ijk)b}$	0	1	1	1	1	σ^2 (not estimable)

Collapsed Model

Factor	Degrees of Freedom	a F i	b R j	n R k	1 R b	Expected Mean Square
τ_i	$a-1$	0	b	n	1	$\sigma^2 + n\sigma^2_{\tau\beta} + \dfrac{bn\sum\tau_i^2}{a-1}$
β_j	$b-1$	a	1	n	1	$\sigma^2 + an\sigma^2_\beta$
$(\tau\beta)_{ij}$	$(a-1)(b-1)$	0	1	n	1	$\sigma^2 + n\sigma^2_{\tau\beta}$
δ_k	$n-1$	a	b	1	1	$\sigma^2 + ab\sigma^2_\delta$
$\epsilon_{(ijk)b}$	$(ab-1)(n-1)$	1	1	1	1	σ^2

sembles that of a factorial design, with the error sum of squares reduced by the sum of squares for blocks. Computationally, we find the sum of squares for blocks as the sum of squares between the n block totals $\{y_{..k}\}$.

The expected mean squares for this design are derived in Table 12-2, assuming that factor A is fixed, and that factor B and blocks are random. In the left-hand side of this table we have expanded the model to include the interactions of treatments with blocks, so that one may more clearly see that experimental error is composed of these interactions. In using the expected mean square algorithm, we have denoted the observations by y_{ijkh}, where the subscript h represents replication. In this design we have only one replicate, so $h = 1$. Because each block contains only one replicate, the error variance σ^2 is not estimable. If we could run r replicates in each block, an estimate of σ^2 could be computed.

As discussed above, it is relatively standard to assume the interactions of blocks and treatments to be negligible, and to pool their mean squares to obtain an estimate of error. The right-hand side of Table 12-2 shows how the expanded model is collapsed to produce the expected mean squares.

In the previous example, the randomization was restricted to within a batch of raw material. In practice, a variety of phenomena may cause randomization restrictions, such as time, operators, and so on. For example, if we could not run the entire factorial experiment on one day, then the experimenter could run a complete replicate on day one, a second replicate on day two, and so on. Consequently, each day would be a block.

■**EXAMPLE 12-1.** An engineer is studying methods for improving the ability to detect targets on a radar scope. Two factors he considers to be important are the amount of background noise on the scope, or "ground clutter," and the type of filter placed over the screen. An experiment is designed using three levels of ground clutter and two filter types. The experiment is performed by randomly selecting a treatment combination (ground clutter level and filter type) and then introducing a signal representing the target into the scope. The intensity of this target is increased until the operator observes it. The intensity level at detection is then measured as the response variable. Because of operator availability, it is convenient to select an operator and keep him at the scope until all necessary runs have been made. Furthermore, operators differ in their skill and abilities to use the scope. Consequently, it seems logical to use the operators as blocks. Four operators were randomly selected. Once an operator is chosen, the order in which the six treatment combinations are run is randomly determined. Thus, we have a 3 X 2 factorial experiment run in a randomized complete block. The data are shown in Table 12-3.

The linear model for this experiment is

$$y_{ijk} = \mu + \tau_i + \beta_j + (\tau\beta)_{ij} + \delta_k + \epsilon_{ijk} \quad \begin{cases} i = 1, 2, 3 \\ j = 1, 2 \\ k = 1, 2, 3, 4 \end{cases}$$

Table 12-3 Intensity Level at Target Detection

Operators (Blocks)	1		2		3		4	
Filter Type	1	2	1	2	1	2	1	2
Ground Clutter								
Low	90	86	96	84	100	92	92	81
Medium	102	87	106	90	105	97	96	80
High	114	93	112	91	108	95	98	83

where τ_i represents the ground clutter effect (fixed), β_j represents the filter type effect (fixed), $(\tau\beta)_{ij}$ is the interaction, δ_k is the block effect (random), and ϵ_{ijk} is the NID$(0, \sigma^2)$ error component. The sums of squares for ground clutter, filter type, and their interaction are computed in the usual manner. The sum of squares due to blocks is found from the operator totals $\{y_{..k}\}$ as follows:

$$SS_{\text{Blocks}} = \sum_{k=1}^{n} \frac{y_{..k}^2}{ab} - \frac{y_{...}^2}{abn}$$

$$= \frac{(572)^2 + (579)^2 + (597)^2 + (530)^2}{(3)(2)} - \frac{(2278)^2}{(3)(2)(4)}$$

$$= 402.17$$

The complete analysis of variance for this experiment is summarized in Table 12-4. A column has been added to this table listing the expected mean squares. These expected mean squares may be derived by the algorithm in Chapter 7. Since both factors are fixed, all effects are tested by dividing their mean squares by mean squares error. Both ground clutter level and filter type are significant at the one percent level, while their interaction would only be significant at ten percent. Thus, we conclude

Table 12-4 Analysis of Variance for Example 12-1

Source of Variation	Sum of Square	Degrees of Freedom	Mean Square	Expected Mean Square	F_0
Ground Clutter (G)	335.58	2	167.79	$\sigma^2 + 8 \sum \tau_i^2/2$	15.13[a]
Filter Type (F)	1066.67	1	1066.67	$\sigma^2 + 12 \sum \beta_j^2$	96.19[a]
GF	77.08	2	38.54	$\sigma^2 + 4 \sum\sum (\tau\beta)_{ij}^2/6$	3.48
Blocks	402.17	3	134.06	$\sigma^2 + 6\sigma_\delta^2$	
Error	166.33	15	11.09	σ^2	
Total	2047.83	23			

[a]Significant at 1 percent.

that both ground clutter level and the type of scope filter used affect the operator's ability to detect the target, and there is some evidence of mild interaction between these factors.   ■

In the case of two randomization restrictions, each with p levels, if the number of treatment combinations in a k-factor factorial design exactly equals the number of restriction levels, that is, if $p = ab \ldots m$, then the factorial design may be run in a $p \times p$ Latin square. For example, consider a modification of the radar target detection experiment of Example 12-1. The factors in this experiment are filter type (two levels) and ground clutter (three levels), and operators are considered as blocks. Suppose now that because of the setup time required, only six runs can be made per day. Thus days become a second randomization restriction, resulting in the 6 × 6 Latin square design shown in Table 12-5. In this table we have used the lower-case letters f_i and g_j to represent the ith and jth levels of filter type and ground clutter, respectively. That is, $f_1 g_2$ represents filter type 1 and medium ground clutter. Note that now six operators are required, rather than 4 as in the original experiment, so that the number of treatment combinations in the 3 × 2 factorial design exactly equals the number of restriction levels. Furthermore, in this design, each operator would be used only once on each day. The Latin letters A, B, C, D, E, and F represent the 3 × 2 = 6 factorial treatment combinations as follows: $A = f_1 g_1$, $B = f_1 g_2$, $C = f_1 g_3$, $D = f_2 g_1$, $E = f_2 g_2$, and $F = f_2 g_3$.

The five degrees of freedom between the six Latin letters correspond to the main effects of filter type (one degree of freedom), ground clutter (two degrees of freedom), and their interaction (two degrees of freedom). The linear statistical model for this design is

$$y_{ijkl} = \mu + \alpha_i + \tau_j + \beta_k + (\tau\beta)_{jk} + \theta_l + \epsilon_{ijkl} \begin{cases} i = 1, 2, \ldots, 6 \\ j = 1, 2, 3 \\ k = 1, 2 \\ l = 1, 2, \ldots, 6 \end{cases} \tag{12-3}$$

where τ_j and β_k are the effects of ground clutter and filter type, respectively, while α_i and θ_l represent the randomization restrictions days and operators, respectively. To compute the sums of squares, the following two-way table of treatment totals is helpful:

Ground Clutter	Filter Type 1	Filter Type 2	$y_{.j..}$
low	560	512	1072
medium	607	528	1135
high	646	543	1189
$y_{..k.}$	1813	1583	3396 = $y_{....}$

Table 12-5 Radar Detection Experiment Run in a 6 × 6 Latin Square

Day	Operator					
	1	2	3	4	5	6
1	$A(f_1g_1 = 90)$	$B(f_1g_2 = 106)$	$C(f_1g_3 = 108)$	$D(f_2g_1 = 81)$	$F(f_2g_3 = 90)$	$E(f_2g_2 = 88)$
2	$C(f_1g_3 = 114)$	$A(f_1g_1 = 96)$	$B(f_1g_2 = 105)$	$F(f_2g_3 = 83)$	$E(f_2g_2 = 86)$	$D(f_2g_1 = 84)$
3	$B(f_1g_2 = 102)$	$E(f_2g_2 = 90)$	$F(f_2g_3 = 95)$	$A(f_1g_1 = 92)$	$D(f_2g_1 = 85)$	$C(f_1g_3 = 104)$
4	$E(f_2g_2 = 87)$	$D(f_2g_1 = 84)$	$A(f_1g_1 = 100)$	$B(f_1g_2 = 96)$	$C(f_1g_3 = 110)$	$F(f_2g_3 = 91)$
5	$F(f_2g_3 = 93)$	$C(f_1g_3 = 112)$	$D(f_2g_1 = 92)$	$E(f_2g_2 = 80)$	$A(f_1g_1 = 90)$	$B(f_1g_2 = 98)$
6	$D(f_2g_1 = 86)$	$F(f_2g_3 = 91)$	$E(f_2g_2 = 97)$	$C(f_1g_3 = 98)$	$B(f_1g_2 = 100)$	$A(f_1g_1 = 92)$

Furthermore the row and column totals are

$$\text{Rows} (y_{.jkl}): \quad 563 \quad 568 \quad 568 \quad 568 \quad 565 \quad 564$$
$$\text{Columns} (y_{ijk.}): \quad 572 \quad 579 \quad 597 \quad 530 \quad 561 \quad 557$$

The analysis of variance is summarized in Table 12-6. We have added a column to this table indicating how the number of degrees of freedom for each sum of squares is determined.

Table 12-6   Analysis of Variance for the Radar Detection Experiment Run as a 3 X 2 Factorial in a Latin Square

Source of Variation	Sum of Squares	Degrees of Freedom	General Formula for Degrees of Freedom	Mean Square	F_0
Ground Clutter, G	571.50	2	$a - 1$	285.75	28.86[a]
Filter Type, F	1469.44	1	$b - 1$	1469.44	148.43[a]
GF	126.73	2	$(a - 1)(b - 1)$	63.37	6.40[b]
Days (rows)	4.33	5	$ab - 1$	0.87	
Operators (columns)	428.00	5	$ab - 1$	85.60	
Error	198.00	20	$(ab - 1)(ab - 2)$	9.90	
Total	2798.00	35	$(ab)^2 - 1$		

[a] Significant at 1 percent.
[b] Significant at 5 percent.

We have emphasized running factorial experiments in randomized blocks or Latin squares. However, other multifactor designs could be treated similarly. The statistical analysis of such designs is straightforward, and follows the general methods outlined above. One may also use incomplete block designs as the basis of multifactor experiments. An interesting example of this is discussed by John (1971, pg. 240).

12-2   THE SPLIT-PLOT DESIGN

In some multifactor designs involving randomized blocks, we may be unable to completely randomize the order of the runs within the block. This often results in a generalization of the randomized block design called a *split-plot design*. As an example, consider a paper manufacturer who is interested in three different pulp preparation methods and four different cooking temperatures for the pulp, and wishes to study the effect of these two factors on the tensile strength of the paper. Each replicate of a factorial experiment requires 12 observations, and the experimenter has decided to run three replicates. However, the pilot plant is only capable of making 12 runs per day, so the experimenter will run one replicate on each of the three days, and consider the days or replicates as blocks. On any day, he conducts the experiment as follows. A batch of pulp is produced by one of the three methods under study. Then this batch is divided into four samples, and each sample is cooked at one of the four temperatures. Then a second batch of pulp is made up using another of the three methods. This second batch is also divided into four samples that are tested at the four temperatures. The process is then repeated, using a batch of pulp produced by the third method. The data are shown in Table 12-7.

Table 12-7 Data for Tensile Strength of Paper

Blocks		1			2			3	
Pulp Preparation Method	1	2	3	1	2	3	1	2	3
Temperature									
200	30	34	29	28	31	31	31	35	32
225	35	41	26	32	36	30	37	40	34
250	37	38	33	40	42	32	41	39	39
275	36	42	36	41	40	40	40	44	45

Initially, we might consider this to be a factorial experiment with three levels of preparation methods (factor A), and four levels of temperature (factor B) in a randomized block. If this is the case then the order of experimentation within the block should have been completely randomized. That is, within a block, we should randomly select a treatment combination (a preparation method and a temperature) and obtain an observation, then another treatment combination should be randomly selected and a second observation obtained, and so on, until the 12 observations in the block have been taken. However, the experimenter did not collect the data this way. He made up a batch of pulp and obtained observations for all four temperatures from that batch. Because of the economics of preparing the batches and the size of the batches, this is the only feasible way to collect the data.

Each block in the design is divided into three parts called *whole plots*, and preparation methods are called the whole plot or *main* treatments. Each whole plot is divided into four parts called subplots (or split-plots), and one temperature is assigned to each. Temperature is called the *subplot* treatment. Note that if there are other uncontrolled or undesigned factors present, and if these uncontrolled factors vary as the pulp preparation methods are changed, then any effect of the undesigned factors on the response will be completely confounded with the effect of the pulp preparation methods. Since the whole plot treatments in a split-plot design are confounded with the whole plots and the subplot treatments are not confounded, it is best to assign the factor we are most interested in to the subplots, if possible.

The linear model for the split-plot design is

$$y_{ijk} = \mu + \tau_i + \beta_j + (\tau\beta)_{ij} + \gamma_k + (\tau\gamma)_{ik} + (\beta\gamma)_{jk} + (\tau\beta\gamma)_{ijk} \begin{cases} i = 1, 2, \ldots, a \\ j = 1, 2, \ldots, b \quad (12\text{-}4) \\ k = 1, 2, \ldots, c \end{cases}$$

Table 12-8 Expected Mean Square Derivation for Split-Plot Design

	Factor	a R i	b F j	c F k	1 R b	Expected Mean Square
	τ_i	1	b	c	1	$\sigma^2 + bc\sigma_\tau^2$
Whole plot	β_j	a	0	c	1	$\sigma^2 + c\sigma_{\tau\beta}^2 + \dfrac{ac\sum\beta_j^2}{b-1}$
	$(\tau\beta)_{ij}$	1	0	c	1	$\sigma^2 + c\sigma_{\tau\beta}^2$
	γ_k	a	b	0	1	$\sigma^2 + b\sigma_{\tau\gamma}^2 + \dfrac{ab\sum\gamma_k^2}{(c-1)}$
	$(\tau\gamma)_{ik}$	1	b	0	1	$\sigma^2 + b\sigma_{\tau\gamma}^2$
Subplot	$(\beta\gamma)_{jk}$	a	0	0	1	$\sigma^2 + \sigma_{\tau\beta\gamma}^2 + \dfrac{a\sum\sum(\beta\gamma)_{jk}^2}{(b-1)(c-1)}$
	$(\tau\beta\gamma)_{ijk}$	1	0	0	1	$\sigma^2 + \sigma_{\tau\beta\gamma}^2$
	$\epsilon_{(ijk)b}$	1	1	1	1	σ^2 (not estimable)

where τ_i, β_j, and $(\tau\beta)_{ij}$ represent the whole plot and correspond respectively to blocks (factor A), main treatments (factor B), and whole plot error (AB); while γ_k, $(\tau\gamma)_{ik}$, $(\beta\gamma)_{jk}$, and $(\tau\beta\gamma)_{ijk}$ represent the subplot and correspond respectively to the subplot treatment (factor C), and the AC and BC interactions, and the subplot error. Note that the whole plot error is the AB interaction and the subplot error is the three-factor interaction ABC. Sums of square for these factors are computed as in the three-way analysis of variance without replication.

The expected mean squares for the split-plot design with blocks random and main treatments and subplot treatments fixed are derived in Table 12-8. Note that the main factor (B) in the whole plot is tested against the whole plot error, while the sub-treatment (C) is tested against the block $\times$ subtreatment (AC) interaction. The BC interaction is tested against the subplot error. Notice that there are no tests for the block effect (A) or the block $\times$ subtreatment (AC) interaction. Some authors believe that the subplot error should be found by pooling the ABC *and* AC interactions, and then testing both subtreatment and the BC interaction by this pooled estimate. If the experimenter has previous experience in the area and is reasonably sure that blocks do not interact with the various treatments, this may be appropriate. On the other hand, if he cannot be reasonably sure that the interactions are negligible, then pooling is inadvisable.

The analysis of variance for the tensile strength data in Table 12-7 is summarized in Table 12-9. Since both preparation methods and temperatures are fixed and blocks are

Table 12-9 Analysis of Variance for Split-Plot Design, Tensile Strength Data of Table 12-8

Source of Variation	Sum of Squares	Degrees of Freedom	Mean Square	F_0
Blocks (A)	77.55	2	38.78	
Preparation method (B)	128.39	2	64.20	7.08[a]
AB (whole plot error)	36.28	4	9.07	
Temperature (C)	434.08	3	144.69	41.94[b]
AC	20.67	6	3.45	
BC	75.17	6	12.53	2.96
ABC (subplot error)	50.83	12	4.24	
Total	822.97	35		

[a]Significant at 5 percent
[b]Significant at 1 percent

random, the expected mean squares in Table 12-8 apply. The mean square for prepara tion methods is compared to the whole plot error mean square, and the mean square for temperatures is compared to the block X temperature or AC mean square. Finally, the preparation method X temperature mean square is tested against the subplot error. Both preparation methods and temperature have a significant effect on strength.

Note from Table 12-9 the subplot error (4.24) is less than the whole plot error (9.07). This is the usual case in split-plot designs, since the subplots are generally more homogeneous than the whole plots. Because the subplot treatments are compared with greater precision, it is preferable to assign the treatment we are most interested in to the subplots, if possible.

The split-plot design has an agricultural heritage, with the whole plots usually being large areas of land and the subplots being small areas of land. For example, several varieties of a crop could be planted in different fields (whole plots), one variety to a field. Then each field could be divided into four subplots, for example, and each sub plot treated with a different type of fertilizer. Here the crop varieties are the main treatments and the different fertilizers are the subtreatments. The split-plot design is also useful in many scientific or industrial experiments where some factors require large experimental units and other factors require small experimental units, such as the tensile strength problem described above.

John (1971) notes that we must carefully consider how the data will be collected and incorporate all restrictions on randomization into the analysis. We illustrate his point with a modification of the eye focus time experiment in Example 10-2. Suppose there are only two factors, visual acuity (A) and illumination level (B). A factorial experiment with a levels of acuity, b levels of illumination, and n replicates would require that all abn observations be taken in random order. However, in the test apparatus, it is fairly difficult to adjust these two factors to different levels, so the experimenter decides to obtain the n replicates by adjusting the device to one of the a acuities and one of the b illumination levels and running all n observations at once. In the factorial design, the error actually represents the scatter or noise in the system plus the ability of the subject to reproduce the same focus time. The model for the factorial design could be written as

$$y_{ijk} = \mu + \tau_i + \beta_j + (\tau\beta)_{ij} + \phi_{ijk} + \theta_{ijk} \begin{cases} i = 1, 2, \ldots, a \\ j = 1, 2, \ldots, b \\ k = 1, 2, \ldots, n \end{cases} \tag{12-5}$$

where ϕ_{ijk} represents the scatter or noise in the system due to "experimental error," that is, our failure to exactly duplicate the same levels of acuity and illumination on different runs, variability in environmental conditions, and the like, and θ_{ijk} represents the "reproducibility error" of the subject. Usually, we combine these components into one overall error term, say $\epsilon_{ijk} = \phi_{ijk} + \theta_{ijk}$. Assume that $V(\epsilon_{ijk}) = \sigma^2 = \sigma_\phi^2 + \sigma_\theta^2$. Now in the factorial design, the error mean square has expectation $\sigma^2 = \sigma_\phi^2 + \sigma_\theta^2$, with $ab(n - 1)$ degrees of freedom.

If we restrict the randomization as in the second design above, then the "error" mean square in the analysis of variance provides an estimate of the "reproducibility error" σ_θ^2 with $ab(n - 1)$ degrees of freedom, but yields no information on the "experimental error" σ_ϕ^2. Thus, the mean square for error in this second design is too small, and consequently we will wrongly reject the null hypothesis very frequently. As pointed out by John (1971), this design is similar to a split-plot design with ab whole plots, each divided into n subplots, with no subtreatment. The situation is also similar to *subsampling*, as described by Ostle (1963). Assuming that A and B are fixed, the expected mean squares in this case are

$$E(MS_A) = \sigma_\theta^2 + n\sigma_\phi^2 + \frac{bn \sum \tau_i^2}{a - 1}$$

$$E(MS_B) = \sigma_\theta^2 + n\sigma_\phi^2 + \frac{an \sum \beta_j^2}{b - 1}$$

$$E(MS_{AB}) = \sigma_\theta^2 + n\sigma_\phi^2 + \frac{n \sum\sum (\tau\beta)_{ij}^2}{(a - 1)(b - 1)}$$

$$E(MS_E) = \sigma_\theta^2$$

$$\tag{12-6}$$

Thus, there are no tests on main effects unless interaction is negligible. The situation is exactly that of a two-way analysis of variance with one observation per cell. If both factors are random, then main effects may be tested against the AB interaction. If only one factor is random, then the fixed factor can be tested against the AB interaction.

John (1971) also states that, in general, if one analyzes a factorial design and all main effects and interactions are significant, then examine carefully *how* the data were collected. There may be randomization restrictions in the model not accounted for in the analysis, and consequently the data should not be analyzed as a factorial. Anderson and McLean (1974, p. 191) present an example where the experimenter has confused the factorial and split-plot layouts.

12-3 THE SPLIT-SPLIT-PLOT DESIGN

The concept of split-plot designs can be extended to situations in which randomization restrictions may occur at any number of levels within a block. If there are two levels of randomization restrictions within a block, then the layout is called a split-split-plot design. The following example illustrates such a design.

■EXAMPLE 12-2.   A medical researcher is studying the absorption times of a particular type of antibiotic capsule. There are three technicians, three dosage strengths, and four capsule wall thicknesses of interest to the researcher. Each replicate of a factorial experiment would require 36 observations. The experimenter has decided on four replicates, and it is necessary to run each replicate on a different day. Thus the days are *blocks*. Within a block (day), the experiment is performed by assigning a unit of antibiotic to a technician who conducts the experiment on the three dosage strengths and the four wall thicknesses. Once a particular dosage strength was formulated, all four wall thicknesses were tested at that strength. Then another dosage strength was selected and all four wall thickness were tested. Finally, the third dosage strength and the four wall thicknesses were tested. Meanwhile, two other laboratory technicians followed this plan, each starting with a unit of antibiotic.

Note that there are *two* randomization restrictions within a block: technician and dosage strength. The whole plots correspond to the technician. The order in which the technicians are assigned the units of antibiotic is randomly determined. The dosage strengths form three subplots. Dosage strength may be randomly assigned to a subplot. Finally, within a particular dosage strength, the four capsule wall thicknesses are tested in random order, forming four sub-subplots. The wall thicknesses are usually called sub-subtreatments. Because there are *two* randomization restrictions within a block (some authors say two "splits" in the design) the design is called a split-split-plot design. Figure 12-1 illustrates the randomization restrictions and experimental layout in this design.   ■

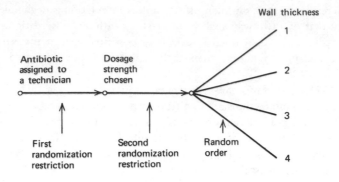

Blocks	Dosage strength	Technician								
		1			2			3		
		1	2	3	1	2	3	1	2	3
1	Wall thicknesses	1	1	1	1	1	1	1	1	1
		2	2	2	2	2	2	2	2	2
		3	3	3	3	3	3	3	3	3
		4	4	4	4	4	4	4	4	4
2	Wall thicknesses	1	1	1	1	1	1	1	1	1
		2	2	2	2	2	2	2	2	2
		3	3	3	3	3	3	3	3	3
		4	4	4	4	4	4	4	4	4
3	Wall thicknesses	1	1	1	1	1	1	1	1	1
		2	2	2	2	2	2	2	2	2
		3	3	3	3	3	3	3	3	3
		4	4	4	4	4	4	4	4	4
4	Wall thicknesses	1	1	1	1	1	1	1	1	1
		2	2	2	2	2	2	2	2	2
		3	3	3	3	3	3	3	3	3
		4	4	4	4	4	4	4	4	4

Figure 12-1. A split-split-plot design.

Table 12-10 Expected Mean Square Derivation for the Split-Split-Plot Design

	Factor	a R i	b F j	c F k	d F h	1 R l	Expected Mean Square
	τ_i	1	b	c	d	1	$\sigma^2 + bcd\sigma_\tau^2$
Whole plot	β_j	a	0	c	d	1	$\sigma^2 + cd\sigma_{\tau\beta}^2 + \dfrac{acd\sum\beta_j^2}{(b-1)}$
	$(\tau\beta)_{ij}$	1	0	c	d	1	$\sigma^2 + cd\sigma_{\tau\beta}^2$
	γ_k	a	b	0	d	1	$\sigma^2 + bd\sigma_{\tau\gamma}^2 + \dfrac{abd\sum\gamma_k^2}{(c-1)}$
Subplot	$(\tau\gamma)_{ik}$	1	b	0	d	1	$\sigma^2 + bd\sigma_{\tau\gamma}^2$
	$(\beta\gamma)_{jk}$	a	0	0	d	1	$\sigma^2 + d\sigma_{\tau\beta\gamma}^2 + \dfrac{ad\sum\sum(\beta\gamma)_{jk}^2}{(b-1)(c-1)}$
	$(\tau\beta\gamma)_{ijk}$	1	0	0	d	1	$\sigma^2 + d\sigma_{\tau\beta\gamma}^2$
	δ_h	a	b	c	0	1	$\sigma^2 + bc\sigma_{\tau\delta}^2 + \dfrac{abc\sum\gamma_k^2\,\delta_h^2}{(c-1)}$
	$(\tau\delta)_{ih}$	1	b	c	0	1	$\sigma^2 + bc\sigma_{\tau\delta}^2$
	$(\beta\delta)_{jh}$	a	0	c	0	1	$\sigma^2 + c\sigma_{\tau\beta\delta}^2 + \dfrac{ac\sum\sum(\beta\delta)_{jh}^2}{(b-1)(d-1)}$
	$(\tau\beta\delta)_{ijh}$	1	0	c	0	1	$\sigma^2 + c\sigma_{\tau\beta\delta}^2$
Sub-subplot	$(\gamma\delta)_{kh}$	a	b	0	0	1	$\sigma^2 + b\sigma_{\tau\gamma\delta}^2 + \dfrac{ab\sum\sum(\gamma\delta)_{kh}^2}{(c-1)(d-1)}$
	$(\tau\gamma\delta)_{ikh}$	1	b	0	0	1	$\sigma^2 + b\sigma_{\tau\gamma\delta}^2$
	$(\beta\gamma\delta)_{jkh}$	a	0	0	0	1	$\sigma^2 + \sigma_{\tau\beta\gamma\delta}^2 + \dfrac{a\sum\sum\sum(\beta\gamma\delta)_{jkh}^2}{(b-1)(c-1)(d-1)}$
	$(\tau\beta\gamma\delta)_{ijkh}$	1	0	0	0	1	$\sigma^2 + \sigma_{\tau\beta\gamma\delta}^2$
	$\epsilon_{l(ijkh)}$	1	1	1	1	1	σ^2 (not estimable)

The linear statistical model for the split-split-plot design is

$$y_{ijkh} = \mu + \tau_i + \beta_j + (\tau\beta)_{ij} + \gamma_k + (\tau\gamma)_{ik} + (\beta\gamma)_{jk} + (\tau\beta\gamma)_{ijk} + \delta_h + (\tau\delta)_{ih} + (\beta\delta)_{jh}$$

$$+ (\tau\beta\delta)_{ijh} + (\gamma\delta)_{kh} + (\tau\gamma\delta)_{ikh} + (\beta\gamma\delta)_{jkh}$$

$$+ (\tau\beta\gamma\delta)_{ijkh} \begin{cases} i = 1, 2, \ldots, a \\ j = 1, 2, \ldots, b \\ k = 1, 2, \ldots, c \\ h = 1, 2, \ldots, d \end{cases} \qquad (12\text{-}7)$$

where τ_i, β_j, and $(\tau\beta)_{ij}$ represent the whole plot and correspond to blocks (factor A), main treatments (factor B) and whole plot error (AB), respectively; while γ_k, $(\tau\gamma)_{ik}$, $(\beta\gamma)_{jk}$, and $(\tau\beta\gamma)_{ijk}$ represent the subplot and correspond to the subplot treatment (factor C), the AC and BC interactions, and the subplot error, respectively; and δ_h and the remaining parameters correspond to the sub-subplot, and represent respectively the sub-subplot treatment (factor D) and the remaining interactions. The four-factor interaction $(\tau\beta\gamma\delta)_{ijkh}$ is called the sub-subplot error.

Assuming that blocks are random and that the other factors are fixed, we may derive the expected mean squares as shown in Table 12-10. Tests on main treatments, subtreatments, sub-subtreatments and their interactions are obvious from inspection of this table. Note that no tests on blocks or interactions involving the blocks exist.

The statistical analysis of a split-split-plot design is like that of a single replicate of a four-factor factorial. The number of degrees of freedom for each test are determined in the usual manner. To illustrate, in Example 12-2, where we had four blocks, three technicians, three dosage strengths, and four wall thicknesses, we would have only $(a - 1)(b - 1) = (4 - 1)(3 - 1) = 6$ whole plot error degrees of freedom for testing technicians. This is a relatively small number of degrees of freedom, and the experimenter might consider using additional blocks to increase the precision of the test. If there are a blocks, then we will have $2(a - 1)$ degrees of freedom for whole plot error. Thus, five blocks yields $2(5 - 1) = 8$ error degrees of freedom, six blocks yields $2(6 - 1) = 10$ error degrees of freedom, seven blocks yields $2(7 - 1) = 12$ error degrees of freedom, and so on. Consequently, we would probably not want to run fewer than four blocks since this would yield only 4 error degrees of freedom. Each additional block allows us to gain 2 degrees of freedom for error. If the experimenter could afford to run five blocks, he could increase the precision of the test by one-third (from 6 to 8 degrees of freedom). Also, in going from five to six blocks, there is an additional 25 percent gain in precision. If resources permit, the experimenter should run five or six blocks.

12-4   PROBLEMS

12-1.   The yield of a chemical process is being studied. The two factors under study are the temperature and the pressure. Three levels of each factor are selected, and only nine observations can be run in one day. The experimenter runs a complete replicate of the design on each day. The data are shown below. Analyze the data, assuming that the days are blocks.

	Day 1 Pressure			Day 2 Pressure		
Temperature	250	260	270	250	260	270
Low	86.3	84.0	85.8	86.1	85.2	87.3
Medium	88.5	87.3	89.0	89.4	89.9	90.3
High	89.1	90.2	91.3	91.7	93.2	93.7

12-2. Consider the data in Problem 8-2. Analyze the data assuming that replicates are blocks. Assume that all factors are fixed.

12-3. Consider the data in Problem 8-18. Analyze the data assuming that replicates are blocks. Assume that men are randomly selected, and that bottle type and shelf type are fixed.

12-4. Rework Problem 12-3, computing all interactions of the three factors with blocks.

12-5. Rework Example 11-1, assuming that each replicate represents a block. Derive the expected mean squares.

12-6. Rework Example 11-2, assuming that each replicate represents a block. Derive the expected mean squares.

12-7. Outline the analysis of variance for a 2^3 factorial experiment run in a Latin square.

12-8. Steel is normalized by heating above the critical temperature, soaking, and then air cooling. This process increases the strength of the steel, refines the grain, and homogenizes the structure. An experiment is performed to deter-

Shift	Time (minutes)	Temperature (°F) 1500	1600
1	10	63	89
	20	54	91
	30	61	62
2	10	50	80
	20	52	72
	30	59	69
3	10	48	73
	20	74	81
	30	71	69
4	10	54	88
	20	48	92
	30	59	64

mine the effect of temperature and heat treatment time on the strength of normalized steel. Two temperatures and three times are selected. The experiment is performed by heating the oven to a randomly selected temperature, and inserting three specimens. After 10 minutes one specimen is removed, after 20 minutes the second is removed, and after 30 minutes the final specimen is removed. Then the temperature is changed to the other level and the process repeated. Four shifts were required to collect the data, which are shown on page 301. Analyze the data and draw conclusions, assuming both factors are fixed.

12-9. Repeat Problem 12-8, assuming that temperature and time were randomly selected.

12-10. An experiment was designed to study pigment dispersion in paint. Four different mixes of a particular pigment were studied. The procedure consisted of preparing a particular mix, and then applying that mix to a panel by three application methods (brushing, spraying, and rolling). The response measured is the percentage reflectance of pigment. Three days were required to run the experiment, and the data obtained is shown below. Analyze the data and draw conclusions, assuming that mixes and application methods are fixed.

Day	Application Method	Mix 1	2	3	4
	1	64.5	66.3	74.1	66.5
1	2	68.3	69.5	73.8	70.0
	3	70.3	73.1	78.0	72.3
	1	65.2	65.0	73.8	64.8
2	2	69.2	70.3	74.5	68.3
	3	71.2	72.8	79.1	71.5
	1	66.2	66.5	72.3	67.7
3	2	69.0	69.0	75.4	68.6
	3	70.8	74.2	80.1	72.4

12-11. Repeat Problem 12-10 assuming that mixes are random and application methods are fixed.

12-12. Consider the split-split-plot design described in Example 12-2. Suppose this experiment is conducted as described and the data on page 303 obtained. Analyze this data and draw conclusions.

Blocks	Dosage Strengths	Technician								
		1			2			3		
		1	2	3	1	2	3	1	2	3
	Wall Thickness									
	1	95	71	108	96	70	108	95	70	100
1	2	104	82	115	99	84	100	102	81	106
	3	101	85	117	95	83	105	105	84	113
	4	108	85	116	97	85	109	107	87	115
	1	95	78	110	100	72	104	92	69	101
2	2	106	84	109	101	79	102	100	76	104
	3	103	86	116	99	80	108	101	80	109
	4	109	84	110	112	86	109	108	86	113
	1	96	70	107	94	66	100	90	73	98
3	2	105	81	106	100	84	101	97	75	100
	3	106	88	112	104	87	109	100	82	104
	4	113	90	117	121	90	117	110	91	112
	1	90	68	109	98	68	106	98	72	101
4	2	100	84	112	102	81	103	102	78	105
	3	102	85	115	100	85	110	105	80	110
	4	114	88	118	118	85	116	110	95	120

12-13. Rework Problem 12-12 assuming that dosage strengths are chosen at random.

12-14. Suppose that in Problem 12-12 four technicians had been used. Assuming that all factors are fixed, how many blocks should be run to obtain an adequate number of degrees of freedom on the test for differences between technician?

12-15. Consider the experiment described in Example 12-2. Demonstrate how the order in which the treatment combinations are run would be determined if this experiment were run as (a) a split-split-plot, (b) a split-plot, (c) a factorial design in a randomized block, and (d) a completely randomized factorial design.

Chapter Thirteen
Regression Analysis

13-1 INTRODUCTION

In many problems there are two or more variables that are inherently related, and it is necessary to explore the nature of this relationship. For example, in a chemical process the yield of product is related to the operating temperature. It may be of interest to build a model relating yield to temperature, and then use the model for prediction, process optimization, or control purposes.

In general, suppose that there is a single *dependent variable* or response y which depends on k *independent variables*, for example, $x_1, x_2, \ldots, x_k$. The relationship between these variables is characterized by a mathematical model called a *regression equation*. More precisely, we speak of the regression of y on $x_1, x_2, \ldots, x_k$. The regression model is fit to a set of experimental data. In some instances, the experimenter will know the exact form of the true functional relationship between y and $x_1, x_2, \ldots, x_k$, say $y = \phi(x_1, x_2, \ldots, x_k)$. However, in most cases, the true functional relationship is unknown, and the experimenter will choose an appropriate function to approximate ϕ. Polynomial models are widely used as approximating functions, and for that reason we discuss the fitting of polynomials to data in some detail in this chapter.

Regression methods are frequently used to analyze data from *unplanned experiments*, such as might arise from observation of uncontrolled phenomena, or historical records. However, designed experiments may also be used to great advantage in regression analysis, as will be illustrated later.

13-2 SIMPLE LINEAR REGRESSION

We wish to determine the relationship between a single independent variable x and a dependent variable y. The independent variable x is usually assumed to be a continuous variable, controllable by the experimenter. Then if the experiment is *designed*, we choose the values of x and observe the corresponding value of y.

Suppose the true relationship between y and x is a straight line, and that the observation y at each level of x is a random variable. Now, the expected value of y for each value of x is

$$E(y \mid x) = \beta_0 + \beta_1 x \tag{13-1}$$

where the parameters of the straight line, β_0 and β_1, are unknown constants. We assume that each observation, y, can be described by the model

$$y = \beta_0 + \beta_1 x + \epsilon \tag{13-2}$$

where ϵ is a random error with mean zero and variance σ^2. The $\{\epsilon\}$ are also assumed to be uncorrelated random variables. The regression model (13-2) involving only a single independent variable x is often called the *simple* linear regression model.

If we have n pairs of data $(y_1, x_1), (y_2, x_2), \ldots, (y_n, x_n)$ we may estimate the model parameters β_0 and β_1 by least squares. By Equation 13-2 we may write

$$y_j = \beta_0 + \beta_1 x_j + \epsilon_j \qquad j = 1, 2, \ldots, n$$

and the least squares function is

$$L = \sum_{j=1}^{n} \epsilon_j^2 = \sum_{j=1}^{n} (y_j - \beta_0 - \beta_1 x_j)^2 \tag{13-3}$$

Minimizing the least squares function is simplified if we rewrite the model, Equation 13-2, as

$$y = \beta_0' + \beta_1 (x - \bar{x}) + \epsilon \tag{13-4}$$

where $\bar{x} = (1/n) \sum_{j=1}^{n} x_j$ and $\beta_0' = \beta_0 + \beta_1 \bar{x}$. In (13-4) we have simply corrected the independent variable for its mean, resulting in a transformation on the intercept. Equation 13-4 is frequently called the transformed simple linear regression model, or more simply, the transformed model. In addition to simplifying the estimation problem the use of the transformed model allows other inference tasks to be performed more easily.

Employing the transformed model, the least squares function is

$$L = \sum_{j=1}^{n} \{y_j - \beta_0' - \beta_1 (x_j - \bar{x})\}^2 \tag{13-5}$$

The least squares estimators of β_0 and β_1, say $\hat{\beta}_0$ and $\hat{\beta}_1$, must satisfy

$$\left.\frac{\partial L}{\partial \beta_0'}\right|_{\hat{\beta}_0', \hat{\beta}_1} = -2 \sum_{j=1}^{n} [y_j - \hat{\beta}_0' - \hat{\beta}_1(x_j - \bar{x})] = 0$$

$$\left.\frac{\partial L}{\partial \beta_1}\right|_{\hat{\beta}_0', \hat{\beta}_1} = -2 \sum_{j=1}^{n} [y_j - \hat{\beta}_0 - \hat{\beta}_1(x_j - \bar{x})] (x_j - \bar{x}) = 0$$

Simplifying these two equations yields

$$n\hat{\beta}_0' = \sum_{j=1}^{n} y_j$$

$$\hat{\beta}_1 \sum_{j=1}^{n} (x_j - \bar{x})^2 = \sum_{j=1}^{n} y_j (x_j - \bar{x})$$

(13-6)

Equations 13-6 are called the *least squares normal equations*. Their solution is

$$\hat{\beta}_0' = \frac{1}{n} \sum_{j=1}^{n} y_j = \bar{y}$$

(13-7)

$$\hat{\beta}_1 = \frac{\sum_{j=1}^{n} y_j (x_j - \bar{x})}{\sum_{j=1}^{n} (x_j - \bar{x})^2}$$

(13-8)

Thus, $\hat{\beta}_0'$ and $\hat{\beta}_1$ are the least squares estimators of the intercept and slope, respectively.
The fitted simple linear regression model is

$$\hat{y} = \hat{\beta}_0' + \hat{\beta}_1 (x - \bar{x})$$

(13-9)

If the analyst wishes to present his results in terms of the *original* intercept, β_0, then

$$\hat{\beta}_0 = \hat{\beta}_0' - \hat{\beta}_1 \bar{x}$$

and the fitted model is

$$\hat{y} = \hat{\beta}_0 + \hat{\beta}_1 x$$

(13-10)

Notationally, it is convenient to give special symbols to the numerator and denominator of Equation 13-8. That is, let

$$S_{xx} = \sum_{j=1}^{n} (x_j - \bar{x})^2 = \sum_{j=1}^{n} x_j^2 - \frac{\left(\sum_{j=1}^{n} x_j\right)^2}{n}$$

(13-11)

and

$$S_{xy} = \sum_{j=1}^{n} y_j (x_j - \bar{x}) = \sum_{j=1}^{n} x_j y_j - \frac{\left(\sum_{j=1}^{n} x_j\right)\left(\sum_{j=1}^{n} y_j\right)}{n}$$

(13-12)

We call S_{xx} the corrected sum of squares of x and S_{xy} the corrected sum of cross-products of x and y. The extreme right-hand sides of (13-11) and (13-12) are the usual computational formulas. Using this new notation, the least squares estimator of the slope is

$$\hat{\beta}_1 = \frac{S_{xy}}{S_{xx}} \qquad (13\text{-}13)$$

■**EXAMPLE 13-1.** A study was made to determine the effect of stirring rate on the amount of impurity in paint produced by a chemical process. The study yielded the following data:

Stirring rate, RPM (x)	20	22	24	26	28	30	32	34	36	38	40	42
Impurity, % (y)	8.4	9.5	11.8	10.4	13.3	14.8	13.2	14.7	16.4	16.5	18.9	18.5

The straight-line model $y = \beta_0 + \beta_1 x + \epsilon$ was proposed, and the following quantities computed:

$$n = 12, \quad \sum_{j=1}^{12} x_j = 372, \quad \sum_{j=1}^{12} y_j = 166.4, \quad \sum_{j=1}^{12} x_j^2 = 12{,}104$$

$$\bar{y} = 13.87, \quad \bar{x} = 31, \quad \sum_{j=1}^{12} y_j^2 = 2435.14, \quad \sum_{j=1}^{12} y_j x_j = 5419.60$$

From (13-11) and (13-12) we find

$$S_{xx} = 12{,}104 - \frac{(372)^2}{12} = 572.00$$

$$S_{xy} = 5419.60 - \frac{(372)(166.4)}{12} = 261.20$$

Thus,

$$\hat{\beta}_1 = \frac{S_{xy}}{S_{xx}} = \frac{261.20}{572.00} = 0.46$$

$$\hat{\beta}_0' = \bar{y} = 13.87$$

and the fitted model is

$$\hat{y} = \hat{\beta}_0' + \hat{\beta}_1 (x - \bar{x})$$

or

$$\hat{y} = 13.87 + 0.46 (x - 31)$$

If we wish to express the model in terms of the original intercept, then

$$\hat{\beta}_0 = \hat{\beta}_0' - \hat{\beta}_1 \bar{x}$$

$$= 13.87 - (0.46)(31) = -0.29$$

and since $\hat{y} = \hat{\beta}_0 + \hat{\beta}_1 x$, we have

$$\hat{y} = -0.29 + 0.46x$$ ∎

The statistical properties of the least squares estimators are of considerable value in judging the adequacy of the fitted model. The estimators $\hat{\beta}_0'$ and $\hat{\beta}_1$ are random variables, since they are just linear combinations of the y_j, and the y_j are random variables. We will investigate the bias and variance properties of these estimators. Consider first $\hat{\beta}_1$. The expected value of $\hat{\beta}_1$ is

$$E(\hat{\beta}_1) = E\left(\frac{S_{xy}}{S_{xx}}\right)$$

$$= \frac{1}{S_{xx}} E\left[\sum_{j=1}^{n} y_j (x_j - \bar{x})\right]$$

$$= \frac{1}{S_{xx}} E\left[\sum_{j=1}^{n} (\beta_0' + \beta_1 (x_j - \bar{x}) + \epsilon_j)(x_j - \bar{x})\right]$$

$$= \frac{1}{S_{xx}} \left\{E\left[\beta_0' \sum_{j=1}^{n} (x_j - \bar{x})\right] + E\left[\beta_1 \sum_{j=1}^{n} (x_j - \bar{x})^2\right] + E\left[\sum_{j=1}^{n} \epsilon_j(x_j - \bar{x})\right]\right\}$$

$$= \frac{1}{S_{xx}} \beta_1 S_{xx}$$

$$= \beta_1$$

since $\sum_{j=1}^{n} (x_j - \bar{x}) = 0$, and by assumption $E(\epsilon_j) = 0$. Thus $\hat{\beta}_1$ is an *unbiased* estimator of the true slope β_1. Now consider the variance of $\hat{\beta}_1$. Since we have assumed that $V(\epsilon_j) = \sigma^2$, it follows that $V(y_j) = \sigma^2$, and

$$V(\hat{\beta}_1) = V\left(\frac{S_{xy}}{S_{xx}}\right)$$

$$= \frac{1}{S_{xx}^2} V\sum_{j=1}^{n} y_j (x_j - \bar{x}) \tag{13-14}$$

The random variables y_j are uncorrelated because the ϵ_j are uncorrelated. Therefore, the variance of the sum in (13-14) is just the sum of the variances, and the variance of each term in the sum, say $V[y_j(x_j - \bar{x})]$, is $\sigma^2 (x_j - \bar{x})^2$. Thus,

$$V(\hat{\beta}_1) = \frac{1}{S_{xx}^2} \sigma^2 \sum_{j=1}^{n} (x_j - \bar{x})^2$$

$$= \frac{\sigma^2}{S_{xx}} \tag{13-15}$$

By using a similar approach, we can show that

$$E(\hat{\beta}_0') = \beta_0' \qquad V(\hat{\beta}_0') = \frac{\sigma^2}{n}, \tag{13-16}$$

and

$$E(\hat{\beta}_0) = \beta_0 \qquad V(\hat{\beta}_0) = \sigma^2 \left[\frac{1}{n} + \frac{\bar{x}^2}{S_{xx}} \right] \tag{13-17}$$

To find $V(\hat{\beta}_0)$, we must make use of the result $\text{Cov}(\hat{\beta}_0', \hat{\beta}_1) = 0$. However, the covariance of $\hat{\beta}_0$ and $\hat{\beta}_1$ is not zero; in fact $\text{Cov}(\hat{\beta}_0, \hat{\beta}_1) = -\sigma^2 \bar{x}/S_{xx}$. Refer to Problems 13-14 and 13-15. Note that $\hat{\beta}_0'$ and $\hat{\beta}_0$ are unbiased estimators of β_0' and β_0, respectively.

It is usually necessary to obtain an estimate of σ^2. The difference between the observation y_j and the corresponding predicted value $\hat{y}_j$, say $e_j = y_j - \hat{y}_j$, is called a *residual*. The sum of the squares of the residuals, or the error sum of squares, would be

$$SS_E = \sum_{j=1}^{n} e_j^2$$

$$= \sum_{j=1}^{n} (y_j - \hat{y}_j)^2 \tag{13-18}$$

A more convenient computing formula for SS_E may be found by substituting the estimated model $\hat{y}_j = \bar{y} + \hat{\beta}_1 (x_j - \bar{x})$ into (13-18) and simplifying as follows:

$$SS_E = \sum_{j=1}^{n} [y_j - \bar{y} - \hat{\beta}_1 (x_j - \bar{x})]^2$$

$$= \sum_{j=1}^{n} [y_j^2 + \bar{y}^2 + \hat{\beta}_1^2 (x_j - \bar{x})^2 - 2\bar{y}y_j - 2\hat{\beta}_1 y_j (x_j - \bar{x}) - 2\hat{\beta}_1 \bar{y} (x_j - \bar{x})]$$

$$= \sum_{j=1}^{n} y_j^2 + n\bar{y}^2 + \hat{\beta}_1^2 S_{xx} - 2\bar{y} \sum_{j=1}^{n} y_j - 2\hat{\beta}_1 S_{xy} - 2\hat{\beta}_1 \bar{y} \sum_{j=1}^{n} (x_j - \bar{x}) \tag{13-19}$$

The last term in (13-19) is zero, $2\bar{y} \sum_{j=1}^{n} y_j = 2n\bar{y}^2$, and $\hat{\beta}_1^2 S_{xx} = \hat{\beta}_1 (S_{xy}/S_{xx}) S_{xx} = \hat{\beta}_1 S_{xy}$. Therefore, (13-19) becomes

$$SS_E = \sum_{j=1}^{n} y_j^2 - n\bar{y}^2 - \hat{\beta}_1 S_{xy}.$$

But $\sum_{j=1}^{n} y_j^2 - n\bar{y}^2 = \sum_{j=1}^{n} (y_j - \bar{y})^2 \equiv S_{yy}$, say, so we may write SS_E as

$$SS_E = S_{yy} - \hat{\beta}_1 S_{xy} \tag{13-20}$$

By taking expectation of SS_E we may show that $E(SS_E) = (n - 2) \sigma^2$. Therefore,

$$\hat{\sigma}^2 = \frac{SS_E}{n - 2} \equiv MSE \tag{13-21}$$

is an unbiased estimator of σ^2.

Regression analysis is widely used, and frequently *misused*. There are several common abuses of regression that should be briefly mentioned. Care should be taken in selecting variables with which to construct regression models, and in determining the form of the approximating function. It is quite possible to develop relationships among variables that are completely meaningless in a practical sense. For example, one might attempt to relate the shear strength of spot welds with the number of boxes of punched cards used by the data processing department. A straight line may even appear to provide a good fit to the data, but the relationship is unreasonable and cannot be relied on. The analyst familiar with the process under study must be the final judge of such functional relationships.

Regression relationships are valid only for values of the independent variable within the range of the original data. The linear relationship that we have tentatively assumed may be valid over the original range of x but it is unlikely to remain so as we encounter x values beyond that range. In other words, as we move beyond the original range of x we become less certain about the validity of the assumed model. Regression models should never be used for extrapolation.

Finally, one occasionally feels that the model $y = \beta x + \epsilon$ is appropriate. The omission of the intercept from this model implies, of course, that $y = 0$ when $x = 0$. This is a very strong assumption that often is unjustified. Even when two variables, such as the height and weight of men, would seem to qualify for the use of this model, we would usually obtain a better fit by including the intercept because of the limited range of data concerning the independent variable.

13-3 HYPOTHESIS TESTING IN SIMPLE LINEAR REGRESSION

To test hypotheses about the slope and intercept of the regression model, we must make an additional assumption about the error term, that is, that the ϵ_j are normally distributed. Thus, we are assuming that the errors ϵ_j are NID($0, \sigma^2$). Later we will discuss how these assumptions can be checked through *residual analysis*.

Suppose the experimenter wishes to test the hypothesis that the slope equals some value, for example, β_{10}. The appropriate hypotheses are

$$H_0: \beta_1 = \beta_{10}$$
$$H_1: \beta_1 \neq \beta_{10}$$

(13-22)

where we have specified a two-sided alternative. Now since the ϵ_j are NID($0, \sigma^2$) it follows directly that the observations y_j are NID($\beta_0 + \beta_1 x_j, \sigma^2$). Thus $\hat{\beta}_1$ is a linear combination of independent normal random variables, and consequently $\hat{\beta}_1$ is $N(\beta_1, \sigma^2/S_{xx})$, using the bias and variance properties of $\hat{\beta}_1$ from Section 13-2. Furthermore, $\hat{\beta}_1$ is independent of MS_E. Then as a result of the normality assumption, the statistic

$$t_0 = \frac{\hat{\beta}_1 - \hat{\beta}_{10}}{\sqrt{MS_E/S_{xx}}}$$

(13-23)

follows the t distribution with $n - 2$ degrees of freedom under H_0: $\beta_1 = \beta_{10}$. We would reject H_0: $\beta_1 = \beta_{10}$ if

$$|t_0| > t_{\alpha/2, n-2} \tag{13-24}$$

where t_0 is computed from Equation 13-23.

A similar procedure can be used to test hypotheses about the intercept. To test

$$H_0: \beta_0 = \beta_{00}$$
$$H_1: \beta_0 \neq \beta_{00} \tag{13-25}$$

we would use the statistic

$$t_0 = \frac{\hat{\beta}_0 - \beta_{00}}{\sqrt{MS_E \left(\dfrac{1}{n} + \dfrac{\bar{x}^2}{S_{xx}} \right)}} \tag{13-26}$$

and reject the null hypothesis if $|t_0| > t_{\alpha/2, n-2}$.

A very important special case of the hypotheses (13-22) is

$$H_0: \beta_1 = 0$$
$$H_1: \beta_1 \neq 0 \tag{13-27}$$

The hypothesis H_0: $\beta_1 = 0$ relates to the *significance* of regression. Failing to reject H_0: $\beta_1 = 0$ is equivalent to concluding that there is no *linear* relationship between x and y, that is, the best estimator of y_j for any x_j is $\hat{y}_j = \bar{y}$. In many instances this may mean that there is no *causal* relationship between x and y, or that the true relationship is not linear.

The test procedure for H_0: $\beta_1 = 0$ may be developed from two approaches. The first approach starts with the following partitioning of the total corrected sum of squares for y:

$$S_{yy} \equiv \sum_{j=1}^{n} (y_j - \bar{y})^2 = \sum_{j=1}^{n} (\hat{y}_j - \bar{y})^2 + \sum_{j=1}^{n} (y_j - \hat{y}_j)^2 \tag{13-28}$$

The two components of S_{yy} measure, respectively, the amount of variability in the y_j accounted for by the regression line and the residual variation left unexplained by the regression line. We usually define $SS_E = \sum_{j=1}^{n} (y_j - \hat{y}_j)^2$ as the *error* sum of squares and $SS_R = \sum_{j=1}^{n} (\hat{y}_j - \bar{y})^2$ as the *regression* sum of squares. Thus (13-28) may be written as

$$S_{yy} = SS_R + SS_E \tag{13-29}$$

From Equation 13-20, we obtain the computing formula for SS_R as

$$SS_R = \hat{\beta}_1 S_{xy} \tag{13-30}$$

S_{yy} has $n - 1$ degrees of freedom, and SS_R and SS_E have 1 and $n - 2$ degrees of freedom, respectively.

We may show that $E[SS_E/(n-2)] = \sigma^2$, that $E[SS_R/1] = \sigma^2 + \beta_1^2 S_{xx}$, and that SS_E and SS_R are independent. Thus, if $H_0: \beta_1 = 0$ is true, the statistic

$$F_0 = \frac{SS_R/1}{SS_E/(n-2)} = \frac{MS_R}{MS_E} \qquad (13\text{-}31)$$

follows the $F_{1,n-2}$ distribution, and we would reject H_0 if $F_0 > F_{\alpha,1,n-2}$. The test procedure is usually arranged in an analysis of variance table, such as Table 13-1.

The test for significance of regression may also be developed from Equation 13-23 with $\beta_{10} = 0$, say

$$t_0 = \frac{\hat{\beta}_1}{\sqrt{MS_E/S_{xx}}} \qquad (13\text{-}32)$$

Squaring both sides of (13-32) we obtain

$$t_0^2 = \frac{\hat{\beta}_1^2 S_{xx}}{MS_E} = \frac{\hat{\beta}_1 S_{xy}}{MS_E} = \frac{MS_R}{MS_E} \qquad (13\text{-}33)$$

Table 13-1 Analysis of Variance for Testing Significance of Regression

Source of Variation	Sum of Squares	Degrees of Freedom	Mean Square	F_0
Regression	$SS_R = \hat{\beta}_1 S_{xy}$	1	MS_R	MS_R/MS_E
Error or residual	$SS_E = S_{yy} - \hat{\beta}_1 S_{xy}$	$n-2$	MS_E	
Total	S_{yy}	$n-1$		

Note that t_0^2 in (13-33) is identical to F_0 in (13-31). It is true in general that the square of a t random variable with f degrees of freedom is an F random variable with one and f degrees of freedom in the numerator and denominator, respectively. Thus the test using t_0^2 is identical to the test based on F_0.

■**EXAMPLE 13-2.** For the data given in Example 13-1, we will test for significance of regression. The fitted model is $\hat{y} = -0.29 + 0.46x$, and S_{yy} is computed as

$$S_{yy} = \sum_{j=1}^{n} y_j^2 - \frac{\left(\sum_{j=1}^{n} y_j\right)^2}{12}$$

$$= 2435.14 - \frac{(166.4)^2}{12} = 127.73$$

The regression sum of squares is

$$SS_R = \hat{\beta}_1 S_{xy} = (0.46)(261.20) = 120.15$$

Thus, the error sum of squares is

$$SS_E = S_{yy} - SS_R$$

$$= 127.73 - 120.15 = 7.58$$

The analysis of variance for testing H_0: $\beta_1 = 0$ is summarized in Table 13-2. Since $F_{.01,1,10} = 10.0$ we reject H_0 and conclude that $\beta_1 \neq 0$.   ∎

Table 13-2   Analysis of Variance for Example 13-2

Source of Variation	Sum of Squares	Degrees of Freedom	Mean Square	F_0
Regression	120.15	1	120.15	158.51
Error	7.58	10	0.758	
Total	127.73	11		

13-4   INTERVAL ESTIMATION IN SIMPLE LINEAR REGRESSION

In addition to point estimates of the slope and intercept, it is possible to obtain interval estimates of these parameters. If the c_j are normally and independently distributed, then

$$(\hat{\beta}_1 - \beta_1)/\sqrt{MS_E/S_{xx}} \quad \text{and} \quad (\hat{\beta}_0 - \beta_0)\Big/\sqrt{MS_E\left(\frac{1}{n} + \frac{\bar{x}^2}{S_{xx}}\right)}$$

are both distributed as t with $n - 2$ degrees of freedom. Therefore, a $100(1 - \alpha)$ percent confidence interval on β_1 is given by

$$\left[\hat{\beta}_1 \pm t_{\alpha/2,n-2}\sqrt{\frac{MS_E}{S_{xx}}}\right] \tag{13-34}$$

Similarly, a $100(1 - \alpha)$ percent confidence interval on β_0 is

$$\left[\hat{\beta}_0 \pm t_{\alpha/2,n-2}\sqrt{MS_E\left(\frac{1}{n} + \frac{\bar{x}^2}{S_{xx}}\right)}\right] \tag{13-35}$$

As an illustration, a 95 percent confidence interval for β_1 for the data in Example 13-1 would be found from (13-34) as

$$\left[0.46 \pm (2.228)\sqrt{\frac{0.758}{572.00}}\right]$$

or

$$[0.46 \pm 0.08]$$

Thus, the 95 percent confidence interval for β_1 is $0.38 \leqslant \beta_1 \leqslant 0.54$.

A confidence interval may be constructed for the mean response at a specified x, for example, x_0. This is a confidence interval about $E(y \mid x_0)$ and is often called a confidence interval about the regression line. Since $E(y \mid x_0) = \beta'_0 + \beta_1 (x_0 - \bar{x})$, we may obtain a point estimate of $E(y \mid x_0)$ from the fitted model as

$$\widehat{E(y \mid x_0)} \equiv \hat{y}_0 = \hat{\beta}'_0 + \hat{\beta}_1 (x_0 - \bar{x})$$

Now it is clear that $E(\hat{y}_0) = \beta'_0 + \beta_1 (x_0 - \bar{x})$, since $\hat{\beta}'_0$ and $\hat{\beta}_1$ are unbiased, and furthermore

$$V(\hat{y}_0) = \sigma^2 \left[\frac{1}{n} + \frac{(x_0 - \bar{x})^2}{S_{xx}} \right]$$

since $\text{Cov}(\hat{\beta}'_0, \hat{\beta}_1) = 0$. Also, $\hat{y}_0$ is normally distributed, as $\hat{\beta}_0$ and $\hat{\beta}_1$ are normally distributed. Therefore, a $100(1 - \alpha)$ percent confidence interval about the true regression line at $x = x_0$ may be computed from

$$\left[\hat{y}_0 \pm t_{\alpha/2, n-2} \sqrt{MS_E \left(\frac{1}{n} + \frac{(x_0 - \bar{x})^2}{S_{xx}} \right)} \right] \tag{13-36}$$

Notice that the width of the confidence interval for $E(y \mid x_0)$ is a function of x_0. The interval width is a minimum for $x_0 = \bar{x}$ and widens as $|x_0 - \bar{x}|$ increases.

■**EXAMPLE 13-3.** We shall construct a 95 percent confidence interval about the regression line for the data in Example 13-1. Since $\hat{y}_0 = -0.29 + 0.46 x_0$, the 95 percent confidence interval is given by

$$\left[\hat{y}_0 \pm 2.228 \sqrt{(0.758) \left(\frac{1}{12} + \frac{(x_0 - 31)^2}{572.00} \right)} \right]$$

The predicted values and the 95 percent confidence limits for $x_0 = x_i, i = 1, 2, \ldots, 12$ are displayed in Table 13-3. To demonstrate the use of this table, we could find the 95 percent confidence interval on the true regression line at $x_0 = 26$ (say) as

$$11.67 - 0.69 \leqslant E(y \mid x_0 = 26) \leqslant 11.67 + 0.69$$

Table 13-3 Confidence Limits for Example 13-3

x_0	20	22	24	26	28	30	32	34	36	38	40	42
$\hat{y}_0$	8.91	9.83	10.75	11.67	12.59	13.51	14.43	15.35	16.27	17.19	18.11	19.03
95 percent confidence limits	±1.05	±0.92	±0.80	±0.69	±0.61	±0.57	±0.57	±0.61	±0.69	±0.80	±0.92	±1.05

Figure 13-1. Fitted model and the 95 percent confidence interval for Example 13-3.

or

$$10.98 \leqslant E(y \mid x_0 = 26) \leqslant 12.35$$

The fitted model and the 95 percent confidence interval about the true regression line is shown in Figure 13-1.   ∎

Another useful concept in simple linear regression is the *prediction interval*, an interval estimate on the mean of k *future* observations at a particular value of x, say x_0. To illustrate the prediction interval, suppose that in Example 13-1 the analyst wishes to construct an interval estimate on the mean impurity of the next four batches of paint processed with stirring rate $x_0 = 34$. A confidence interval is inadequate, because it refers to the true mean impurity (an unknown constant) and not future determinations of this quantity.

Let the jth future observation on the response at x_0 be denoted by y_{0j}, and denote the mean of these values by

$$\bar{y}_0 = \frac{1}{k} \sum_{j=1}^{k} y_{0j}.$$

The best point estimator of $\bar{y}_0$ is $\hat{y}_0 = \hat{\beta}'_0 + \hat{\beta}_1 (x_0 - \bar{x})$. The random variable

$$\psi = \bar{y}_0 - \hat{y}_0$$

is normally distributed with mean zero and variance

$$V(\psi) = V(\bar{y}_0 - \hat{y}_0)$$

$$= \sigma^2 \left[\frac{1}{k} + \frac{1}{n} + \frac{(x_0 - \bar{x})^2}{S_{xx}} \right]$$

because $\bar{y}_0$ is independent of $\hat{y}_0$. Thus the $100 (1 - \alpha)$ percent prediction interval on the mean of k future observations at x_0 is

$$\left[\hat{y}_0 \pm t_{\alpha/2, n-2} \sqrt{MS_E \left(\frac{1}{k} + \frac{1}{n} + \frac{(x_0 - \bar{x})^2}{S_{xx}} \right)} \right]. \qquad (13\text{-}37)$$

Notice that the prediction interval is of minimum width at $x_0 = \bar{x}$, and widens as $|x_0 - \bar{x}|$ increases. Furthermore, if $k = 1$, then (13-38) yields a prediction interval on a single future observation at x_0. By comparing (13-37) with (13-36), we observe that the prediction interval at x_0 is always wider than the confidence interval at x_0. This results from the prediction interval depending on both the error from the fitted model and the error associated with future observations.

To illustrate the construction of a prediction interval, we may use the data in Example 13-1 and find a 95 percent prediction interval on the mean impurity of the next two batches of paint produced at $x_0 = 34$ from

$$\left[15.35 \pm 2.228 \sqrt{(0.758) \left(\frac{1}{2} + \frac{1}{12} + \frac{(34-31)^2}{572.00} \right)} \right].$$

This calculation yields $[15.35 \pm 1.50]$. Thus, the 95 percent prediction interval for $k = 2$ at $x_0 = 34$ is $13.85 \leqslant \bar{y}_0 \leqslant 16.85$.

13-5 MULTIPLE LINEAR REGRESSION

Many regression problems involve more than one independent variable. For example, the yield of a chemical process may depend on temperature, pressure, and the concentration of catalyst. In this case, at least three independent variables will be necessary.

The general problem of fitting the model

$$y = \beta_0 + \beta_1 x_1 + \beta_2 x_2 + \cdots + \beta_k x_k + \epsilon \qquad (13\text{-}38)$$

is called the multiple linear regression problem. The unknown parameters $\{\beta_i\}$ are usually called regression coefficients. Note that the model (13-38) describes a hyperplane in the k-dimensional space of the independent variables $\{x_i\}$.

The method of least squares is used to estimate the regression coefficients in (13-38). Suppose that $n > k$ observations are available, and let x_{ij} denote the jth observation or level of variable x_i. The data will appear as in Table 13-4. The estimation procedure requires that the random error component have $E(\epsilon) = 0$, $V(\epsilon) = \sigma^2$, and that the $\{\epsilon\}$ are uncorrelated.

Table 13-4 Data For Multiple Linear Regression

y	x_1	x_2	$\cdots$	x_k
y_1	x_{11}	x_{21}	$\cdots$	x_{k1}
y_2	x_{12}	x_{22}	$\cdots$	x_{k2}
$\vdots$	$\vdots$	$\vdots$		$\vdots$
y_n	x_{1n}	x_{2n}	$\cdots$	x_{kn}

We may write the model in terms of the data as

$$
\begin{aligned}
y_j &= \beta_0 + \beta_1 x_{1j} + \beta_2 x_{2j} + \cdots + \beta_k x_{kj} + \epsilon_j \\
&= \beta_0 + \sum_{i=1}^{k} \beta_i x_{ij} + \epsilon_j \qquad j = 1, 2, \ldots, n
\end{aligned}
\tag{13-39}
$$

As in the simple linear regression case, the intercept is redefined as

$$
\beta_0' = \beta_0 + \beta_1 \bar{x}_1 + \beta_2 \bar{x}_2 + \cdots + \beta_k \bar{x}_k
\tag{13-40}
$$

where $\bar{x}_i = (1/n) \sum_{j=1}^{n} x_{ij}$ is the average level for the ith variable. The model now becomes

$$
y_j = \beta_0' + \sum_{i=1}^{k} \beta_i (x_{ij} - \bar{x}_i) + \epsilon_j \qquad j = 1, 2, \ldots, n
\tag{13-41}
$$

and the least squares function is

$$
L = \sum_{j=1}^{n} \left[y_j - \beta_0' - \sum_{i=1}^{k} \beta_i (x_{ij} - \bar{x}_i) \right]^2
\tag{13-42}
$$

It will be convenient to define

$$
S_{ii} = \sum_{j=1}^{n} (x_{ij} - \bar{x}_i)^2 = \sum_{j=1}^{n} x_{ij}^2 - \frac{\left(\sum_{j=1}^{n} x_{ij} \right)^2}{n} \qquad i = 1, 2, \ldots, k
$$

$$
S_{rs} = S_{sr} = \sum_{j=1}^{n} (x_{rj} - \bar{x}_r)(x_{sj} - \bar{x}_s) = \sum_{j=1}^{n} x_{rj} x_{sj} - \frac{\left(\sum_{j=1}^{n} x_{rj} \right)\left(\sum_{j=1}^{n} x_{sj} \right)}{n} \qquad r \neq s \tag{13-43}
$$

$$
S_{iy} = \sum_{j=1}^{n} y_j (x_{ij} - \bar{x}_i) = \sum_{j=1}^{n} y_j x_{ij} - \frac{\left(\sum_{j=1}^{n} y_j \right)\left(\sum_{j=1}^{n} x_{ij} \right)}{n} \qquad i = 1, 2, \ldots, k
$$

Note that S_{ii} is the corrected sum of squares of the ith independent variable, S_{rs} is the corrected sum of cross-products between x_r and x_s, and S_{iy} is the corrected sum of cross-products between x_i and y.

The least squares estimators of $\beta_0', \beta_1, \ldots, \beta_k$ must satisfy

$$\left. \frac{\partial L}{\partial \beta_0'} \right|_{\hat{\beta}_0', \hat{\beta}_1, \ldots, \hat{\beta}_k} = -2 \left[\sum_{j=1}^{n} y_j - \hat{\beta}_0' - \sum_{u=1}^{k} \hat{\beta}_u (x_{uj} - \bar{x}_u) \right] = 0 \qquad (13\text{-}44)$$

and

$$\left. \frac{\partial L}{\partial \beta_i} \right|_{\hat{\beta}_0', \hat{\beta}_1, \ldots, \hat{\beta}_k} = -2 \left[\sum_{j=1}^{n} y_j - \hat{\beta}_0' - \sum_{u=1}^{k} \hat{\beta}_u (x_{uj} - \bar{x}_u) \right] (x_{ij} - \bar{x}_i)$$

$$= 0, \quad i = 1, 2, \ldots, k \qquad (13\text{-}45)$$

Simplifying (13-44) and (13-45), and invoking (13-43), we obtain the least squares normal equations

$$n\hat{\beta}_0' = \sum_{j=1}^{n} y_j$$

$$\hat{\beta}_1 S_{i1} + \hat{\beta}_2 S_{i2} + \cdots + \hat{\beta}_k S_{ik} = S_{iy} \qquad i = 1, 2, \ldots, k \qquad (13\text{-}46)$$

Note that there are $p = k + 1$ normal equations, one for each of the unknown regression coefficients. The solution to the normal equations will be the least squares estimators $\hat{\beta}_0', \hat{\beta}_1, \ldots, \hat{\beta}_k$.

It is simpler to solve the normal equations if they are first expressed in matrix notation. We present a matrix development of the normal equations that parallels the development of (13-46). The model in terms of the observations, Equation 13-39, may be written in matrix notation as

$$\mathbf{y} = \mathbf{X}\boldsymbol{\beta} + \boldsymbol{\epsilon}$$

where

$$\mathbf{y} = \begin{bmatrix} y_1 \\ y_2 \\ \vdots \\ y_n \end{bmatrix} \qquad \mathbf{X} = \begin{bmatrix} 1 & (x_{11} - \bar{x}_1) & (x_{21} - \bar{x}_2) & \cdots & (x_{k1} - \bar{x}_k) \\ 1 & (x_{12} - \bar{x}_1) & (x_{22} - \bar{x}_2) & \cdots & (x_{k2} - \bar{x}_k) \\ \vdots & \vdots & \vdots & & \vdots \\ 1 & (x_{1n} - \bar{x}_1) & (x_{2n} - \bar{x}_2) & \cdots & (x_{kn} - \bar{x}_k) \end{bmatrix}$$

$$\boldsymbol{\beta} = \begin{bmatrix} \beta_0' \\ \beta_1 \\ \vdots \\ \beta_k \end{bmatrix} \qquad \text{and} \qquad \boldsymbol{\epsilon} = \begin{bmatrix} \epsilon_1 \\ \epsilon_2 \\ \vdots \\ \epsilon_n \end{bmatrix}$$

In general, $\mathbf{y}$ is an $(n \times 1)$ vector of the observations, $\mathbf{X}$ is an $(n \times p)$ matrix of the levels of the independent variables, $\boldsymbol{\beta}$ is a $(p \times 1)$ vector of the regression coefficients, and $\boldsymbol{\epsilon}$ is an $(n \times 1)$ vector of random errors.

We wish to find the vector of least squares estimators, $\hat{\boldsymbol{\beta}}$, which minimizes

$$L = \sum_{j=1}^{n} \epsilon_j^2 = \boldsymbol{\epsilon}'\boldsymbol{\epsilon} = (\mathbf{y} - \mathbf{X}\boldsymbol{\beta})' \, (\mathbf{y} - \mathbf{X}\boldsymbol{\beta})$$

Note that L may be expressed as

$$L = \mathbf{y}'\mathbf{y} - \boldsymbol{\beta}'\mathbf{X}'\mathbf{y} - \mathbf{y}'\mathbf{X}\boldsymbol{\beta} + \boldsymbol{\beta}'\mathbf{X}'\mathbf{X}\boldsymbol{\beta}$$
$$= \mathbf{y}'\mathbf{y} - 2\boldsymbol{\beta}'\mathbf{X}'\mathbf{y} + \boldsymbol{\beta}'\mathbf{X}'\mathbf{X}\boldsymbol{\beta} \tag{13-47}$$

since $\boldsymbol{\beta}'\mathbf{X}'\mathbf{y}$ is a (1×1) matrix, or a scalar, and its transpose $(\boldsymbol{\beta}'\mathbf{X}'\mathbf{y})' = \mathbf{y}'\mathbf{X}\boldsymbol{\beta}$ is the same scalar. The least squares estimators must satisfy

$$\left. \frac{\partial L}{\partial \boldsymbol{\beta}} \right|_{\hat{\boldsymbol{\beta}}} = -2\mathbf{X}'\mathbf{y} + 2\mathbf{X}'\mathbf{X}\hat{\boldsymbol{\beta}} = 0$$

which simplifies to

$$\mathbf{X}'\mathbf{X}\hat{\boldsymbol{\beta}} = \mathbf{X}'\mathbf{y} \tag{13-48}$$

Equations 13-48 are the least squares normal equations. They are identical to Equations 13-46. To solve the normal equations multiply both sides of (13-48) by the inverse of $\mathbf{X}'\mathbf{X}$. Thus the least squares estimator of $\boldsymbol{\beta}$ is

$$\hat{\boldsymbol{\beta}} = (\mathbf{X}'\mathbf{X})^{-1}\mathbf{X}'\mathbf{y} \tag{13-49}$$

It is easy to see that the matrix form of the normal equations is identical to the scalar form. Writing out Equation 13-48 in detail we obtain, in general,

$$
\begin{bmatrix}
n & 0 & 0 & \cdots & 0 \\
0 & S_{11} & S_{12} & \cdots & S_{1k} \\
0 & S_{12} & S_{22} & \cdots & S_{2k} \\
\cdot & \cdot & \cdot & & \cdot \\
\cdot & \cdot & \cdot & & \cdot \\
\cdot & \cdot & \cdot & & \cdot \\
0 & S_{1k} & S_{2k} & \cdots & S_{kk}
\end{bmatrix}
\begin{bmatrix}
\hat{\beta}_0 \\
\hat{\beta}_1 \\
\hat{\beta}_2 \\
\cdot \\
\cdot \\
\cdot \\
\hat{\beta}_k
\end{bmatrix}
=
\begin{bmatrix}
\sum_{j=1}^{n} y_j \\
S_{1y} \\
S_{2y} \\
\cdot \\
\cdot \\
\cdot \\
S_{ky}
\end{bmatrix}
$$

If the indicated matrix multiplication is carried out, the scalar form of the normal equations will result. In this form it is easy to see that $\mathbf{X}'\mathbf{X}$ is a $(p \times p)$ symmetric matrix and $\mathbf{X}'\mathbf{y}$ is a $(p \times 1)$ column vector. Note the special structure of $\mathbf{X}'\mathbf{X}$. The diagonal elements of $\mathbf{X}'\mathbf{X}$ are the sums of squares of the columns of $\mathbf{X}$, and the off-diagonal elements are the sums of cross-products of columns of $\mathbf{X}$.

The statistical properties of the least squares estimator $\hat{\boldsymbol{\beta}}$ may be easily investigated. Consider first bias:

$$
\begin{aligned}
E(\hat{\boldsymbol{\beta}}) &= E[(\mathbf{X}'\mathbf{X})^{-1}\mathbf{X}'\mathbf{y}] \\
&= E[(\mathbf{X}'\mathbf{X})^{-1}\mathbf{X}'(\mathbf{X}\boldsymbol{\beta}+\boldsymbol{\epsilon})] \\
&= E[(\mathbf{X}'\mathbf{X})^{-1}\mathbf{X}'\mathbf{X}\boldsymbol{\beta}+(\mathbf{X}'\mathbf{X})^{-1}\mathbf{X}'\boldsymbol{\epsilon}] \\
&= \boldsymbol{\beta}
\end{aligned}
$$

since $E(\boldsymbol{\epsilon}) = 0$ and $(\mathbf{X}'\mathbf{X})^{-1}\mathbf{X}'\mathbf{X} = \mathbf{I}$. Thus, $\hat{\boldsymbol{\beta}}$ is an unbiased estimator of $\boldsymbol{\beta}$.

The variance property of $\hat{\boldsymbol{\beta}}$ is expressed in the *covariance matrix*

$$
\text{Cov}\,(\hat{\boldsymbol{\beta}}) \equiv E\{[\hat{\boldsymbol{\beta}} - E(\hat{\boldsymbol{\beta}})]\,[\hat{\boldsymbol{\beta}} - E(\hat{\boldsymbol{\beta}})]'\} \tag{13-50}
$$

which is just a symmetric matrix whose iith element is the variance of $\hat{\beta}_i$ and whose (i, j)th element is the covariance between $\hat{\beta}_i$ and $\hat{\beta}_j$. The covariance matrix of $\hat{\boldsymbol{\beta}}$ is

$$
\text{Cov}\,(\hat{\boldsymbol{\beta}}) = \sigma^2 (\mathbf{X}'\mathbf{X})^{-1} \tag{13-51}
$$

■**EXAMPLE 13-4.** A beer distributor is analyzing the delivery system for his product. Specifically, he is interested in predicting the amount of time required to service a retail outlet. The industrial engineer in charge of the study has suggested that the two most important factors influencing the delivery time are the number of cases of beer delivered and the maximum distance the deliveryman must travel. The engineer has collected some delivery time data, which are shown in Table 13-5.

Table 13-5 Delivery Time Data for Example 13-4

Number of Cases x_1	Distance x_2	Time (minutes) y
10	30	24
15	25	27
10	40	29
20	18	31
25	22	25
18	31	33
12	26	26
14	34	28
16	29	31
22	37	39
24	20	33
17	25	30
13	27	25
30	23	42
24	33	40

We fit the multiple linear regression model

$$y = \beta_0' + \beta_1 (x_1 - \bar{x}_1) + \beta_2 (x_2 - \bar{x}_2) + \epsilon$$

to these data. Since $\bar{x}_1 = 18$ and $\bar{x}_2 = 28$, we may write the model in matrix terms as

$$
\begin{bmatrix}
24 \\ 27 \\ 29 \\ 31 \\ 25 \\ 33 \\ 26 \\ 28 \\ 31 \\ 39 \\ 33 \\ 30 \\ 25 \\ 42 \\ 40
\end{bmatrix}
=
\begin{matrix}
\beta_0' \quad \beta_1 \quad \beta_2 \\
\begin{bmatrix}
1 & -8 & 2 \\
1 & -3 & -3 \\
1 & -8 & 12 \\
1 & 2 & -10 \\
1 & 7 & -6 \\
1 & 0 & 3 \\
1 & -6 & -2 \\
1 & -4 & 6 \\
1 & -2 & 1 \\
1 & 4 & 9 \\
1 & 6 & -8 \\
1 & -1 & -3 \\
1 & -5 & -1 \\
1 & 12 & -5 \\
1 & 6 & 5
\end{bmatrix}
\end{matrix}
\begin{bmatrix}
\beta_0' \\ \beta_1 \\ \beta_2
\end{bmatrix}
+
\begin{bmatrix}
\epsilon_1 \\ \epsilon_2 \\ \epsilon_3 \\ \epsilon_4 \\ \epsilon_5 \\ \epsilon_6 \\ \epsilon_7 \\ \epsilon_8 \\ \epsilon_9 \\ \epsilon_{10} \\ \epsilon_{11} \\ \epsilon_{12} \\ \epsilon_{13} \\ \epsilon_{14} \\ \epsilon_{15}
\end{bmatrix}
$$

where for clarity we have labeled each column of the **X** matrix to indicate the independent variable associated with that column. It is easy to verify that

$$
\mathbf{X'X} = \begin{bmatrix} 15 & 0 & 0 \\ 0 & 504 & -213 \\ 0 & -213 & 548 \end{bmatrix}
\qquad
\mathbf{X'y} = \begin{bmatrix} 463 \\ 345 \\ 63 \end{bmatrix}
$$

and

$$
\mathbf{(X'X)^{-1}} = \begin{bmatrix} (1/15) & 0 & 0 \\ 0 & 0.0024 & 0.0009 \\ 0 & 0.0009 & 0.0022 \end{bmatrix}
$$

Thus the vector of least squares estimators is

$$
\hat{\boldsymbol{\beta}} = \begin{bmatrix} \hat{\beta}_0' \\ \hat{\beta}_1 \\ \hat{\beta}_2 \end{bmatrix} = \mathbf{(X'X)^{-1} X'y} = \begin{bmatrix} 30.8667 \\ 0.8847 \\ 0.4491 \end{bmatrix}
$$

The final regression model is

$$\hat{y} = 30.8667 + 0.8847\,(x_1 - 18) + 0.4491\,(x_2 - 28)$$ ■

■**EXAMPLE 13-5.** **A Designed Experiment in Regression Analysis.** The yield of a chemical process is being studied. The chemical engineer has chosen three factors (temperature, pressure, and percent concentration), each at a high and low level. Eight runs corresponding to the treatment combinations associated with a 2^3 design are made, and the resulting yields are shown in Table 13-6. Note that the high and low levels of each factor have been coded to an arbitrary $[-1, +1]$ interval.

Table 13-6   Data For Example 13-5

y (% yield)	x_1 (temp)	x_2 (pressure)	x_3 (% concentration)
32	-1	-1	-1
36	-1	-1	1
57	-1	1	-1
46	1	-1	-1
65	1	1	-1
57	-1	1	1
48	1	-1	1
68	1	1	1

We fit the regression model

$$y = \beta_0 + \beta_1 x_1 + \beta_2 x_2 + \beta_3 x_3 + \epsilon$$

Notice that in this problem

$$\mathbf{X'X} = \begin{bmatrix} 8 & 0 & 0 & 0 \\ 0 & 8 & 0 & 0 \\ 0 & 0 & 8 & 0 \\ 0 & 0 & 0 & 8 \end{bmatrix} = 8\mathbf{I}_4 \qquad \mathbf{X'y} = \begin{bmatrix} 409 \\ 45 \\ 85 \\ 9 \end{bmatrix}$$

Since $\mathbf{X'X}$ is *diagonal*, the required universe is just $(\mathbf{X'X})^{-1} = (1/8)\mathbf{I}_4$, and the least squares estimators are

$$\hat{\boldsymbol{\beta}} = \begin{bmatrix} \hat{\beta}_0 \\ \hat{\beta}_1 \\ \hat{\beta}_2 \\ \hat{\beta}_3 \end{bmatrix} = \begin{bmatrix} 51\frac{1}{8} \\ 5\frac{5}{8} \\ 10\frac{5}{8} \\ 1\frac{1}{8} \end{bmatrix}$$

Thus the fitted model is

$$\hat{y} = 51\tfrac{1}{8} + 5\tfrac{5}{8}\,x_1 + 10\tfrac{5}{8}\,x_2 + 1\tfrac{1}{8}\,x_3$$ ■

In this example the required inverse is very easy to obtain, since $X'X$ is diagonal. Intuitively, this seems to be advantageous, not only because of the computational simplicity, but because the estimators of all regression coefficients are uncorrelated; that is, Cov $(\hat{\beta}_i, \hat{\beta}_j) = 0$. If we can choose the levels of x_{ij} before the data are collected, we might wish to design the experiment so that a diagonal $X'X$ will result.

In practice it may be relatively easy to do this. We know that the off-diagonal elements in $X X$ are the sums of cross-products of columns in X. Therefore, we must make the dot product of the columns of X equal zero; that is, these columns must be *orthogonal*. Experimental designs that have this property for fitting a regression model such as Equation 13-38 are called orthogonal designs. In general, the 2^k factorial design is an orthogonal design for fitting the multiple linear regression model. Other designs for fitting polynomials are discussed in Chapter 14.

13-6 HYPOTHESIS TESTING IN MULTIPLE LINEAR REGRESSION

In multiple linear regression the experimenter frequently wishes to test hypotheses about the model parameters. This will require the additional assumption that the errors are NID $(0, \sigma^2)$. As a direct result of this assumption the observations y_j are NID $(\beta_0 + \Sigma_{i=1}^{k} \beta_i x_{ij}, \sigma^2)$.

Consider testing for significance of regression. In multiple linear regression this is accomplished by testing

$$H_0: \beta_1 = \beta_2 = \cdots = \beta_k = 0$$
$$H_1: \beta_i \neq 0 \text{ at least one } i$$

(13-52)

Rejection of H_0 in (13-52) implies that at least one variable in the model contributes significantly to the fit. The test procedure for (13-52) is a generalization of the procedure used in simple linear regression. The total sum of squares S_{yy} is partitioned into regression and error sums of squares,

$$S_{yy} = SS_R + SS_E$$

and if $H_0: \beta_i = 0$ is true, then $SS_R/\sigma^2 \sim \chi_k^2$, where the number of degrees of freedom for χ^2 are equal to the number of independent variables in the model. Also, we can show that $SS_E/\sigma^2 \sim \chi_{n-k-1}^2$, and SS_E and SS_R are independent. Therefore, the test procedure for $H_0: \beta_i = 0$ is to compute

$$F_0 = \frac{SS_R/k}{SS_E/(n-k-1)} = \frac{MS_R}{MS_E}$$

(13-53)

and to reject H_0 if $F_0 > F_{\alpha, k, n-k-1}$. The procedure is usually summarized in an analysis of variance table such as Table 13-7.

Table 13-7 Analysis of Variance for Significance of Regression in Multiple Linear Regression

Source of Variation	Sum of Squares	Degrees of Freedom	Mean Square	F_0
Regression	SS_R	k	MS_R	MS_R/MS_E
Error or residual	SS_E	$n - k - 1$	MS_E	
Total	S_{yy}	$n - 1$		

A computational formula for the regression sum of squares SS_R may be obtained as follows:

$$SS_E = \sum_{j=1}^{n} (y_j - \hat{y}_j)^2$$

$$= (\mathbf{y} - \mathbf{X}\hat{\boldsymbol{\beta}})'(\mathbf{y} - \mathbf{X}\hat{\boldsymbol{\beta}})$$

$$= \mathbf{y}'\mathbf{y} - \hat{\boldsymbol{\beta}}'\mathbf{X}'\mathbf{y}$$

$$= \sum_{j=1}^{n} y_j^2 - [\hat{\beta}_0', \hat{\beta}_1, \ldots, \hat{\beta}_k] \begin{bmatrix} \sum_{j=1}^{n} y_j \\ S_{1y} \\ \cdot \\ \cdot \\ \cdot \\ S_{ky} \end{bmatrix}$$

$$= \sum_{j=1}^{n} y_j^2 - \frac{\left(\sum_{j=1}^{n} y_j\right)^2}{n} - \sum_{i=1}^{k} \hat{\beta}_i S_{iy}$$

$$= S_{yy} - \sum_{i=1}^{k} \hat{\beta}_i S_{iy}$$

Thus, since $S_{yy} = SS_E + SS_R$, we see that the regression sum of squares is

$$SS_R = \sum_{i=1}^{k} \hat{\beta}_i S_{iy} \tag{13-54}$$

■**EXAMPLE 13-6.** Consider the data in Example 13-4. The fitted model is $\hat{y} = 30.8667 + 0.8847 (x_1 - 18) + 0.4491 (x_2 - 28)$, and from $\mathbf{X}'\mathbf{y}$ we see that $S_{1y} = 345$ and $S_{2y} = 63$. The total sum of squares is

$$S_{yy} = \sum_{j=1}^{15} y_j^2 - \frac{\left(\sum_{j=1}^{15} y_j\right)^2}{15} = 14{,}741 - \frac{(463)^2}{15} = 449.73$$

Table 13-8 Analysis of Variance for Significance of Regression in Example 13-6

Source of Variation	Sum of Squares	Degrees of Freedom	Mean Square	F_0
Regression	333.51	2	166.76	17.23
Error	116.22	12	9.68	
Total	449.73	14		

and from Equation 13-54 the regression sum of squares is

$$SS_R = \sum_{i=1}^{2} \hat{\beta}_i S_{iy} = (0.8847)(345) + (0.4491)(63) = 333.51$$

The analysis of variance is summarized in Table 13-8. Since $F_{.01, 2, 12} = 6.93$, we conclude that at least one variable contributes significantly to the regression. ∎

We are frequently interested in testing hypotheses on the individual regression coefficients. Such tests would be useful in determining the value of each of the independent variables in the regression. For example, the model might be more effective with the inclusion of additional variables, or perhaps with the deletion of one or more of the variables already in the model.

Adding a variable to a regression model always causes the sum of squares for regression to increase and the error sum of squares to decrease. We must decide whether the increase in the regression sum of squares is sufficient to warrant using the additional variable in the model. Further, adding an unimportant variable to the model can actually *increase* the mean square error, thereby decreasing the usefulness of the model.

The least squares estimator $\hat{\boldsymbol{\beta}}$ is a random variable. Furthermore, since $\hat{\boldsymbol{\beta}}$ is a linear combination of the observations y_j, the distribution of $\hat{\boldsymbol{\beta}}$ is

$$\hat{\boldsymbol{\beta}} \sim N(\boldsymbol{\beta}, \sigma^2 (\mathbf{X'X})^{-1})$$

Thus the variance of the regression coefficient $\hat{\beta}_i$ is σ^2 times the $(i + 1)$st diagonal element of $(\mathbf{X'X})^{-1}$, for example, C_{ii}. Thus, each regression coefficient has the distributional property

$$\hat{\beta}_i \sim N(\beta_i, \sigma^2 C_{ii})$$

The hypotheses for testing the significance of any individual coefficient, for example, β_i, are

$$H_0: \beta_i = 0$$
$$H_1: \beta_i \neq 0 \qquad (13\text{-}55)$$

The appropriate test statistic for (13-55) is

$$t_0 = \frac{\hat{\beta}_i}{\sqrt{MS_E C_{ii}}} \tag{13-56}$$

and H_0: $\beta_i = 0$ is rejected if $|t_0| > t_{\alpha/2,\, n-k-1}$. This test procedure can be misleading, however, because the $\hat{\beta}_i$ are not usually independent. That is, there will usually be elements C_{ij} that are not zero. This implies that the $\hat{\beta}_i$ are not independent, and consequently, the t tests in Equation 13-56 will not be independent. The results of this may be that β_j *appears* to be significant only because its estimator $\hat{\beta}_j$ is not independent of $\hat{\beta}_i$, and β_i really *is* significant.

We need a procedure whereby the contribution to the regression sum of squares for a parameter (e.g., $\hat{\beta}_i$) can be determined *given* that other parameters $\hat{\beta}_j$ ($j \neq i$) are already in the model. That is, we wish to measure the value of adding a term β_i to a model that originally did not include such a term. The general regression significance test can be used to do this.

While we have discussed the general regression significance test before, we will give a statement of the procedure in the present context. Suppose the model is $y = X\beta + \epsilon$, and the β vector is partitioned as follows:

$$\beta = \begin{bmatrix} \beta_1 \\ \hline \beta_2 \end{bmatrix}$$

where β_1 is $(r \times 1)$ and β_2 is $[(p-r) \times 1]$. We wish to test the hypotheses

$$\begin{aligned} H_0: \beta_1 &= 0 \\ H_1: \beta_1 &\neq 0 \end{aligned} \tag{13-57}$$

The model may be written as

$$y = X\beta + \epsilon = X_1\beta_1 + X_2\beta_2 + \epsilon \tag{13-58}$$

where X_1 represents the columns of X associated with β_1 and X_2 represents the columns of X associated with β_2.

For the *full* model (including both β_1 and β_2), we know that $\hat{\beta} = (X'X)^{-1}X'y$. Also,

$$SS_R(\beta) = \hat{\beta}'X'y \quad (p \text{ degrees of freedom})$$

and

$$MS_E = \frac{\sum_{j=1}^{n} y_j^2 - \hat{\beta}'X'y}{n-p}$$

$SS_R(\beta)$ is called the regression sum of squares *due to* β. To find the contribution of the terms in β_1 to the regression, fit the model assuming the null hypothesis H_0: $\beta_1 = 0$ to be true. The *reduced* model is found from (13-58) as

$$y = X_2\beta_2 + \epsilon. \tag{13-59}$$

The least squares estimator of $\boldsymbol{\beta}_2$ is $\hat{\boldsymbol{\beta}}_2 = (X_2'X_2)^{-1}X_2'y$, and

$$SS_R(\boldsymbol{\beta}_2) = \hat{\boldsymbol{\beta}}_2'X_2'y. \quad (p - r \text{ degrees of freedom}) \tag{13-60}$$

The regression sum of squares due to $\boldsymbol{\beta}_1$ *adjusted for the presence of* $\boldsymbol{\beta}_2$ *already in the model is*

$$SS_R(\boldsymbol{\beta}_1 | \boldsymbol{\beta}_2) = SS_R(\boldsymbol{\beta}) - SS_R(\boldsymbol{\beta}_2) \tag{13-61}$$

This sum of squares has r degrees of freedom. It is sometimes called the "extra sum of squares" due to $\boldsymbol{\beta}_1$. Now $SS_R(\boldsymbol{\beta}_1 | \boldsymbol{\beta}_2)$ is independent of MS_E and the null hypothesis $\boldsymbol{\beta}_1 = 0$ may be tested by the statistic

$$F_0 = \frac{SS_R(\boldsymbol{\beta}_1 | \boldsymbol{\beta}_2)/r}{MS_E} \tag{13-62}$$

If $F_0 > F_{\alpha, r, n-p}$ we reject H_0 and conclude that at least one of the parameters in $\boldsymbol{\beta}_1$ is not zero.

This procedure is extremely useful. If we have several variables in a regression model we can measure the value of β_i as though it were added to the model last by computing

$$SS_R(\beta_i | \beta_0, \beta_1, \ldots, \beta_{i-1}, \beta_{i+1}, \ldots, \beta_k) \quad i = 1, 2, \ldots, k$$

■**EXAMPLE 13-7.** We use the data in Example 13-4 to illustrate the general regression significance test. The regression model is

$$\hat{y} = 30.8667 + 0.8847(x_1 - 18) + 0.4491(x_2 - 28).$$

We would like to test the hypothesis $H_0: \beta_1 = 0$ against $H_1: \beta_1 \neq 0$. In the general regression significance test notation, this implies that

$$\boldsymbol{\beta} = \begin{bmatrix} \beta_1 \\ -- \\ \beta_2 \end{bmatrix} = \begin{bmatrix} \beta_1 \\ -- \\ \beta_0' \\ \beta_2 \end{bmatrix}$$

Note that $\boldsymbol{\beta}_1 = \beta_1$ is a (1×1) vector (thus $r = 1$) and $\boldsymbol{\beta}_2$ is a (2×1) vector. For the full model, the regression sum of squares is

$$SS_R(\boldsymbol{\beta}) = SS_R(\beta_0', \beta_1, \beta_2) = \hat{\boldsymbol{\beta}}'x'y$$

$$= \hat{\beta}_0' \sum_{j=1}^{15} y_j + \sum_{i=1}^{2} \hat{\beta}_i S_{iy}$$

$$= 14,625.80$$

Note that this is not the same regression sum of squares computed in Example 13-4, because $SS_R(\boldsymbol{\beta})$ includes contribution to regression due to the intercept. Also, $SS_R(\boldsymbol{\beta})$ has $p = 3$ degrees of freedom. The error mean square, with $n - p = 15 - 3 = 12$ degrees of freedom, is

$$MS_E = \frac{\sum_{j=1}^{15} y_j^2 - \hat{\beta} X'y}{n - p} = \frac{14,741 - 14,625.80}{15 - 3} = 9.60$$

which agrees closely with the results of Example 13-4.

To test H_0: $\beta_1 = 0$ we must find $SS_R(\beta_1 | \beta_2) = SS_R(\beta_1 | \beta_0', \beta_2)$. This is obtained by fitting the reduced model $y = \beta_0' + \beta_2(x_2 - \bar{x}) + \epsilon$. We find that $\hat{y} = 30.8667 + 0.1150$ $(x_2 - 28)$ and, as a result,

$$SS_R(\beta_2) = \hat{\beta}_2' X_2' y$$

$$= \hat{\beta}_0' \sum_{j=1}^{15} y_j + \hat{\beta}_2 S_{2y}$$

$$= 14,298.51$$

with $p - r = 2$ degrees of freedom. Thus, the regression sum of squares for β_1 *adjusted* for β_2 is

$$SS_R(\beta_1 | \beta_2) = SS_R(\beta_1 | \beta_0', \beta_2)$$

$$= SS_R(\beta) - SS_R(\beta_2)$$

$$= 14,625.80 - 14,298.51 = 327.29$$

with $r = 1$ degree of freedom. From (13-62), the test statistic is

$$F_0 = \frac{SS_R(\beta_1 | \beta_0', \beta_2)/1}{MS_E} = \frac{327.29}{9.60} = 34.09$$

and since $F_{.01, 1, 12} = 9.33$, we conclude that $\beta_1 \neq 0$. Therefore, x_1 contributes significantly to the model. ∎

13-7   OTHER LINEAR REGRESSION MODELS

The linear model $y = X\beta + \epsilon$ is a general model. It can be used to fit any relationship that is *linear* in the unknown parameters β. An example would be the kth degree polynomial in one variable

$$y = \beta_0 + \beta_1 x + \beta_2 x^2 + \cdots + \beta_k x^k + \epsilon$$

which we used previously in estimating the polynomial effects of a quantitative factor. Other examples include the second-degree polynomial in two variables

$$y = \beta_0 + \beta_1 x_1 + \beta_2 x_2 + \beta_{11} x_1^2 + \beta_{22} x_2^2 + \beta_{12} x_1 x_2 + \epsilon$$

and the trignometric model

$$y = \beta_0 + \beta_1 \sin x + \beta_2 \cos x + \epsilon$$

Use of the *general linear model* $y = X\beta + \epsilon$ to fit models such as these is illustrated in the following example.

■**EXAMPLE 13-8.** Suppose we have the data shown below and we wish to fit the model $y = \beta_0 + \beta_1 x + \beta_2 x^2 + \epsilon$.

x	1.0	1.2	1.4	1.6	1.8	2.0
y	6.15	7.90	9.40	10.50	11.00	14.00

In matrix notation the model is

$$
y = \begin{bmatrix} 6.15 \\ 7.90 \\ 9.40 \\ 10.50 \\ 11.00 \\ 14.00 \end{bmatrix} \quad
X = \begin{matrix} \beta_0 \; \beta_1 \; \beta_2 \\ \begin{bmatrix} 1 & 1.0 & 1.00 \\ 1 & 1.2 & 1.44 \\ 1 & 1.4 & 1.96 \\ 1 & 1.6 & 2.56 \\ 1 & 1.8 & 3.24 \\ 1 & 2.0 & 4.00 \end{bmatrix} \end{matrix} \quad
\beta = \begin{bmatrix} \beta_0 \\ \beta_1 \\ \beta_2 \end{bmatrix}
$$

The least squares estimator of β is

$$
\hat{\beta} = (X'X)^{-1} X'y = \begin{bmatrix} 6.0000 & 9.0000 & 14.2000 \\ 9.0000 & 14.2000 & 23.4000 \\ 14.2000 & 23.4000 & 39.9664 \end{bmatrix}^{-1} \begin{bmatrix} 58.95 \\ 93.39 \\ 154.47 \end{bmatrix}
$$

or

$$
\hat{\beta} = \begin{bmatrix} 79.5702 & -109.2841 & 35.7137 \\ -109.2841 & 152.0959 & -50.2225 \\ 35.7137 & -50.2225 & 16.7408 \end{bmatrix} \begin{bmatrix} 58.95 \\ 93.39 \\ 154.47 \end{bmatrix} = \begin{bmatrix} 1.3284 \\ 4.0798 \\ 1.0044 \end{bmatrix}
$$

Therefore, the regression model would be

$$
\hat{y} = 1.3284 + 4.0798 + 1.0044 x^2
$$

■

This general method can be employed if one wishes to fit a response curve to data arising from a designed experiment with unequal spacing of the design variables. If the levels of the design variables are equally-spaced, then the use of orthogonal polynomials greatly simplifies the computational effort, as illustrated in Sections 3-6 and 6-6. A derivation of the formulas used in those sections is given below.

Orthogonal Polynomials

In Chapters 3 and 6 we illustrated the use of orthogonal polynomials to compute the polynomial effects of quantitative factors in an experimental design. The procedure involves fitting the model

$$y_j = \alpha_0 + \alpha_1 P_1(x_j) + \alpha_2 P_2(x_j) + \cdots + \alpha_k P_k(x_j) + \epsilon_j$$

where $P_u(x_j)$ is a uth-order orthogonal polynomial; that is, $P_0(x_j) = 1$, $\sum_{j=1}^{n} P_u(x_j) = 0$, and $\sum_{j=1}^{n} P_u(x_j) P_s(x_j) = 0$ for $u \neq s$. The first five orthogonal polynomials are given in Section 3-6. The least-squares normal equations for this model are

$$\begin{bmatrix} n & 0 & 0 & \cdots & 0 \\ 0 & \sum_{j=1}^{n} \{P_1(x_j)\}^2 & 0 & \cdots & 0 \\ 0 & 0 & \sum_{j=1}^{n} \{P_2(x_j)\}^2 & & \\ \vdots & & & \ddots & \vdots \\ 0 & & & & \sum_{j=1}^{n} \{P_k(x_j)\}^2 \end{bmatrix} \begin{bmatrix} \hat{\alpha}_0 \\ \hat{\alpha}_1 \\ \vdots \\ \hat{\alpha}_k \end{bmatrix} = \begin{bmatrix} \sum_{j=1}^{n} y_j \\ \sum_{j=1}^{n} y_j P_1(x_j) \\ \vdots \\ \sum_{j=1}^{n} y_j P_k(x_j) \end{bmatrix}$$

Note that all off-diagonal elements in the $\mathbf{X'X}$ matrix are zero because of the orthogonality property of the polynomials $P_u(x_j)$. Therefore, the least squares estimators of the model parameters are

$$\hat{\alpha}_0 = \bar{y}$$

$$\hat{\alpha}_i = \frac{\sum_{j=1}^{n} y_j P_i(x_j)}{\sum_{j=1}^{n} \{P_i(x_j)\}^2} \qquad i = 1, 2, \ldots, k$$

and the regression sum of squares for any model parameter adjusted for the presence of other model parameters is

$$SS_R(\alpha_i) = \hat{\alpha}_i \sum_{j=1}^{n} y_j P_i(x_j) = \frac{\left\{\sum_{j=1}^{n} y_j P_i(x_j)\right\}^2}{\sum_{j=1}^{n} \{P_i(x_j)\}^2} \qquad i = 1, 2, \ldots, k$$

with $SS_R(\alpha_0) = (\sum_{j=1}^{n} y_j)^2/n$. These results were used previously in Section 3-6.

13-8 EXAMINING THE REGRESSION EQUATION

Fitting a regression model requires several assumptions. To estimate the model parameters we usually assume that the errors are uncorrelated random variables with mean zero and constant variance. Tests of hypotheses and interval estimation require that the errors are normally distributed. Additionally, we assume that the *order* of the fitted model is correct; that is, if we fit a first-order polynomial then we are assuming that the phenomena actually behaves in a first-order manner.

The experimenter should always consider the validity of these assumptions to be doubtful, and conduct analyses to examine the adequacy of the model he has tentatively entertained. In this section we discuss methods useful in this respect.

13-8.1 Residual Analysis

We define the residuals as $e_j = y_j - \hat{y}_j, j = 1, 2, \ldots, n$, where y_j is an observation and $\hat{y}_j$ is the corresponding fitted value from the regression model. Analysis of the residuals is frequently helpful in checking the assumption that the errors are NID $(0, \sigma^2)$ and in determining if additional terms in the model would be useful.

As an approximate check of normality, the experimenter can construct a frequency histogram of the residuals or plot them on normal probability paper. It requires judgment to assess the abnormality of such plots. One may also *standardize* the residuals by computing $d_j = e_j/\sqrt{MS_E}, j = 1, 2, \ldots, n$. If the errors are NID $(0, \sigma^2)$ then approximately 95 percent of the standardized residuals should fall in the interval $(-2, +2)$. Residuals far outside this interval may indicate the presence of an *outlier*; that is, an observation that is not typical of the rest of the data. Various rules have been proposed for discarding outliers. However, sometimes outliers provide important information about unusual circumstances of interest to the experimenter, and should not be discarded. For further discussion of outliers, see Anscombe (1960).

It is frequently helpful to plot the residuals (1) in time sequence (if known), (2) against the $\hat{y}_j$, and (3) against the independent variables $x_{ij}, i = 1, 2, \cdots k$. These

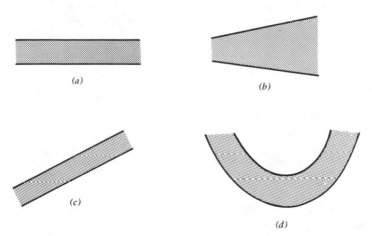

Figure 13-2. Patterns for residual plots. (Adapted with permission from *Applied Regression Analysis*, by N. R. Draper and H. Smith, John Wiley and Sons, 1966.)

graphs will usually look like one of the four general patterns in Figure 13-2. Pattern (*a*) in Figure 13-2 represents the normal situation, while patterns (*b*), (*c*), and (*d*) represent anomalies. If the residuals appear as in (*b*), then the variance of the observations may be increasing with time, or with the magnitude of the y_j or x_{ij}. If a plot of the residuals against time has the appearance of (*c*), then a linear term or time should be added to the model. Plots against $\hat{y}_j$ and x_{ij} that look like (*c*) indicate an error in calculation or analysis. Residual plots that look like (*d*) indicate model inadequacy; that is, higher-order terms should be added to the model.

13-8.2   The Lack-of-Fit Test

Regression models are often fit to data when the true functional relationship is unknown. Naturally, we would like to know whether the order of the model tentatively assumed is correct. This section will describe a test for the validity of this assumption.

   The danger of using a regression model that is a poor approximation of the true functional relationship is illustrated in Figure 13-3. Obviously, a polynomial of degree two or greater should have been used for this hypothetical situation. The result is that a very poor model has been obtained.

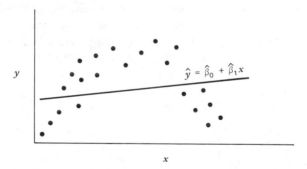

Figure 13-3.   A regression model displaying lack of fit.

   We present a test for the "goodness of fit" of a regression model. While the case of only one independent variable is explicitly treated, the procedure will generalize to k independent variables easily. The hypotheses we wish to test are

H_0: The model adequately fits the data

H_1: The model does not fit the data

The test involves partitioning the error or residual sum of squares into the following two components:

$$SS_E = SS_{PE} + SS_{LOF}$$

where SS_{PE} is the sum of squares attributable to "pure" experimental error, and SS_{LOF} is the sum of squares attributable to the lack of fit of the model. To compute SS_{PE} we will require repeated observations on y for at least one level of x. Suppose that we have n observations such that

$$y_{11}, y_{12}, \ldots, y_{1n_1} \quad \text{repeated observations at } x_1$$

$$y_{21}, y_{22}, \ldots, y_{2n_2} \quad \text{repeated observations at } x_2.$$

$$\vdots$$

$$y_{m1}, y_{m2}, \ldots, y_{mn_m} \quad \text{repeated observations at } x_m$$

We see that there are m distinct levels of x. The contribution to the pure-error sum of squares at x_1 (say) would be

$$\sum_{u=1}^{n_1} (y_{1u} - \bar{y}_1)^2 \tag{13-63}$$

The total sum of squares for pure error would be obtained by summing (13-63) over all levels of x as

$$SS_{PE} = \sum_{i=1}^{m} \sum_{u=1}^{n_i} (y_{iu} - \bar{y}_i)^2$$

There are $n_e = \sum_{i=1}^{m} (n_i - 1) = n - m$ degrees of freedom associated with the pure-error sum of squares. The sum of squares for lack of fit is simply

$$SS_{LOF} = SS_E - SS_{PE}$$

with $n - 2 - n_e = m - 2$ degrees of freedom. The test statistic for lack of fit would then be

$$F_0 = \frac{SS_{LOF}/(m-2)}{SS_{PE}/(n-m)} = \frac{MS_{LOF}}{MS_{PE}} \tag{13-64}$$

and we would reject if $F_0 > F_{\alpha, m-2, n-m}$.

This test procedure may be easily introduced into the analysis of variance conducted for the significance of regression. If the null hypothesis of model adequacy is rejected then the model must be abandoned and attempts made to find a more appropriate model. If H_0 is not rejected, then there is no apparent reason to doubt the adequacy of the model, and MS_{PE} and MS_{LOF} are often combined to estimate σ^2.

■**EXAMPLE 13-9.** Suppose we have the data

x	1.0	1.0	2.0	3.3	3.3	4.0	4.0	4.0	4.7	5.0
y	2.3	1.8	2.8	1.8	3.7	2.6	2.6	2.2	3.2	2.0
x	5.6	5.6	5.6	6.0	6.0	6.5	6.9			
y	3.5	2.8	2.1	3.4	3.2	3.4	5.0			

We may compute $S_{yy} = 10.97$, $S_{xy} = 13.62$, $S_{xx} = 52.53$, $\bar{y} = 2.847$, and $\bar{x} = 4.382$. The regression model is $\hat{y} = 1.708 + 0.260x$, and the regression sum of squares is $SS_R = \hat{\beta}_1 S_{xy} = (0.260)(13.62) = 3.541$. The pure-error sum of squares is computed as follows:

Level of x	$\Sigma(y_j - \bar{y})^2$	Degrees of Freedom
1.0	0.1250	1
3.3	1.8050	1
4.0	0.1066	2
5.6	0.9800	2
6.0	0.0200	1
Totals:	3.0366	7

The analysis of variance is summarized in Table 13-9. Since $F_{.25,8,7} = 1.70$, we cannot reject the hypothesis that the tentative model adequately describes the data. We will pool lack-of-fit and pure-error mean squares to form the denominator mean square in the test for significance of regression. Also, since $F_{.05,1,15} = 4.54$, we must conclude that $\beta_1 \neq 0$.   ∎

Table 13-9 Analysis of Variance for Example 13-10

Source of Variation	Sum of Squares	Degrees of Freedom	Mean Square	F_0
Regression	3.541	1	3.541	/.15
Residual	7.429	15	0.4952	
(Lack of fit)	4.3924	8	0.5491	1.27
(Pure error)	3.0366	7	0.4338	
Total	10.970	16		

In fitting a regression model to experimental data, a good practice is to use the lowest-degree model which adequately describes the data. The lack-of-fit test may be useful in this respect. However, it is always possible to fit a polynomial of degree $n - 1$ to n data points, and the experimenter should not consider using a model that is "saturated," that is, which has very nearly as many independent variables as observations on y.

13-8.3   The Coefficient of Multiple Determination

The quantity

$$R^2 = \frac{SS_R}{S_{yy}} = \frac{\sum\limits_{j=1}^{n} (\hat{y}_j - \bar{y})^2}{\sum\limits_{j=1}^{n} (y_j - \bar{y})^2} \tag{13-65}$$

is called the coefficient of multiple determination, and it is often used to judge the adequacy of a regression model. Clearly $0 < R^2 \leqslant 1$. We often refer loosely to R^2 as the amount of variability in the data explained or accounted for by the regression model. For the data in Example 13-1, we have $R^2 = SS_R/S_{yy} = 120.15/127.73 = 0.9407$; that is, 94.07 percent of the variability in the data is accounted for by the model.

The statistic R^2 should be used with caution, since it is always possible to make R^2 unity by simply adding enough terms to the model. For example, we can obtain a "perfect" fit to n data points with a polynomial of degree $n - 1$. Also, R^2 will always increase if we add a variable to the model, but this does not necessarily mean the new model is superior to the old one. Unless the error sum of squares in the new model is reduced by an amount equal to the original error mean square, the new model will have a *larger* error mean square than the old one because of the loss of one degree of freedom. Thus the new model will actually be worse than the old one.

13-9 PROBLEMS

13-1. The tensile strength of a paper product is related to the amount of hardwood in the pulp. Ten samples are produced in the pilot plant, and the data obtained are shown below. Fit a simple linear regression model expressing strength as a function of hardwood concentration using these data.

Strength	Percent Hardwood	Strength	Percent Hardwood
160	10	181	20
171	15	188	25
175	15	193	25
182	20	195	28
184	20	200	30

13-2. Using the results of Problem 13-1, test the regression model for lack of fit and significance of regression.

13-3. Using the results of Problem 13-1, construct 95 percent confidence intervals on the slope and intercept.

13-4. A plant distills liquid air to produce oxygen, nitrogen, and argon. The percentage of impurity in the oxygen is thought to be linearly related to the amount of impurities in the air, as measured by the "pollution count" in parts per million (ppm). Fit a linear regression model to the data given below.

Purity (%)	93.3	92.0	92.4	91.7	94.0	94.6	93.6
Pollution Count (ppm)	1:10	1.45	1.36	1.59	1.08	0.75	1.20

	93.1	93.2	92.9	92.2	91.3	90.1	91.6	91.9
	0.99	0.83	1.22	1.47	1.81	2.03	1.75	1.68

13-5. Using the results of Problem 13-4, does a linear relationship between purity and the pollution count seem reasonable?

13-6. Construct 95 percent confidence intervals on the slope and intercept of the linear regression model in Problem 13-4.

13-7. The yield of a chemical process is related to the concentration of reactant and the operating temperature. Fit a multiple linear regression model to the data shown below.

Yield	Concentration	Temperature
81	1.00	150
89	1.00	180
83	2.00	150
91	2.00	180
79	1.00	150
87	1.00	180
84	2.00	150
90	2.00	180

13-8. Test the multiple linear regression model in Problem 13-7 for significance of regression.

13-9. Suppose that in Problem 13-7 we had the additional data points shown below. How do these additional observations change the estimates of the regression coefficients found in Problem 13-7. Test the new model for lack of fit and significance of regression.

Yield	Concentration	Temperature
86	1.50	165
85	1.50	165
87	1.50	165
86	1.50	165
85	1.50	165
85	1.50	165

13-10. The brake horsepower developed by an automobile engine on a dynomometer is thought to be a function of the engine speed in RPM, the road octane number of the fuel, and the engine compression. An experiment is run in the laboratory and the data shown on page 337 collected. Fit a multiple linear regression model to these data.

Brake Horsepower	RPM	Octane Number	Compression
225	2000	90	100
212	1800	94	95
229	2400	88	110
222	1900	91	96
219	1600	86	100
278	2500	96	110
246	3000	94	98
237	3200	90	100
233	2800	88	105
224	3400	86	97
223	1800	90	100
230	2500	89	104

13-11. Test the model in Problem 13-10 for significance of regression. Use the general regression significance test to determine the importance of the variable "octane number" in this model.

13-12. Given the data below, fit the second-order polynomial regression model

$$\hat{y} = \hat{\beta}_0 + \hat{\beta}_1 x_1 + \hat{\beta}_2 x_2 + \hat{\beta}_{11} x_1^2 + \hat{\beta}_{22} x_2^2 + \hat{\beta}_{12} x_1 x_2$$

y	x_1	x_2
26	1.0	1.0
24	1.0	1.0
175	1.5	4.0
160	1.5	4.0
163	1.5	4.0
55	0.5	2.0
62	1.5	2.0
100	0.5	3.0
26	1.0	1.5
30	0.5	1.5
70	1.0	2.5
71	0.5	2.5

13-13. Fit a quadratic model to the data below using orthogonal polynomials.

y	10.0	9.7	8.5	6.8	4.5	3.6	-1.7	-5.4	-9.8
x	1.0	1.5	2.0	2.5	3.0	3.5	4.0	4.5	5.0

13-14. Prove that in simple linear regression Cov $(\hat{\beta}_0', \hat{\beta}_1) = 0$.

13-15. Verify the bias and variance properties of $\hat{\beta}_0'$ and $\hat{\beta}_0$ given in Equations 13-16 and 13-17. Show that Cov $(\hat{\beta}_0, \hat{\beta}_1) = -\sigma^2 \bar{x}/S_{xx}$.

13-16. Derive the expected value of mean square regression for the simple linear regression model.

13-17. Find the residuals for Example 13-1. Analyze them by the methods discussed in Section 13-8.

13-18. *Relationship Between Analysis of Variance and Regression.* Any analysis of variance model can be expressed in terms of the general linear model $y = X\beta + \epsilon$, where the X matrix consists of zeros and ones. Show that the one-way classification model $y_{ij} = \mu + \tau_i + \epsilon_{ij}$, $i = 1, 2, 3, j = 1, 2, 3, 4$ can be written in general linear model form. Then
  (a) Write the normal equations $(X'X)\hat{\beta} = X'y$ and compare them with the normal equations found for this model in Chapter 3, page 43.
  (b) Find the rank of $X'X$. Can $(X'X)^{-1}$ be obtained?
  (c) Suppose the first normal equation is deleted and the restriction $\Sigma_{i=1}^{3} n\hat{\tau}_i = 0$ is added. Can the resulting system of equations be solved? If so, find the solution. Find the regression sum of squares $\hat{\beta}'X'y$, and compare it to the treatment sum of squares in the one-way classification.

13-19. Suppose that we are fitting a straight line and we desire to make the variance of $\hat{\beta}_1$ as small as possible. Restricting ourselves to an even number of experimental points, where should we place these points so as to minimize $V(\hat{\beta}_1)$? (Note: Use the design called for in this exercise with *great* caution, because even though it minimizes $V(\hat{\beta}_1)$ it has some undesirable properties; for example, see Myers (1971). Only if you are *very sure* the true functional relationship is linear should you consider using this design).

13-20. *Weighted Least Squares.* Suppose that we are fitting the straight line $y = \beta_0 + \beta_1 x + \epsilon$, but the variance of the y's now depend on the level of x; that is,

$$V(y \mid x_i) = \sigma_i^2 = \frac{\sigma^2}{w_i} \qquad i = 1, 2, \ldots, n$$

where the w_i are known constants, often called weights. Show that the resulting least squares normal equations are

$$\hat{\beta}_0 \sum_{i=1}^{n} w_i + \hat{\beta}_1 \sum_{i=1}^{n} w_i x_i = \sum_{i=1}^{n} w_i y_i$$

$$\hat{\beta}_0 \sum_{i=1}^{n} w_i x_i + \hat{\beta}_1 \sum_{i=1}^{n} w_i x_i^2 = \sum_{i=1}^{n} w_i x_i y_i$$

13-21. *Nonlinear Regression.* This chapter showed that polynomial models could be fit by multiple linear regression techniques. Actually, any model that is linear in the unknown parameters can be treated in such a fashion. Regression models that are not linear in the parameters are called *nonlinear* models and cannot be handled by the methods of this chapter. Sometimes models that appear nonlinear can be made linear by the use of some appropriate transformation. Could you fit the model

$$y = \beta_0 e^{\beta_1 x} \epsilon$$

by the methods of this chapter? Suppose the model were

$$y = \beta_0 e^{\beta_1 x} + \epsilon$$

instead. Obtain the least squares normal equations for this latter case and try to solve them.

Chapter Fourteen
Response Surface Methodology

14-1 INTRODUCTION

Response surface methodology, or RSM, is a collection of mathematical and statistical techniques useful for analyzing problems where several independent variables influence a dependent variable or response, and the goal is to optimize this response. We denote the independent variables by $x_1, x_2, \ldots, x_k$. It is assumed that these variables are continuous and controllable by the experimenter with negligible error. The response (e.g., y) is assumed to be a random variable.

For example, suppose that a chemical engineer wishes to find the temperature (x_1) and pressure (x_2) that maximize the yield of his process. We may write the observed response y as a function of the levels of temperature and pressure as

$$y = f(x_1, x_2) + \epsilon$$

where ϵ is a random error component. If we denote the expected response by $E(y) = \eta$, then the surface represented by $\eta = f(x_1, x_2)$ is called a *response surface*. We may represent the two-dimensional response surface graphically by drawing the x_1 and x_2 axes in the plane of the paper and visualizing the $E(y)$ axis perpendicular to the plane of the paper. Then plotting contours of constant expected response yields the response surface, as in Figure 14-1.

340

x_2

$\eta = 80$

$\eta = 70$

$\eta = 60$

$\eta = 50$

x_1

Figure 14-1. A response surface.

In most RSM problems, the form of the relationship between the response and the independent variables is unknown. Thus, the first step in RSM is to find a suitable approximation for f. Usually, a low-order polynomial in some region of the independent variables is employed. If the response is well modeled by a linear function of the independent variables, then the approximating function is the first-order model

$$y = \beta_0 + \beta_1 x_1 + \beta_2 x_2 + \cdots + \beta_k x_k + \epsilon \qquad (14\text{-}1)$$

If there is curvature in the system, then a polynomial of higher degree, such as the second-order model

$$y = \beta_0 + \sum_{i=1}^{k} \beta_i x_i + \sum_{i=1}^{k} \beta_{ii} x_i^2 + \sum_i \sum_j \beta_{ij} x_i x_j + \epsilon \qquad (14\text{-}2)$$
$$\scriptstyle i<j$$

must be used. Almost all RSM problems utilize one or both of these approximating polynomials. Of course, it is unlikely that a polynomial model will be a reasonable approximation of f over the entire space of the independent variables, but in a relatively small region they usually work quite well.

The method of least squares, discussed in Chapter 13, is used to estimate the parameters in the approximating polynomials. The RSM analysis is then done in terms of the fitted surface. If the fitted surface is an adequate approximation of f, then analysis of the fitted surface will be approximately equivalent to analysis of the actual system. The model parameters can be estimated most effectively if proper experimental designs are used to collect the data. Designs for fitting response surfaces are often called *response surface designs*. These designs will be discussed in Section 14-4.

RSM is a *sequential* procedure. Often when we are at a point on the response surface that is remote from the optimum there is little curvature in the system and the first-order model will be appropriate. Our objective here is to rapidly and efficiently lead the experimenter to the general vicinity of the optimum. Once the region of the optimum has been found, a more elaborate model such as the second-order response surface may be employed, and an analysis performed to locate the optimum. From Figure 14-1, we see that the analysis of a response surface can be thought of as "climbing a hill," where the top of the hill represents the point of maximum response. If the true optimum is a point of minimum response, then we may think of descending into a valley.

The eventual objective of RSM is to determine the optimum operating conditions for the system, or to determine a region of the factor space in which operating specifications are satisfied. RSM is not used primarily to gain understanding of the *physical mechanism* of the system, although RSM may assist in gaining such knowledge. Furthermore, note that "optimum" in RSM is used in a special sense. The "hill climbing" procedures of RSM guarantee convergence to a local optimum only.

14-2 THE METHOD OF STEEPEST ASCENT

Frequently, the initial estimate of the optimum operating conditions for the system will be far from the actual optimum. In such circumstances, the experimenter's objective is to move rapidly to the general vicinity of the optimum. He wishes to use a simple and economically efficient experimental procedure. When we are remote from the optimum we usually assume that a first-order model is an adequate approximation to the true surface in a small region of the x's.

The *method of steepest ascent* is a procedure for moving sequentially along the path of steepest ascent, that is, the direction of maximum increase in response. Of course, if minimization is desired, then we are talking about the method of steepest *descent*. The fitted first-order model is

$$\hat{y} = \hat{\beta}_0 + \sum_{i=1}^{k} \hat{\beta}_i x_i \tag{14-3}$$

and the first-order response surface, that is, the contours of $\hat{y}$, is a series of parallel lines such as shown in Figure 14-2. The direction of steepest ascent is the direction in which $\hat{y}$ increases most rapidly. This direction is parallel to the normal to the fitted response surface. We usually take as the *path of steepest ascent* the line through the center of the region of interest and normal to the fitted surface. Thus, the steps along the path are proportional to the regression coefficients $\{\hat{\beta}_i\}$. The actual step size is determined by the experimenter based on experience.

Experiments are conducted along the path of steepest ascent until no further increase in response is observed. Then a new first-order model may be fit, a new path of steepest ascent determined, and the procedure continued. Eventually, the experimen-

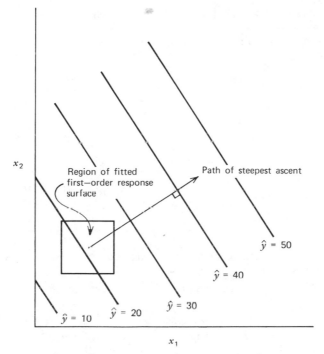

Figure 14-2.   First-order response surface and path of steepest ascent.

ter will arrive in the vicinity of the optimum. This is usually indicated by lack of fit of the first-order model. At that time additional experiments, to be described in Section 14-3, are conducted to obtain a more precise estimate of the optimum.

■**EXAMPLE 14-1.** A chemical engineer is interested in determining the operating conditions that maximize the yield of his process. Two controllable variables influence process yield, reaction time, and reaction temperature. He is currently operating the process with reaction time of 35 minutes and temperature of 155°F, resulting in yields around 40 percent. Since it is unlikely that this region contains the optimum, we will fit a first-order model and apply the method of steepest ascent.

The engineer decides that the region of exploration for fitting the first-order model should be (30, 40) minutes of reaction and (150, 160) °F. To simplify the calculations, the independent variables will be coded to a (−1, 1) interval. Thus, if ξ_1 denotes the *natural* variable time and ξ_2 denotes the *natural* variable temperature, then the *coded* variables are

$$x_1 = \frac{\xi_1 - 35}{5} \qquad x_2 = \frac{\xi_2 - 155}{5}$$

The data are shown in Table 14-1.

Table 14-1 Process Data for Fitting the First-Order Model

Natural Variables		Coded Variables		Response
ξ_1	ξ_2	x_1	x_2	y
30	150	-1	-1	39.3
30	160	-1	1	40.0
40	150	1	-1	40.9
40	160	1	1	41.5
35	155	0	0	40.3
35	155	0	0	40.5
35	155	0	0	40.7
35	155	0	0	40.2
35	155	0	0	40.6

Note that the design used to collect this data is a 2^2 factorial augmented by five center points. Repeat observations at the center can be used to estimate the experimental error. Also, the design is centered about the current operating conditions for the process.

A first-order model may be fit to these data by least squares. Following the methods of Chapter 13, we obtain the following model in the coded variables:

$$\hat{y} = 40.44 + 0.775x_1 + 0.325x_2$$

The analysis of variance for this model is shown in Table 14-2. From inspection of this table, we conclude that the first-order model adequately approximates the data.

Table 14-2 Analysis of Variance

Source of Variation	Sum of Squares	Degrees of Freedom	Mean Square	F_0
Regression	2.8250	2	1.4125	47.83[a]
Error	0.1772	6	0.0295	
(Lack of fit)	(0.0052)	2	0.0026	0.06
(Pure error)	(0.1720)	4	0.0430	
Total	3.0022	8		

[a]Significant at 1 percent.

To move away from the design center [the point $(x_1 = 0, x_2 = 0)$] along the path of steepest ascent we would move 0.775 units in the x_1 direction for every 0.325 units in the x_2 direction. Thus, the path of steepest ascent passes through the point $(x_1 = 0, x_2 = 0)$ and has slope 0.325/0.775. The engineer decides to use five minutes of reaction time as the basic step size. Using the relationship between ξ_1 and x_1 we see that five minutes of reaction time is equivalent to a step in the *coded* variable x_1 of $\Delta x_1 = 1$. Therefore, the steps along the path of steepest ascent are $\Delta x_1 = 1.0000$ and $\Delta x_2 = (0.325/0.775)\Delta x_1 = 0.4194$.

The engineer computes points along this path and observes yields at these points until a decrease in response is noted. The results are shown in Table 14-3. The steps are shown in both coded and natural variables. While the coded variables are easier to manipulate mathematically, the natural variables must be used in running the process. Increases in response are observed through the tenth step; however, the eleventh step produces a decrease in yield. Therefore, another first-order model must be fit in the general vicinity of the point ($\xi_1 = 85, \xi_2 = 175.970$).

Table 14-3 Steepest Ascent Experiment for Example 14-1

	Coded Variables		Natural Variables		Response
	x_1	x_2	ξ_1	ξ_2	y
Origin	0	0	35	155	
Δ	1.0000	0.4194			
Origin + Δ	1.0000	0.4194	40	157.097	41.0
Origin + 2Δ	2.0000	0.8388	45	159.194	41.9
Origin + 3Δ	3.0000	1.2573	50	161.287	43.1
.	.	.	.	.	.
.	.	.	.	.	.
Origin + 10Δ	10.0000	4.1940	85	175.970	80.3
Origin + 11Δ	11.0000	4.6134	90	178.067	79.2

A new first-order model is fit about the point ($\xi_1 = 85, \xi_2 = 175$). The region of exploration for ξ_1 is [80, 90] and for ξ_2 it is [170, 180]. Thus the coded variables are

$$x_1 = \frac{\xi_1 - 85}{5} \qquad x_2 = \frac{\xi_2 - 175}{5}$$

Once again, a 2^2 design with five center points is used. The data are shown in Table 14-4.

The first-order model fit to the coded data in Table 14-4 is

$$\hat{y} = 78.97 + 1.00x_1 + 0.50x_2$$

The analysis of variance for this model is shown in Table 14-5. The test statistic for lack of fit is significant at one percent, so we conclude that the first-order model is not an adequate approximation. This curvature in the true surface may indicate that we are near the optimum. At this point, additional analysis must be done to locate the optimum more precisely. ∎

Table 14-4 Data for Second First-Order Model

Natural Variables		Coded Variables		Response
ξ_1	ξ_2	x_1	x_2	y
80	170	-1	-1	76.5
80	180	-1	1	77.0
90	170	1	-1	78.0
90	180	1	1	79.5
85	175	0	0	79.9
85	175	0	0	80.3
85	175	0	0	80.0
85	175	0	0	79.7
85	175	0	0	79.8

Table 14-5 Analysis of Variance

Source of Variation	Sum of Squares	Degrees of Freedom	Mean Square	F_0
Regression	5.0000	2		
Error	11.1200	6		
(Lack of fit)	(10.9080)	2	5.4540	102.91[a]
(Pure error)	(0.2120)	4	0.0530	
Total	16.1200	8		

[a]Significant at 1 percent.

14-3 ANALYSIS OF QUADRATIC MODELS

When the experimenter is relatively close to the optimum, a model of degree 2 or higher is usually required to approximate the response because of curvature in the true surface. In most cases, the second-order model

$$\hat{y} = \hat{\beta}_0 + \sum_{i=1}^{k} \hat{\beta}_i x_i + \sum_{i=1}^{k} \hat{\beta}_{ii} x_i^2 + \sum_i \sum_{\substack{j \\ i<j}} \hat{\beta}_{ij} x_i x_j \tag{14-4}$$

is an adequate approximation. The analysis of the fitted second-order response surface is often called the *canonical analysis*.

Suppose we wish to find the levels of $x_1, x_2, \ldots, x_k$ that maximize the predicted response. This maximum point, if it exists, will be the set of $x_1, x_2, \ldots, x_k$ such that the partial derivatives $\partial \hat{y}/\partial x_1 = \partial \hat{y}/\partial x_2 = \cdots = \partial \hat{y}/\partial x_k = 0$. This point, say $x_{1,0}$,

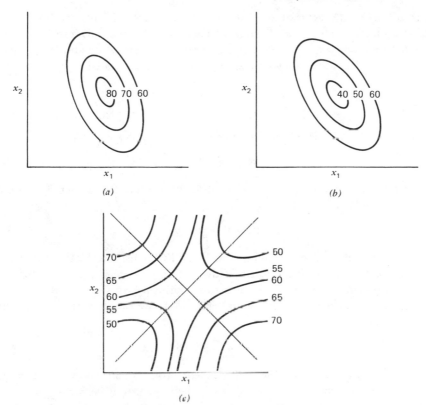

Figure 14-3. Stationary points in a fitted second-order response surface. (*a*) Maximum response. (*b*) Minimum response. (*c*) Saddle point.

$x_{2,0}, \ldots, x_{k,0}$, is called the *stationary point*. The stationary point could represent (1) a point of maximum response, (2) a point of minimum response, or (3) a *saddle point*. These three possibilities are shown in Figure 14-3. Part of the canonical analysis is characterizing the nature of the stationary point.

We may obtain a general solution for the stationary point. Writing the second-order model in matrix notation we have

$$\hat{y} = \hat{\beta}_0 + \mathbf{x}'\mathbf{b} + \mathbf{x}'\mathbf{B}\mathbf{x} \tag{14-5}$$

where

$$\mathbf{x} = \begin{bmatrix} x_1 \\ x_2 \\ \cdot \\ \cdot \\ \cdot \\ x_k \end{bmatrix} \qquad \mathbf{b} = \begin{bmatrix} \hat{\beta}_1 \\ \hat{\beta}_2 \\ \cdot \\ \cdot \\ \cdot \\ \hat{\beta}_k \end{bmatrix} \qquad \text{and} \qquad \mathbf{B} = \begin{bmatrix} \hat{\beta}_{11}, \hat{\beta}_{12}/2, \ldots, \hat{\beta}_{1k}/2 \\ \hat{\beta}_{22}, \ldots, \hat{\beta}_{2k}/2 \\ \cdot \\ \cdot \\ \text{sym.} \qquad \hat{\beta}_{kk} \end{bmatrix}$$

That is, **b** is a $(k \times 1)$ vector of the first-order regression coefficients and **B** is a $(k \times k)$ symmetric matrix whose main diagonal elements are the *pure* quadratic coefficients $(\hat{\beta}_{ii})$ and whose off-diagonal elements are one-half the *mixed* quadratic coefficients $(\hat{\beta}_{ij}, i \neq j)$. The derivative of $\hat{y}$ with respect to the vector **x** equated to **0** is

$$\frac{\partial \hat{y}}{\partial x} = b + 2Bx = 0 \qquad (14\text{-}6)$$

The stationary point is the solution to (14-6), or

$$x_0 = -\tfrac{1}{2} B^{-1} b \qquad (14\text{-}7)$$

Furthermore, by substituting (14-7) into (14-5) we can find the predicted response at the stationary point as

$$\hat{y}_0 = \hat{\beta}_0 + \tfrac{1}{2} x_0' b \qquad (14\text{-}8)$$

To characterize the stationary point, it is helpful to transform the fitted model to a new coordinate system with the origin at the stationary point x_0, and then rotate the axes of this system until they are parallel to the principal axes of the fitted response surface. This transformation is shown in Figure 14-4. We can show that this results in the fitted model

$$\hat{y} = \hat{y}_0 + \lambda_1 w_1^2 + \lambda_2 w_2^2 + \cdots + \lambda_k w_k^2 \qquad (14\text{-}9)$$

where the $\{w_i\}$ are the transformed independent variables and the $\{\lambda_i\}$ are constants. Equation 14-9 is called the *canonical form* of the model. Furthermore, the $\{\lambda_i\}$ are just the *eigenvalues* or *characteristic roots* of the matrix **B**.

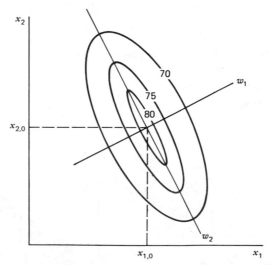

Figure 14-4.   Canonical form of the second-order model.

The nature of the response surface can be determined from the stationary point and the *sign* and *magnitude* of the $\{\lambda_i\}$. First suppose that the stationary point is within the region of exploration for fitting the second-order model. If the λ_i are all positive, then x_0 is a point of minimum response; if the $\{\lambda_i\}$ are all negative, then x_0 is a point of maximum response. If the $\{\lambda_i\}$ have different signs, x_0 is a saddle point. Furthermore, the surface is steepest in the w_i direction for which $|\lambda_i|$ is the greatest. For example, Figure 14-4 depicts a system for which x_0 is a maximum (λ_1 and λ_2 negative) and with $|\lambda_1| > |\lambda_2|$. Should one or more of the λ_i be very small (e.g., $\lambda_i \simeq 0$), then the system is insensitive to the variable w_i. This type of surface is often called a *stationary ridge*, and is illustrated in Figure 14-5.

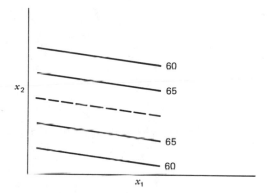

Figure 14-5. A stationary ridge system.

If the stationary point is far outside the region of exploration for fitting the second-order model, and one (or more) λ_i is near zero, then the surface may be a *rising ridge*. Figure 14-6 illustrates a rising ridge for $k = 2$ variables with λ_1 near zero and λ_2 negative. In this type of ridge system we cannot draw inferences about the true surface or the stationary point, since x_0 is outside the region where we have fit the model. How-

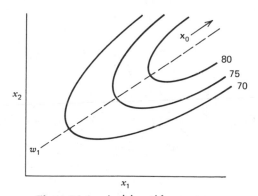

Figure 14-6. A rising ridge system.

ever, further exploration is warranted in the w_1 direction. If λ_2 had been positive, we would call this system a falling ridge.

■**EXAMPLE 14-2.** We will continue the analysis of the chemical process in Example 14-1. A second-order model in the variables x_1, x_2 cannot be fit using the data in Table 14-4. The experimenter decides to augment these data with enough points to fit a second-order model. He obtains four observations at $(x_1 = 0, x_2 = \pm 1.414)$ and $(x_1 = \pm 1.414, x_2 = 0)$. The complete data set is shown in Table 14-6, and the design is displayed in Figure 14-7. This design is called a *central composite design*. It will be discussed in more detail in Section 14-4.2.

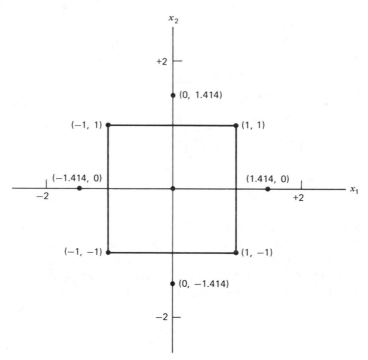

Figure 14-7. Central composite design for Example 14-2.

A second-order model is fit to the coded data by least squares, yielding

$$\hat{y} = 79.9408 + 0.9949x_1 + 0.5151x_2 - 1.3770x_1^2 - 1.0018x_2^2 + 0.2500x_1 x_2$$

The analysis of variance for this model is shown in Table 14-7. Lack of fit is not significant and regression is significant, so we conclude that the second-order model adequately approximates the true surface.

Table 14-6   Central Composite Design for Example 14-2

Natural Variables		Coded Variables		Response
ξ_1	ξ_2	x_1	x_2	y
80	170	-1	-1	76.5
80	180	-1	1	77.0
90	170	1	-1	78.0
90	180	1	1	79.5
85	175	0	0	79.9
85	175	0	0	80.3
85	175	0	0	80.0
85	175	0	0	79.7
85	175	0	0	79.8
92.07	175	1.414	0	78.4
77.93	175	-1.414	0	75.6
85	182.07	0	1.414	78.5
85	167.93	0	-1.414	77.0

Table 14-7   Analysis of Variance for Second-Order Model

Source of Variation	Sum of Squares	Degrees of Freedom	Mean Square	F_0
Regression	28.2560	5	5.6512	81.20[a]
Error	0.4871	7	0.0696	
(Lack of fit)	(0.2751)	3	0.0917	1.73
(pure error)	(0.2120	4	0.0530	
Total	28.7431	12		

[a]Significant at 1 percent.

We now perform the canonical analysis. Note that

$$b = \begin{bmatrix} 0.9949 \\ 0.5151 \end{bmatrix} \qquad B = \begin{bmatrix} -1.3770 & 0.1250 \\ 0.1250 & -1.0018 \end{bmatrix}$$

and from (14-7) the stationary point is

$$x_0 = -\tfrac{1}{2}B^{-1}b$$

$$= -\tfrac{1}{2}\begin{bmatrix} -0.7345 & -0.0917 \\ -0.0917 & -1.0096 \end{bmatrix}\begin{bmatrix} 0.9949 \\ 0.5151 \end{bmatrix} = \begin{bmatrix} 0.3890 \\ 0.3056 \end{bmatrix}$$

That is, $x_{1,0} = 0.3890$ and $x_{2,0} = 0.3056$. In terms of the natural variables, the stationary point is

$$0.3890 = \frac{\xi_1 - 85}{5} \qquad 0.3056 = \frac{\xi_2 - 175}{5}$$

which yields $\xi_1 = 86.9450$ and $\xi_2 = 176.5280$. Using (14-8) we may find the predicted response at the stationary point as $\hat{y}_0 = 80.21$.

To further characterize the stationary point we obtain the canonical form (14-9). The eigenvalues λ_1 and λ_2 are the roots of the determinantal equation

$$|\mathbf{B} - \lambda \mathbf{I}| = 0$$

$$\begin{vmatrix} -1.3770 - \lambda & 0.1250 \\ 0.1250 & -1.0018 - \lambda \end{vmatrix} = 0$$

which reduces to

$$\lambda^2 + 2.3788\lambda + 1.3639 = 0$$

The roots of this quadratic equation are $\lambda_1 = -0.9641$ and $\lambda_2 = -1.4147$. Thus the canonical form of the fitted model is

$$\hat{y} = 80.21 - 0.9641w_1^2 - 1.4147w_2^2$$

Since both λ_1 and λ_2 are negative and the stationary point is within the region of exploration, we conclude that the stationary point is a maximum. ■

In some RSM problems it may be necessary to find the relationship between the *canonical* variables $\{w_i\}$ and the *design* variables $\{x_i\}$. This is particularly true if it is impossible to operate the process at the stationary point. As an illustration, suppose that in Example 14-2 we could not operate the process at $\xi_1 = 86.9450$ and $\xi_2 = 176.5280$ because this factor combination results in excessive cost. We now wish to "back away" from the stationary point to a point of lower cost, but without large losses in yield. The canonical form of the model indicates that the surface is less sensitive to yield loss in the w_1 direction. Exploration of the canonical form requires converting points in the (w_1, w_2) space to points in the (x_1, x_2) space.

In general, the variables **x** are related to the canonical variables **w** by

$$\mathbf{w} = \mathbf{M}'(\mathbf{x} - \mathbf{x}_0)$$

where **M** is a $(k \times k)$ orthogonal matrix. The columns of **M** are the normalized eigenvectors associated with the $\{\lambda_i\}$. That is, if $\mathbf{m}_i$ is the ith column of **M**, then $\mathbf{m}_i$ is the solution to

$$(\mathbf{B} - \lambda_i \mathbf{I})\mathbf{m}_i = 0 \qquad\qquad (14\text{-}10)$$

for which $\sum_{j=1}^{k} m_{ji}^2 = 1$.

We illustrate the procedure using the fitted second-order model in Example 14-2. For $\lambda_1 = -0.9641$, Equation (14-10) becomes

$$\begin{bmatrix} (-1.3770 + 0.9641) & 0.1250 \\ 0.1250 & (-1.0018 + 0.9641) \end{bmatrix} \begin{bmatrix} m_{11} \\ m_{21} \end{bmatrix} = \begin{bmatrix} 0 \\ 0 \end{bmatrix}$$

or

$$-0.4129 m_{11} + 0.1250 m_{21} = 0$$

$$0.1250 m_{11} - 0.0377 m_{21} = 0$$

We wish to obtain the normalized solution to these equations, that is, the one for which $m_{11}^2 + m_{21}^2 = 1$. There is no unique solution to these equations, and it is most convenient to assign an arbitrary value to one unknown, solve the system, and normalize the solution. Letting $m_{21}^* = 1$, we find $m_{11}^* = 0.3027$. To normalize this solution, divide m_{11}^* and m_{21}^* by $\sqrt{(m_{11}^*)^2 + (m_{21}^*)^2} = \sqrt{(0.3027)^2 + (1)^2} = 1.0448$. This yields the normalized solution

$$m_{11} = \frac{m_{11}^*}{1.0448} = \frac{0.3027}{1.0448} = 0.2897, \quad m_{21} = \frac{m_{21}^*}{1.0448} = \frac{1}{1.0448} = 0.9571$$

which is the first column of the M matrix.

Using $\lambda_2 = -1.4147$, we can repeat the above procedure, yielding $m_{12} = -0.9574$ and $m_{22} = 0.2888$ as the second column of M. Thus, we have

$$M = \begin{bmatrix} 0.2897 & -0.9574 \\ 0.9571 & 0.2888 \end{bmatrix}$$

The relationship between the w and x variables is

$$\begin{bmatrix} w_1 \\ w_2 \end{bmatrix} = \begin{bmatrix} 0.2897 & 0.9571 \\ -0.9574 & 0.2888 \end{bmatrix} \begin{bmatrix} x_1 - 0.3890 \\ x_2 - 0.3056 \end{bmatrix}$$

or

$$w_1 = 0.2897(x_1 - 0.3890) + 0.9571(x_2 - 0.3056)$$

$$w_2 = -0.9574(x_1 - 0.3890) + 0.2888(x_2 - 0.3056)$$

If the experimenter wished to explore the response surface in the vicinity of the stationary point, he could determine appropriate points to take observations in the (w_1, w_2) space, and then use the above relationship to convert these points into the (x_1, x_2) space so that the runs may be made.

14-4 RESPONSE SURFACE DESIGNS

A response surface can be most efficiently fit if proper attention is given to the choice of experimental design. In this section we discuss designs useful in response surface methodology.

14-4.1 Designs for Fitting the First-Order Model

Suppose we wish to fit the first-order model in k variables

$$y = \beta_0 + \sum_{i=1}^{k} \beta_i x_i + \epsilon \tag{14-11}$$

There is a unique class of designs that minimize the variance of the regression coefficients $\{\hat{\beta}_i\}$. These are the *orthogonal* first-order designs. A first-order design is orthogonal if the off-diagonal elements of the $(\mathbf{X'X})$ matrix are all zero. This implies that the cross-products of the columns of the $\mathbf{X}$ matrix sum to zero.

The class of orthogonal first-order designs includes the 2^k factorial and fractions of the 2^k series in which main effects are not aliased with each other. In using these designs, we assume that the k factors are coded to the standardized levels ± 1. As an example, suppose we use a 2^3 design to fit the first-order model

$$y = \beta_0 + \beta_1 x_1 + \beta_2 x_2 + \beta_3 x_3 + \epsilon$$

The $\mathbf{X}$ matrix for fitting this model is

$$
\mathbf{X} =
\begin{array}{cccc}
\beta_0 & \beta_1 & \beta_2 & \beta_3 \\
\end{array}
\begin{bmatrix}
1 & -1 & -1 & -1 \\
1 & -1 & -1 & 1 \\
1 & -1 & 1 & -1 \\
1 & -1 & 1 & 1 \\
1 & 1 & -1 & -1 \\
1 & 1 & -1 & 1 \\
1 & 1 & 1 & -1 \\
1 & 1 & 1 & 1 \\
\end{bmatrix}
$$

It is easy to verify that the off-diagonal elements of $(\mathbf{X'X})$ are zero for this design.

The 2^k design does not afford an estimate of the experimental error unless some runs are repeated. A common method of including replication in the 2^k design is to augment the design with several observations at the center (the point $x_i = 0$, $i = 1, 2, \ldots, k$). The addition of center points to the 2^k design does not influence the $\{\hat{\beta}_i\}$ for $i \geq 1$, but the estimate of β_0 becomes the grand average of all observations. Example 14-1 illustrates the use of a 2^2 design augmented with five center points to fit a first-order model.

Another orthogonal first-order design is the *simplex*. The simplex is a regularly sided figure with $k + 1$ vertices in k dimensions. Thus, for $k = 2$ the simplex design is an equilateral triangle, while for $k = 3$ it is a regular tetrahedron. A simplex design in two dimensions is shown in Figure 14-8.

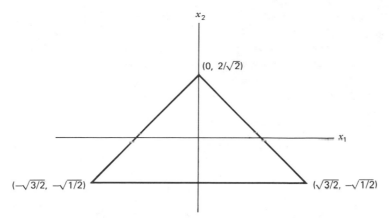

Figure 14-8. Simplex design for $k = 2$.

14-4.2   Designs for Fitting the Second-Order Model

An experimental design for fitting a second-order model must have at least three levels of each factor so that the model parameters can be estimated. The preferred class of second-order response surface designs is the class of *rotatable* designs. An experimental design is said to be rotatable if the variance of the predicted response $\hat{y}$ at some point x is a function only of the *distance* of the point from the design center, and not a function of direction. This implies that the variance contours of $\hat{y}$ are concentric circles. Furthermore, a design with this property will leave the variance of $\hat{y}$ unchanged when the design is rotated about the center $(0, 0, \ldots, 0)$; hence the name *rotatable* design.

The most widely used design for fitting a second-order model is the *central composite design*. These designs consist of a 2^k factorial or fractional factorial (coded to the usual ± 1 notation) augmented by $2k$ *axial* points $(\pm\alpha, 0, 0, \ldots, 0)$, $(0, \pm\alpha, 0, \ldots, 0)$, $(0, 0, \pm\alpha, \ldots, 0)$, $\ldots$, $(0, 0, 0, \ldots, \pm\alpha)$ and n_0 center points $(0, 0, \ldots, 0)$. Central composite designs for $k = 2$ and $k = 3$ are shown in Figure 14-9.

A central composite design is made rotatable by the choice of α. The value of α for rotatability depends on the number of points in the factorial portion of the design; in fact, $\alpha = (F)^{1/4}$ yields a rotatable central composite design where F is the number of points used in the factorial portion of the design. Example 14-2 illustrates the use of a rotatable central composite design. That example has two factors, and the factorial portion contains $F = 2^2 = 4$ points. Thus the value of α for rotatability is $\alpha = (4)^{1/4} = 1.414$. Another useful property of the central composite design is that it may be "built up" from the first-order design (the 2^k) by adding the axial points and perhaps several center points. This was demonstrated in Examples 14-1 and 14-2.

Other properties of the central composite design may be controlled by choice of the number of center points, n_0. With proper choice of n_0 the central composite design may be made *orthogonal*, or it can be made *uniform precision*. In a uniform precision design the variance of $\hat{y}$ at the origin is equal to the variance of $\hat{y}$ at unit distance from the origin. A uniform precision design affords more protection against bias in the regression coefficients due to the presence of third-order and higher terms in the true surface than does an orthogonal design. Table 14-8 provides the design parameters for both orthogonal and uniform precision rotatable central composite designs for various values of k.

Table 14-8 Orthogonal and Uniform Precision Rotatable Central Composite Designs[a]

k	2	3	4	5	5 $\frac{1}{2}$ rep.	6	6 $\frac{1}{2}$ rep.	7 $\frac{1}{2}$ rep.	8 $\frac{1}{2}$ rep.
F	4	8	16	32	16	64	32	64	128
Axial									
points	4	6	8	10	10	12	12	14	16
n_0(up)	5	6	7	10	6	15	9	14	20
n_0(orth)	8	9	12	17	10	24	15	22	33
N(up)	13	20	31	52	32	91	53	92	164
N(orth)	16	23	36	59	36	100	59	100	177
α	1.414	1.682	2.000	2.378	2.000	2.828	2.378	2.828	3.364

[a]Adapted from "Multifactor Experimental Designs for Exploring Response Surfaces" by G. E. P. Box and J. S. Hunter, *Annals of Mathematical Statistics*, Vol. 28, pp. 195–242, 1937, with permission of the publisher.

There are several other rotatable designs that are occasionally useful for problems involving two or three variables. These designs consist of points equally-spaced on a circle ($k = 2$) or a sphere ($k = 3$), and are regular polygons or polyhedrons (see Figure 14-9). Because the design points are equidistant from the origin, these arrangements are often called *equiradial* designs.

For $k = 2$, a rotatable equiradial design is obtained by combining $n_2 \geqslant 5$ points equally spaced on a circle with $n_1 \geqslant 1$ points at the center of the circle. Particularly useful designs for $k = 2$ are the pentagon and hexagon. These designs are shown in Figure 14-10a and b. For $k = 3$ the only equiradial arrangements that contain enough points to allow all parameters in the second-order model to be estimated are the icosahedron (12 points) and the dodecahedron (20 points).

14-5 EVOLUTIONARY OPERATION

Response surface methodology is often applied to pilot plant operations by research and development personnel. When it is applied to a full-scale production process it is usually only done once (or very infrequently), because the experimental procedure is relatively elaborate. However, conditions that were optimum for the pilot plant

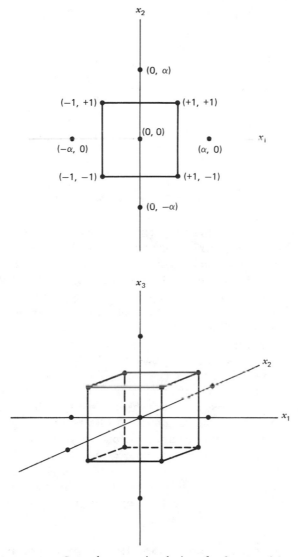

Figure 14-9. Central composite designs for $k = 2$ and $k = 3$.

may not be optimum for the full-scale process. The pilot plant may produce two pounds of product per day while the full-scale process may produce 2000 pounds per day. This "scale-up" of the pilot plant to the full-scale production process usually results in distortion of the optimum conditions. Even if the full-scale plant begins operation at the optimum, it will eventually "drift" away from that point due to variations in raw materials, environmental changes, and operating personnel.

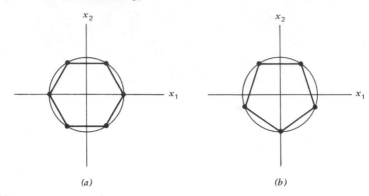

Figure 14-10.   Equiradial designs for two variables. (*a*) Hexagon. (*b*) Pentagon.

A method is needed for continuous operation and monitoring of a full-scale process with the goal of moving the operating conditions towards the optimum or following a "drift." The method should not require large or sudden changes in operating conditions that might disrupt production. Evolutionary operation (EVOP) was proposed by Box (1957) as such an operating procedure. It is designed as a method of routine plant operation that is carried out by operating personnel with minimum assistance from the research and development staff.

EVOP consists of systematically introducing small changes in the levels of the operating variables under consideration. Usually, a 2^k design is employed to do this. The changes in the variables are assumed to be small enough so that serious disturbances in yield, quality, or quantity will not occur, yet large enough so that potential improvements in process performance will eventually be discovered. Data is collected on the response variables of interest at each point of the 2^k design. When one observation has been taken at each design point, a *cycle* is said to have been completed. The effects and interactions of the process variables may then be computed. Eventually, after several cycles, the effect of one or more process variables, or their interactions, may appear to have a significant effect on the response. At this point, a decision may be made to change the basic operating conditions to improve the response. When improved conditions have been detected, a *phase* is said to have been completed.

In testing the significance of process variables and interactions, an estimate of experimental error is required. This is calculated from the cycle data. Also, the 2^k design is usually centered about the current best operating conditions. By comparing the response at this point with the 2^k points in the factorial portion we may check on the *change in mean* (CIM); that is, if the process is really centered at the maximum (say), then the response at the center should be significantly greater than the responses at the 2^k-peripheral points.

In theory, EVOP can be applied to k process variables. In practice, only two or three variables are usually considered. We will give an example of the procedure for two variables. Box and Draper (1969) give a detailed discussion of the three-variable case, including necessary forms and worksheets.

■ **EXAMPLE 14-3.** Consider a chemical process whose yield is a function of temperature (x_1) and pressure (x_2). The current operating conditions are $x_1 = 250°F$ and $x_2 = 145$ psi. The EVOP procedure uses the 2^2 design plus center point shown in Figure 14-11. The cycle is completed by running each design point in numerical order (1, 2, 3, 4, 5). The yields in the first cycle are shown in Figure 14-11.

Table 14-9 EVOP Calculation Sheet—Example 14-3, n = 1

		Cycle: $n = 1$	Phase: 1
		Response: Yield	Date: 1-11-75

```
  5 ┌──┐ 3
    │.1│
  2 └──┘ 4
```

						Calculation of Standard Deviation
	Calculation of Averages					
Operating Conditions	(1)	(2)	(3)	(4)	(5)	
(i) Previous cycle sum						Previous sum S =
(ii) Previous cycle average						Previous average S −
(iii) New observations	84.5	84.2	84.9	84.5	84.3	New S = range × $f_{5,n}$ =
(iv) Differences						
[(ii) − (iii)]						Range of (iv) =
(v) New sums [(i) + (iii)]	84.5	84.2	84.9	84.5	84.3	New sum S =
(vi) New averages	84.5	84.2	84.9	84.5	84.3	New average S = $\dfrac{\text{new sum } S}{n-1}$
$[\bar{y}_i = (v)/n]$						

Calculation of Effects	Calculation of Error Limits
Temperature effect = $\frac{1}{2}(\bar{y}_3 + \bar{y}_4 - \bar{y}_2 - \bar{y}_5) = 0.45$	For new average $\dfrac{2}{\sqrt{n}} S =$
Pressure effect = $\frac{1}{2}(\bar{y}_3 + \bar{y}_5 - \bar{y}_2 - \bar{y}_4) = 0.25$	For new effects $\dfrac{2}{\sqrt{n}} S =$
$T \times P$ interaction effect = $\frac{1}{2}(\bar{y}_2 + \bar{y}_3 - \bar{y}_4 - \bar{y}_5) = 0.15$	
Change-in-mean effect $\frac{1}{5}(\bar{y}_2 + \bar{y}_3 + \bar{y}_4 + \bar{y}_5 - 4\bar{y}_1) = -0.02$	For change in mean $\dfrac{1.78}{\sqrt{n}} S =$

Figure 14-11. 2^2 design for EVOP.

The yields from the first cycle are entered in the EVOP calculation sheet, shown in Table 14-9. At the end of the first cycle, no estimate of the standard deviation can be made. The effects and interaction for temperature and pressure are calculated in the usual manner for a 2^2.

A second cycle is then run and the yield data entered in another EVOP calculation sheet, shown in Table 14-10. At the end of the second cycle, the experimental error can be estimated and the estimates of the effects compared to approximate 95 percent (two standard deviation) limits. Note that the range refers to the range of the differences in row (iv); thus the range is +1.0 – (–1.0) = 2.0. Since none of the effects in Table 14-10 exceed their error limits, the true effect is probably zero, and no changes in operating conditions are contemplated.

The results of a third cycle are shown in Table 14-11. The effect of pressure now exceeds its error limit and the temperature effect is equal to the error limit. A change in operating conditions is now probably justified.

In light of the results, it seems reasonable to begin a new EVOP phase about point (3). Thus $x_1 = 225°F$ and $x_2 = 150$ psi would become the center of the 2^2 design in the second phase.

An important aspect of EVOP is feeding the information generated back to the process operators and supervisors. This is accomplished by a prominently displayed EVOP information board. The information board for this example at the end of cycle three is shown in Table 14-12. ∎

Most of the quantities on the EVOP calculation sheet follow directly from the analysis of the 2^k factorial design. For example, the variance of any effect, such as

Table 14-10 EVOP Calculation Sheet—Example 14-3, n = 2

5 ___ 3		Cycle: $n = 2$		Phase: 1
[.1]		Response: Yield		Date: 1-11-75
2 ___ 4				

Operating Conditions	(1)	(2)	(3)	(4)	(5)	Calculation of Standard Deviation
		Calculation of Averages				
(i) Previous cycle sum	84.5	84.2	84.9	84.5	84.3	Previous sum S =
(ii) Previous cycle average	84.5	84.2	84.9	84.5	84.3	Previous average S =
(iii) New observations	84.9	84.6	85.9	83.5	84.0	New S = range $\times f_{5,n}$ = 0.60
(iv) Differences [(ii) − (iii)]	−0.4	−0.4	−1.0	+1.0	0.3	Range of (iv) = 2.0
(v) New sums [(i) + (iii)]	169.4	168.8	170.8	168.0	168.3	New sum S = 0.60
(vi) New averages [$\bar{y}_i$ = (v)/n]	84.70	84.40	85.40	84.00	84.15	New average S = $\dfrac{\text{New sum } S}{n-1} = 0.60$

Temperature effect = $\frac{1}{2}(\bar{y}_3 + \bar{y}_4 - \bar{y}_2 - \bar{y}_5) = 0.43$

For new average

$$\frac{2}{\sqrt{n}} S = 0.85$$

Pressure effect = $\frac{1}{2}(\bar{y}_3 + \bar{y}_5 - \bar{y}_2 - \bar{y}_4) = 0.58$

For new effects

$$\frac{2}{\sqrt{n}} S = 0.85$$

$T \times P$ interaction effect = $\frac{1}{2}(\bar{y}_2 + \bar{y}_3 - \bar{y}_4 - \bar{y}_5) = 0.83$

Change-in-mean effect = $\frac{1}{5}(\bar{y}_2 + \bar{y}_3 + \bar{y}_4 + \bar{y}_5 - 4\bar{y}_1) = -0.17$

For change in mean

$$\frac{1.78}{\sqrt{n}} S = 0.76$$

Table 14-11 EVOP Calculation Sheet—Example 14-3, n = 3

Cycle: $n = 3$ Phase: 1

Response: Yield   Date: 1-11-75

	Calculation of Averages					Calculation of Standard Deviation
Operating Conditions	(1)	(2)	(3)	(4)	(5)	
(i) Previous cycle sum	169.4	168.8	170.8	168.0	168.3	Previous sum $S = 0.60$
(ii) Previous cycle average	84.70	84.40	85.40	84.00	84.15	Previous average $S = 0.60$
(iii) New observations	85.0	84.0	86.6	84.9	85.2	New S = range $\times f_{5,n}$ = 0.56
(iv) Differences [(ii) − (iii)]	−0.30	+0.40	−1.20	−0.90	−1.05	Range of (iv) = 1.60
(v) New sums [(i) + (iii)]	254.4	252.8	257.4	252.9	253.5	New sum S = 1.16
(vi) New averages $[\bar{y}_i = (v)/n]$	84.80	84.27	85.80	84.30	84.50	New average S $= \dfrac{\text{New sum } S}{n - 1} = 0.58$

Calculation of Effects	Calculations of Error Limits
Temperature effect $= \frac{1}{2}(\bar{y}_3 + \bar{y}_4 - \bar{y}_2 - \bar{y}_5) = 0.67$	For new average $\dfrac{2}{\sqrt{n}} S = 0.67$
Pressure effect $= \frac{1}{2}(\bar{y}_3 + \bar{y}_5 - \bar{y}_2 - \bar{y}_4) = 0.87$	For new effects $\dfrac{2}{\sqrt{n}} S = 0.67$
$T \times P$ interaction effect $= \frac{1}{2}(\bar{y}_2 + \bar{y}_3 - \bar{y}_4 - \bar{y}_5) = 0.64$	
Change-in-mean effect $= \frac{1}{5}(\bar{y}_2 + \bar{y}_3 + \bar{y}_4 + \bar{y}_5 - 4\bar{y}_1) = -0.07$	For change in mean $\dfrac{1.78}{\sqrt{n}} S = 0.60$

$\frac{1}{2}(\bar{y}_3 + \bar{y}_5 - \bar{y}_2 - \bar{y}_4)$, is simply

$$V[\tfrac{1}{2}(\bar{y}_3 + \bar{y}_5 - \bar{y}_2 - \bar{y}_4)] = \tfrac{1}{4}(\sigma^2_{\bar{y}_3} + \sigma^2_{\bar{y}_5} + \sigma^2_{\bar{y}_2} + \sigma^2_{\bar{y}_4})$$

$$= \frac{1}{4}(4\sigma^2_{\bar{y}}) = \frac{\sigma^2}{n}$$

Table 14-12 EVOP Information Board—Cycle Three

Response: Percent Yield

Requirement: Maximize

Error Limits for Averages: ±0.67

Effects with	Temperature	0.67	±0.67
95% error	Pressure	0.87	±0.67
Limits:	$T \times P$	0.64	±0.67
	Change in mean	0.07	±0.60
Standard deviation:	0.58		

where σ^2 is the variance of the observations (y). Thus two standard deviation (corresponding to 95 percent) error limits on any effect would be $\pm 2\sigma/\sqrt{n}$. The variance of the change in mean is

$$V(\text{CIM}) = V[\tfrac{1}{5}(\bar{y}_2 + \bar{y}_3 + \bar{y}_4 + \bar{y}_5 - 4\bar{y}_1)]$$

$$= \frac{1}{25}(4\sigma_{\bar{y}}^2 + 16\sigma_{\bar{y}}^2) = \left(\frac{20}{25}\right)\frac{\sigma^2}{n}$$

Thus, two standard deviation error limits on the CIM are $\pm(2\sqrt{20/25})\sigma/\sqrt{n} = \pm 1.78\sigma/\sqrt{n}$.

The standard deviation σ is estimated by the range method. Let $y_i(n)$ denote the observation at the ith design point in cycle n, and $\bar{y}_i(n)$ is the corresponding average of $y_i(n)$ after n cycles. The quantities in row (iv) of the EVOP calculation sheet are the differences $y_i(n) - \bar{y}_i(n-1)$. The variance of these differences is $V[y_i(n) - \bar{y}_i(n-1)] \equiv \sigma_D^2 = \sigma^2[1 + 1/(n-1)] = \sigma^2[n/(n-1)]$. The range of the differences, say R_D, is related to the estimate of the standard deviation of the differences by $\hat{\sigma}_D = R_D/d_2$. The factor d_2 depends on the number of observations used in computing R_D. Now $R_D/d_2 = \hat{\sigma}\sqrt{n/(n-1)}$, so

$$\hat{\sigma} = \sqrt{\frac{(n-1)}{n}}\,\frac{R_D}{d_2} = (f_{k,n})\,R_D \equiv S$$

can be used to estimate the standard deviation of the observations, where k denotes the number of points used in the design. For a 2^2 with one center point we have $k = 5$, and for a 2^3 with one center point we have $k = 9$. Values of $f_{k,n}$ are given in Table 14-13.

Table 14-13 Values of $f_{k,n}$

$n =$	2	3	4	5	6	7	8	9	10
$k = 5$	0.30	0.35	0.37	0.38	0.39	0.40	0.40	0.40	0.41
9	0.24	0.27	0.29	0.30	0.31	0.31	0.31	0.32	0.32
10	0.23	0.26	0.28	0.29	0.30	0.30	0.30	0.31	0.31

14-6 PROBLEMS

14-1. A chemical plant produces oxygen by liquifying air and separating it into its component gases by fractional distillation. The purity of the oxygen is a function of the main condenser temperature and the pressure ratio between the upper and lower columns. Current operating conditions are temperature $(\xi_1) = -220°C$ and pressure ratio $(\xi_2) = 1.2$. Using the data below, find the path of steepest ascent.

Temperature (ξ_1)	Pressure Ratio (ξ_2)	Purity
-225	1.1	82.8
-225	1.3	83.5
-215	1.1	84.7
-215	1.3	85.0
-220	1.2	84.1
-220	1.2	84.5
-220	1.2	83.9
-220	1.2	84.3

14-2. An industrial engineer has developed a computer simulation model of a two-item inventory system. The decision variables are the order quantity and the

Item 1		Item 2		
Order Quantity (ξ_1)	Reorder Point (ξ_2)	Order Quantity (ξ_3)	Reorder Point (ξ_4)	Total Cost
100	25	250	40	625
140	45	250	40	670
140	25	300	40	663
140	25	250	80	654
100	45	300	40	648
100	45	250	80	634
100	25	300	80	692
140	45	300	80	686
120	35	275	60	680
120	35	275	60	674
120	35	275	60	681

reorder point for each item. The response to be minimized is total inventory cost. The simulation model is used to produce the data shown in the table on page 364. Identify the experimental design. Find the path of steepest decent.

14-3. Verify that the design shown below is a simplex. Fit the first-order model and find the path of steepest ascent.

x_1	x_2	x_3	y
0	$\sqrt{2}$	-1	18.5
$-\sqrt{2}$	0	1	19.8
0	$-\sqrt{2}$	-1	17.4
$\sqrt{2}$	0	1	22.5

14-4. For the first-order model

$$\hat{y} = 60 + 1.5x_1 - 0.8x_2 + 2.0x_3$$

find the path of steepest ascent. The variables are coded as $-1 \leqslant x_i \leqslant 1$.

14-5. Consider the data shown below. Fit a second-order model and perform a canonical analysis.

x_1	x_2	x_3	y
-1	-1	-1	66
-1	-1	1	70
-1	1	-1	78
-1	1	1	60
1	-1	-1	80
1	-1	1	70
1	1	-1	100
1	1	1	75
-1.682	0	0	100
1.682	0	0	80
0	-1.682	0	68
0	1.682	0	63
0	0	-1.682	65
0	0	1.682	82
0	0	0	113
0	0	0	100
0	0	0	118
0	0	0	88
0	0	0	100
0	0	0	85

14-6. Consider the data shown below. Fit a second-order model and perform a canonical analysis.

x_1	x_2	y
-1	-1	54
-1	1	45
1	-1	32
1	1	47
-1.414	0	50
1.414	0	53
0	-1.414	47
0	1.414	51
0	0	41
0	0	39
0	0	44
0	0	42
0	0	40

14-7. Verify that the data in Table 14-1 result in the first-order model shown on page 344. Also verify the analysis of variance for this model in Table 14-2.

14-8. Verify that the data in Table 14-6 result in the second-order model shown on page 350. Also verify the analysis of variance for this model in Table 14-7.

14-9. The hexagon design shown below is used to collect data for fitting a second-order model. Fit the second-order model and perform a canonical analysis.

x_1	x_2	y
1	0	68
0.5	$\sqrt{0.75}$	74
-0.5	$\sqrt{0.75}$	65
-1	0	60
-0.5	$-\sqrt{0.75}$	63
0.5	$-\sqrt{0.75}$	70
0	0	58
0	0	60
0	0	57
0	0	55
0	0	59

14-10. Verify that an orthogonal first-order design is also first-order rotatable.

14-11. Show that augmenting a 2^k design with n_0 center points does not affect the estimates of the β_i ($i = 1, 2, \ldots, k$), but that the estimate of the intercept β_0 is the average of all $2^k + n_0$ observations.

14-12. *An Orthogonal Central Composite Design.* Suppose a central composite design is used to fit the second-order model

$$y = \beta_0 + \sum_{i=1}^{k} \beta_i x_i + \sum_{i=1}^{k} \beta_{ii} (x_i^2 - \overline{x_i^2}) + \sum_i \sum_{i \neq j} \beta_{ij} x_i x_j + \epsilon$$

Show that the design can be made orthogonal by appropriate choice of α.

14-13. *The Rotatable Central Composite Design.* It can be shown that a second-order design is rotatable if $\sum_{u=1}^{n} x_{iu}^a x_{ju}^b = 0$ if a or b (or both) are odd, and if $\sum_{u=1}^{n} x_{iu}^4 = 3 \sum_{u=1}^{n} x_{iu}^2 x_{ju}^2$. Show that for the central composite design these conditions lead to $\alpha = (F)^{1/4}$ for rotatability, where F is the number of points in the factorial portion.

14-14. Second-order designs require relatively many observations, and often blocking is necessary due to the time effects. The necessary and sufficient conditions for constructing a second-order design in b orthogonal blocks are that each block forms an orthogonal first-order design and that $\sum_m x_{im}^2 / \sum_{u=1}^{n} x_{iu}^2 = n_m/n$ ($m = 1, 2, \ldots, b$), where the summation over m implies summation of observations in the mth block only, and n_m is the number of observations in the mth block. Verify that the central composite design below blocks orthogonally.

Block 1			Block 2			Block 3		
x_1	x_2	x_3	x_1	x_2	x_3	x_1	x_2	x_3
0	0	0	0	0	0	-1.633	0	0
0	0	0	0	0	0	1.633	0	0
1	1	1	1	1	-1	0	-1.633	0
1	-1	-1	1	-1	1	0	1.633	0
-1	-1	1	-1	1	1	0	0	-1.633
-1	1	-1	-1	-1	-1	0	0	1.633
						0	0	0
						0	0	0

14-15. *Blocking in the Central Composite Design.* Consider a central composite design in two blocks. Can rotatable designs always be found which block orthogonally?

14-16. How could a hexagon design be run in two orthogonal blocks?

14-17. Yield during the first four cycles from a chemical process is shown below. The variables are percent concentration (x_1) at levels 30, 31, and 32, and temperature (x_2) at $140°$, $142°$, and $144°$F. Analyze by EVOP methods.

	Conditions				
Cycle	(1)	(2)	(3)	(4)	(5)
1	60.7	59.8	60.2	64.2	57.5
2	59.1	62.8	62.5	64.6	58.3
3	56.6	59.1	59.0	62.3	61.1
4	60.5	59.8	64.5	61.0	60.1

14-18. Design an EVOP calculation sheet for three variables. Derive the two standard error limits for the effects and for the change in mean.

14-19. Suppose that we approximate a response surface with a model of order d_1, such as $y = X_1 \beta_1 + \epsilon$, when the true surface is described by a model of order $d_2 > d_1$, that is, $E(y) = X_1 \beta_1 + X_2 \beta_2$.

  (a) Show that the regression coefficients are biased, that is, that $E(\hat{\beta}_1) = \beta_1 + A\beta_2$, where $A = (X_1' X_1)^{-1} X_1' X_2$. A is usually called the alias matrix.

  (b) If $d_1 = 1$ and $d_2 = 2$, and a full 2^k is used to fit the model, use the result in part (a) to determine the alias structure.

  (c) If $d_1 = 1$, $d_2 = 2$, and $k = 3$, find the alias structure assuming that a 2^{3-1} design is used to fit the model.

  (d) If $d_1 = 1$, $d_2 = 2$, $k = 3$, and the simplex design in Problem 14-3 is used to fit the model, determine the alias structure and compare the results with part (c).

Chapter Fifteen
Analysis of Covariance

15-1 INTRODUCTION

In previous chapters we have discussed the use of randomized block and Latin square designs to improve the precision with which comparisons between treatments are made. The analysis of covariance is another technique occasionally useful for improving the precision of an experiment. Suppose that in an experiment with response variable y, there is another variable, such as x, and y is linearly related to x. Furthermore, x cannot be controlled by the experimenter but can be observed along with y. The variable x is called a *covariate* or *concomitant variable*. The analysis of covariance involves adjusting the observed response variable for the linear effect of the concomitant variable. If such an adjustment is not performed, the concomitant variable could inflate the error mean square and make true differences in the response due to treatments harder to detect. Thus, analysis of covariance is a method of adjusting for the effects of an uncontrollable nuisance variable. As we will see, the procedure is a combination of analysis of variance and regression analysis.

As an example of an experiment in which the analysis of covariance may be employed, consider a study performed to determine if there is a difference in the strength of a monofilament fiber produced by three different machines. However, the strength of the fiber is also affected by its denure or thickness, and, consequently, a thicker fiber will generally be stronger than a thinner one. The analysis of covariance could be used to remove the effect of thickness (x) on strength (y) when testing for differences in strength between machines.

15-2 ONE-WAY CLASSIFICATION WITH A SINGLE COVARIATE

The basic procedure for the analysis of covariance will be illustrated for a single-factor experimental design with one covariate. Assuming that there is a linear relationship between the response and the covariate, an appropriate statistical model is

$$y_{ij} = \mu + \tau_i + \beta(x_{ij} - \bar{x}_{..}) + \epsilon_{ij} \begin{cases} i = 1, 2, \ldots, a \\ j = 1, 2, \ldots, n \end{cases} \tag{15-1}$$

where y_{ij} is the jth observation on the response variable taken under the ith treatment or level of the single factor, x_{ij} is the measurement made on the covariate or concomitant variable corresponding to y_{ij} (i.e., the ijth run), $\bar{x}_{..}$ is the mean of the x_{ij} values, μ is an overall mean, τ_i is the effect of the ith treatment, β is a linear regression coefficient indicating the dependency of y_{ij} on x_{ij} and ϵ_{ij} is a random error component. We assume that the errors ϵ_{ij} are NID $(0, \sigma^2)$, that the slope $\beta \neq 0$ and the true relationship between y_{ij} and x_{ij} is linear, that the regression coefficients for each treatment are identical, that the treatment effects sum to zero $(\sum_{i=1}^{a} \tau_i = 0)$, and that the concomitant variable x_{ij} is not affected by the treatments.

Notice immediately from Equation 15-1 that the analysis of covariance model is a combination of the linear models employed in analysis of variance and regression. That is, we have treatment effects $\{\tau_i\}$ as in a one-way classification analysis of variance, and a regression coefficient β as in regression analysis. The concomitant variable in (15-1) is expressed as $(x_{ij} - \bar{x}_{..})$ instead of x_{ij} so that the parameter μ is preserved as the overall mean. The model could have been written as

$$y_{ij} = \mu' + \tau_i + \beta x_{ij} + \epsilon_{ij} \begin{cases} i = 1, 2, \ldots, a \\ j = 1, 2, \ldots, n \end{cases} \tag{15-2}$$

where μ' is a constant not equal to the overall mean, which for this model is $\mu' + \beta\bar{x}_{..}$. Equation 15-1 is more widely found in the literature.

To present the analysis, we introduce the following notation:

$$S_{yy} = \sum_{i=1}^{a}\sum_{j=1}^{n} (y_{ij} - \bar{y}_{..})^2 = \sum_{i=1}^{a}\sum_{j=1}^{n} y_{ij}^2 - \frac{y_{..}^2}{an} \tag{15-3}$$

$$S_{xx} = \sum_{i=1}^{a}\sum_{j=1}^{n} (x_{ij} - \bar{x}_{..})^2 = \sum_{i=1}^{a}\sum_{j=1}^{n} x_{ij}^2 - \frac{x_{..}^2}{an} \tag{15-4}$$

$$S_{xy} = \sum_{i=1}^{a}\sum_{j=1}^{n} (x_{ij} - \bar{x}_{..})(y_{ij} - \bar{y}_{..}) = \sum_{i=1}^{a}\sum_{j=1}^{n} x_{ij}y_{ij} - \frac{(x_{..})(y_{..})}{an} \tag{15-5}$$

$$T_{yy} = \sum_{i=1}^{a} \sum_{j=1}^{n} (\bar{y}_{i.} - \bar{y}_{..})^2 = \sum_{i=1}^{a} \frac{y_{i.}^2}{n} - \frac{y_{..}^2}{an} \tag{15-6}$$

$$T_{xx} = \sum_{i=1}^{a} \sum_{j=1}^{n} (\bar{x}_{i.} - \bar{x}_{..})^2 = \sum_{i=1}^{a} \frac{x_{i.}^2}{n} - \frac{x_{..}^2}{an} \tag{15-7}$$

$$T_{xy} = \sum_{i=1}^{a} \sum_{j=1}^{n} (\bar{x}_{i.} - \bar{x}_{..})(\bar{y}_{i.} - \bar{y}_{..}) = \sum_{i=1}^{a} \frac{(x_{i.})(y_{i.})}{n} - \frac{(x_{..})(y_{..})}{an} \tag{15-8}$$

$$E_{yy} = \sum_{i=1}^{a} \sum_{j=1}^{n} (y_{ij} - \bar{y}_{i.})^2 = S_{yy} - T_{yy} \tag{15-9}$$

$$E_{xx} = \sum_{i=1}^{a} \sum_{j=1}^{n} (x_{ij} - \bar{x}_{i.})^2 = S_{xx} - T_{xx} \tag{15-10}$$

$$E_{xy} = \sum_{i=1}^{a} \sum_{j=1}^{n} (x_{ij} - \bar{x}_{i.})(y_{ij} - \bar{y}_{i.}) = S_{xy} - T_{xy} \tag{15-11}$$

Note that, in general, $S = T + E$, where the symbols S, T, and E are used to denote sums of squares and cross-products for total, treatments, and error, respectively. The sums of squares for x and y must be nonnegative; however, the sums of cross-products (xy) may be negative.

We now show how the analysis of covariance adjusts the response variable for the effect of the covariate. Consider the full model (15-1). The least-squares estimators of μ, τ_i, and β are $\hat{\mu} = \bar{y}_{..}$, $\hat{\tau}_i = \bar{y}_{i.} - \bar{y}_{..} - \hat{\beta}(\bar{x}_{i.} - \bar{x}_{..})$, and

$$\hat{\beta} = \frac{E_{xy}}{E_{xx}} \tag{15-12}$$

The error sum of squares in this model is

$$SS_E = E_{yy} - (E_{xy})^2 / E_{xx} \tag{15-13}$$

with $a(n - 1) - 1$ degrees of freedom. The experimental error variance is estimated by

$$MS_E = \frac{SS_E}{a(n - 1) - 1}$$

Now suppose that there is no treatment effect. The model (15-1) would then be

$$y_{ij} = \mu + \beta(x_{ij} - \bar{x}_{..}) + \epsilon_{ij} \tag{15-14}$$

and the least-squares estimators of μ and β are $\hat{\mu} = \bar{y}_{..}$ and $\hat{\beta} = S_{xy}/S_{xx}$. The sum of squares for error in this reduced model is

$$SS'_E = S_{yy} - (S_{xy})^2 / S_{xx} \tag{15-15}$$

with $an - 2$ degrees of freedom. In (15-15), the quantity $(S_{xy})^2/S_{xx}$ is the reduction in the sum of squares of y obtained through the linear regression of y on x. Furthermore, note that SS_E is smaller than SS'_E [because the model (15-1) contains additional parameters $\{\tau_i\}$], and the quantity $SS'_E - SS_E$ is a reduction in sum of squares due to the $\{\tau_i\}$. Therefore, the difference between SS'_E and SS_E, that is, $SS'_E - SS_E$, provides a sum of squares with $a - 1$ degrees of freedom for testing the hypothesis of no treatment effects. Consequently, to test H_0: $\tau_i = 0$, compute

$$F_0 = \frac{(SS'_E - SS_E)/(a - 1)}{SS_E/[a(n - 1) - 1]} \tag{15-16}$$

which if the null hypothesis is true is distributed as $F_{a-1,\, a(n-1)-1}$. Thus we reject H_0: $\tau_i = 0$ if $F_0 > F_{\alpha,\, a-1,\, a(n-1)-1}$.

It is instructive to examine the display in Table 15-1. In this table we have presented the analysis of covariance as an "adjusted" analysis of variance. In the source of variation column, the total variability is measured by S_{yy} with $an - 1$ degrees of freedom. The source of variation "regression" has sum of squares $(S_{xy})^2/S_{xx}$ with one degree of freedom. If there were no concomitant variable, we would have $S_{xy} = S_{xx} = E_{xy} = E_{xx} = 0$. Then the sum of squares for error would be simply E_{yy} and the sum of squares for treatments would be $S_{yy} - E_{yy} = T_{yy}$. However, because of the presence of the concomitant variable we must "adjust" S_{yy} and E_{yy} for the regression of y on x as shown in Table 15-1. The adjusted error sum of squares has $a(n - 1) - 1$ degrees of freedom instead of $a(n - 1)$ degrees of freedom because an additional parameter (the slope β) is fitted to the data.

The computations are usually displayed in an analysis of covariance table, such as Table 15-2. This layout is employed because it conveniently summarizes all the required sums of squares and cross-products, as well as the sums of squares for testing hypotheses about treatment effects. In addition to testing the hypothesis that there are no differences in the treatment effects, we frequently find it useful in interpreting the data to present the adjusted treatment means. These adjusted means are computed according to

$$\text{adjusted } \bar{y}_{i.} = \bar{y}_{i.} - \hat{\beta}(\bar{x}_{i.} - \bar{x}_{..}) \qquad i = 1, 2, \ldots, a \tag{15-17}$$

where $\hat{\beta} = E_{xy}/E_{xx}$. This adjustment treatment mean is the least-squares estimator of $\mu + \tau_i$, $i = 1, 2, \ldots, a$, in the model (15-1). The standard error of any adjusted treatment mean is

$$S_{\text{adj. }\bar{y}_{i.}} = \left[MS_E \left(\frac{1}{n} + \frac{(\bar{x}_{i.} - \bar{x}_{..})^2}{E_{xx}} \right) \right]^{1/2} \tag{15-18}$$

Finally, we recall that the regression coefficient β in the model (15-1) has been assumed to be nonzero. We may test the hypothesis H_0: $\beta = 0$ by using the test statistic

$$F_0 = \frac{(E_{xy})^2/E_{xx}}{MS_E} \tag{15-19}$$

which under the null hypothesis is distributed as $F_{1,\, a(n-1)-1}$. Thus we reject H_0: $\beta = 0$ if $F_0 > F_{\alpha,\, 1,\, a(n-1)-1}$.

Table 15-1 Analysis of Covariance as an "Adjusted" Analysis of Variance

Source of Variation	Sum of Squares	Degrees of Freedom	Mean Square	F_0
Regression	$(S_{xy})^2/S_{xx}$	1		
Treatments	$SS'_E - SS_E =$ $S_{yy} - (S_{xy})^2/S_{xx} -$ $[E_{yy} - (E_{xy})^2/E_{xx}]$	$a - 1$	$\dfrac{SS'_E - SS_E}{a - 1}$	$\dfrac{(SS'_E - SS_E)/(a - 1)}{MS_E}$
Error	$SS_E = E_{yy} - (E_{xy})^2/E_{xx}$	$a(n - 1) - 1$	$MS_E = \dfrac{SS_E}{a(n - 1) - 1}$	
Total	S_{yy}	$an - 1$		

Table 15-2 Analysis of Covariance for a Single Factor Experiment with One Covariate

Source of Variation	Degrees of Freedom	Sums of Squares and Products			Adjusted for Regression		
		x	xy	y	y	Degrees of Freedom	Mean Square
Treatments	$a - 1$	T_{xx}	T_{xy}	T_{yy}			
Error	$a(n - 1)$	E_{xx}	E_{xy}	E_{yy}	$SS_E = E_{yy} - (E_{xy})^2/E_{xx}$	$a(n - 1) - 1$	$MS_E = \dfrac{SS_E}{a(n - 1) - 1}$
Total	$an - 1$	S_{xx}	S_{xy}	S_{yy}	$SS'_E = S_{yy} - (S_{xy})^2/S_{xx}$	$an - 2$	
Adjusted treatments					$SS'_E - SS_E$	$a - 1$	$\dfrac{SS'_E - SS_E}{a - 1}$

■**EXAMPLE 15-1.** Consider the experiment described in Section 15-1. Three differ-ent machines produce a monofilament fiber for a textile company. The process engineer is interested in determining if there is a difference in the breaking strength of the fiber produced by the three machines. However, the strength of a fiber is related to its diameter, with thicker fibers being generally stronger than thinner ones. A random sample of five fiber specimens is selected from each machine. The fiber strength (y) and corresponding diameter (x) for each specimen is shown in Table 15-3.

Table 15-3 Breaking Strength Data for Example 15-1
(y = strength in pounds and x = diameter in 10^{-3} in.)

Machine 1		Machine 2		Machine 3	
y	x	y	x	y	x
36	20	40	22	35	21
41	25	48	28	37	23
39	24	39	22	42	26
42	25	45	30	34	21
49	32	44	28	32	15
207	126	216	130	180	106

A graph of the breaking strength versus the fiber diameter is shown in Figure 15-1. There is a strong suggestion of a linear relationship between breaking strength and diameter, and it seems appropriate to remove the effect of diameter on strength by an analysis of covariance. Assuming that a linear relationship between breaking strength

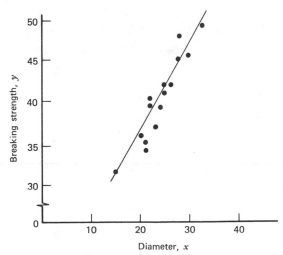

Figure 15-1. Breaking strength (y) versus fiber diameter (x), Example 15-1.

and diameter is appropriate, the model is

$$y_{ij} = \mu + \tau_i + \beta(x_{ij} - \bar{x}_{..}) + \epsilon_{ij} \begin{cases} i = 1, 2, 3 \\ j = 1, 2, \ldots, 5 \end{cases}$$

Using Equations 15-3 through 15-11, we may compute

$$S_{yy} = \sum_{i=1}^{3}\sum_{j=1}^{5} y_{ij}^2 - \frac{y_{..}^2}{an} = (36)^2 + (41)^2 + \cdots + (32)^2 - \frac{(603)^2}{(3)(5)} = 346.40$$

$$S_{xx} = \sum_{i=1}^{3}\sum_{j=1}^{5} x_{ij}^2 - \frac{x_{..}^2}{an} = (20)^2 + (25)^2 + \cdots + (15)^2 - \frac{(362)^2}{(3)(5)} = 261.73$$

$$S_{xy} = \sum_{i=1}^{3}\sum_{j=1}^{5} x_{ij}y_{ij} - \frac{(x_{..})(y_{..})}{an} = (20)(36) + (25)(41) + \cdots + (15)(32)$$

$$- \frac{(362)(603)}{(3)(5)} = 282.60$$

$$T_{yy} = \sum_{i=1}^{3} \frac{y_{i.}^2}{n} - \frac{y_{..}^2}{an} = \frac{(207)^2 + (216)^2 + (184)^2}{5} - \frac{(603)^2}{(3)(5)} = 140.40$$

$$T_{xx} = \sum_{i=1}^{3} \frac{x_{i.}^2}{n} - \frac{x_{..}^2}{an} = \frac{(126)^2 + (130)^2 + (106)^2}{5} - \frac{(362)^2}{(3)(5)} = 66.13$$

$$T_{xy} = \sum_{i=1}^{3} \frac{x_{i.}y_{i.}}{n} - \frac{(x_{..})(y_{..})}{an} = \frac{(126)(207) + (130)(216) + (106)(184)}{5} - \frac{(362)(603)}{(3)(5)}$$

$$= 96.00$$

$$E_{yy} = S_{yy} - T_{yy} = 346.40 - 140.40 = 206.00$$

$$E_{xx} = S_{xx} - T_{xx} = 261.73 - 66.13 = 195.60$$

$$E_{xy} = S_{xy} - T_{xy} = 282.60 - 96.00 = 186.60$$

From (15-15), we find

$$SS_E' = S_{yy} - (S_{xy})^2/S_{xx}$$

$$= 346.40 - (282.60)^2/261.73$$

$$= 41.27,$$

with $an - 2 = (3)(5) - 2 = 13$ degrees of freedom, and from (15-13),

$$SS_E = E_{yy} - (E_{xy})^2/E_{xx}$$

$$= 206.00 - (186.60)^2/195.60$$

$$= 27.99$$

with $a(n - 1) - 1 = 3(5 - 1) - 1 = 11$ degrees of freedom.

Table 15-4 Analysis of Covariance for the Breaking Strength Data

Source of Variation	Degrees of Freedom	Sums of Squares and Products			Adjusted for Regression		
		x	xy	y	y	Degrees of Freedom	Mean Square
Machines	2	66.13	96.00	140.40			
Error	12	195.60	186.60	206.00	27.99	11	2.54
Total	14	261.73	282.60	346.40	41.27	13	
Adjusted machines					13.28	2	6.64

The sum of squares for testing $H_0: \tau_i = 0$ is

$$SS_E' - SS_E = 41.27 - 27.99$$

$$= 13.28,$$

with $a - 1 = 3 - 1 = 2$ degrees of freedom. These calculations are summarized in Table 15-4.

To test the hypothesis that machines differ in the breaking strength of fiber produced, that is, $H_0: \tau_i = 0$, compute the test statistic from (15-16) as

$$F_0 = \frac{(SS_E' - SS_E)/(a - 1)}{SS_E/[a(n - 1) - 1]}$$

$$= \frac{13.28/2}{27.99/11} = \frac{6.64}{2.54} = 2.61$$

Comparing this to $F_{.10, 2, 11} = 2.86$, we find that the null hypothesis cannot be rejected. That is, there is no strong evidence that the fiber produced by the three machines differs in breaking strength.

The estimate of the regression coefficient is computed from (15-12) as

$$\hat{\beta} = \frac{E_{xy}}{E_{xx}} = \frac{186.60}{195.60} = 0.95$$

We may test the hypothesis $H_0: \beta = 0$ by using (15-19). The test statistic is

$$F_0 = \frac{(E_{xy})^2/E_{xx}}{MS_E} = \frac{(186.60)^2/195.60}{2.54} = 70.08$$

and since $F_{.01, 1, 11} = 9.65$, we reject the hypothesis that $\beta = 0$. Therefore, there is a linear relationship between breaking strength and diameter, and the adjustment provided by the analysis of covariance was necessary.

The adjusted treatment means may be computed from Equation 15-17. These adjusted means are

$$\text{adjusted } \bar{y}_{1.} = \bar{y}_{1.} - \hat{\beta}(\bar{x}_{1.} - \bar{x}_{..})$$

$$= 41.40 - (0.95)(25.20 - 24.13) = 40.38$$

$$\text{adjusted } \bar{y}_{2.} = \bar{y}_{2.} - \hat{\beta}(\bar{x}_{2.} - \bar{x}_{..})$$

$$= 43.20 - (0.95)(26.00 - 24.13) = 41.42$$

and

$$\text{adjusted } \bar{y}_{3.} = \bar{y}_{3.} - \hat{\beta}(\bar{x}_{3.} - \bar{x}_{..})$$

$$= 36.00 - (0.95)(21.20 - 24.13) = 38.78$$

Comparing the adjusted treatment means with the unadjusted treatment means (the $\bar{y}_{i.}$), we note that the adjusted means are much closer together, another indication that the covariance analysis was necessary.

A basic assumption in the analysis of covariance is that the treatments do not influence the covariate x because the technique removes the effect of variations in the $\bar{x}_{i.}$. However, if the variability in the $\bar{x}_{i.}$ is due in part to the treatments, then analysis of covariance removes part of the treatment effect. Thus we must be reasonably sure that the treatments do not affect the values x_{ij}. In some experiments this may be obvious from the nature of the covariate, while in others it may be more doubtful. In our example, there may be a difference in fiber diameter (x_{ij}) between the three machines. In such cases, Cochran and Cox (1957) suggest that an analysis of variance on the x_{ij} values may be helpful in determining the validity of this assumption. For our problem, this procedure yields

$$F_0 = \frac{66.13/2}{195.60/12} = \frac{33.07}{16.30} = 2.03$$

which is less than $F_{.10, 2, 12} = 2.81$, so there is no reason to believe that machines produce fiber of different diameter.

It is interesting to note what would have happened in this experiment if an analysis of covariance had not been performed; that is, if the breaking strength data (y) had

Table 15-5 Incorrect Analysis of the Breaking Strength Data as a Single Factor Experiment

Source of Variation	Sum of Squares	Degrees of Freedom	Mean Square	F_0
Machines	140.40	2	70.20	4.09[a]
Error	206.00	12	17.17	
Total	346.40	14		

[a]Significant at 5 percent.

been analyzed as a single-factor experiment. The analysis of variance of the breaking strength data is shown in Table 15-5. Since $F_{.05, 2, 12} = 3.89$ we would conclude that machines differ significantly in the strength of fiber produced, and would ordinarily try to equalize the strength output of the machines. However, after the linear effect of diameter is removed, machines do not differ in the strength of fiber produced. It might be helpful to reduce the within-machine diameter variability. ∎

15-3 DEVELOPMENT BY THE GENERAL REGRESSION SIGNIFICANCE TEST

It is possible to formally develop the procedure for testing H_0: $\tau_i = 0$ in the covariance model

$$y_{ij} = \mu + \tau_i + \beta(x_{ij} - \bar{x}_{..}) + \epsilon_{ij} \begin{cases} i = 1, 2, \ldots, a \\ j = 1, 2, \ldots, n \end{cases} \tag{15-20}$$

using the general regression significance test. Consider estimating the parameters in the model (15-20) by least squares. The least squares function is

$$L = \sum_{i=1}^{a} \sum_{j=1}^{n} [y_{ij} - \mu - \tau_i - \beta(x_{ij} - \bar{x}_{..})]^2 \tag{15-21}$$

and from $\partial L/\partial\mu = \partial L/\partial\tau_i = \partial L/\partial\beta - 0$, we obtain the normal equations

$$\mu: \quad an\hat{\mu} + n \sum_{i=1}^{a} \hat{\tau}_i = y_{..} \tag{15-22a}$$

$$\tau_i: \quad n\hat{\mu} + n\hat{\tau}_i + \hat{\beta} \sum_{j=1}^{n} (x_{ij} - \bar{x}_{..}) = y_{i.} \qquad i = 1, 2, \ldots, a \tag{15-22b}$$

$$\beta: \quad \sum_{i=1}^{a} \hat{\tau}_i \sum_{j=1}^{n} (x_{ij} - \bar{x}_{..}) + \hat{\beta} S_{xx} = S_{xy} \tag{15-22c}$$

Adding the a equations in (15-22b) we obtain (15-22a) since $\sum_{i=1}^{a} \sum_{j=1}^{n} (x_{ij} - \bar{x}_{..}) = 0$, so there is one linear dependency in the normal equations. Therefore, it is necessary to augment (15-22) with a linearly independent equation in order to obtain a solution. A logical side condition is $\sum_{i=1}^{a} \hat{\tau}_i = 0$. Using this condition, we obtain from (15-22a)

$$\hat{\mu} = \bar{y}_{..} \tag{15-23a}$$

and from (15-22b)

$$\hat{\tau}_i = \bar{y}_{i.} - \bar{y}_{..} - \hat{\beta}(\bar{x}_{i.} - \bar{x}_{..}) \tag{15-23b}$$

Equation 15-22c may be rewritten as

$$\sum_{i=1}^{a} (\bar{y}_{i.} - \bar{y}_{..}) \sum_{j=1}^{n} (x_{ij} - \bar{x}_{..}) - \hat{\beta} \sum_{i=1}^{a} (\bar{x}_{i.} - \bar{x}_{..}) \sum_{j=1}^{n} (x_{ij} - \bar{x}_{..}) + \hat{\beta} S_{xx} = S_{xy}$$

after substituting for $\hat{\tau}_i$. But we see that

$$\sum_{i=1}^{a} (\bar{y}_{i.} - \bar{y}_{..}) \sum_{j=1}^{n} (x_{ij} - \bar{x}_{..}) = T_{xy}$$

and

$$\sum_{i=1}^{a} (\bar{x}_{i.} - \bar{x}_{..}) \sum_{j=1}^{n} (x_{ij} - \bar{x}_{..}) = T_{xx}$$

therefore the solution to (15-22c) is

$$\hat{\beta} = \frac{S_{xy} - T_{xy}}{S_{xx} - T_{xx}} = \frac{E_{xy}}{E_{xx}} \tag{15-23c}$$

which was the result given previously in Section 15-2, Equation 15-12.

We may express the reduction in the total sum of squares due to fitting the model (15-20) as

$$R(\mu, \tau, \beta) = \hat{\mu} y_{..} + \sum_{i=1}^{a} \hat{\tau}_i y_{i.} + \hat{\beta} S_{xy}$$

$$= (\bar{y}_{..}) y_{..} + \sum_{i=1}^{a} [\bar{y}_{i.} - \bar{y}_{..} - (E_{xy}/E_{xx})(\bar{x}_{i.} - \bar{x}_{..})] y_{i.} + (E_{xy}/E_{xx}) S_{xy}$$

$$= y_{..}^2/an + \sum_{i=1}^{a} (\bar{y}_{i.} - \bar{y}_{..}) y_{i.} - (E_{xy}/E_{xx}) \sum_{i=1}^{a} (\bar{x}_{i.} - \bar{x}_{..}) y_{i.} + (E_{xy}/E_{xx}) S_{xy}$$

$$= y_{..}^2/an + T_{yy} - (E_{xy}/E_{xx})(T_{xy} - S_{xy})$$

$$= y_{..}^2/an + T_{yy} + (E_{xy})^2/E_{xx}$$

This sum of squares has $a + 1$ degrees of freedom, since the rank of the normal equations is $a + 1$. The error sum of squares for this model is

$$SS_E = \sum_{i=1}^{a} \sum_{j=1}^{n} y_{ij}^2 - R(\mu, \tau, \beta)$$

$$= \sum_{i=1}^{a} \sum_{j=1}^{n} y_{ij}^2 - y_{..}^2/an - T_{yy} - (E_{xy})^2/E_{xx} \tag{15-24}$$

$$= S_{yy} - T_{yy} - (E_{xy})^2/E_{xx}$$

$$= E_{yy} - (E_{xy})^2/E_{xx}$$

with $an - (a + 1) = a(n - 1) - 1$ degrees of freedom. This quantity was obtained previously as Equation 15-13.

Now consider the model restricted to the null hypothesis, that is, to H_0: $\tau_i = 0$. This reduced model is

$$y_{ij} = \mu + \beta(x_{ij} - \bar{x}_{..}) + \epsilon_{ij} \quad \begin{cases} i = 1, 2, \ldots, a \\ j = 1, 2, \ldots, n \end{cases} \tag{15-25}$$

We recognize this as a simple linear regression model, and by analogy with Equation 13-6 the normal equations for this model are

$$an\hat{\mu} = y_{..} \tag{15-26a}$$

$$\hat{\beta} S_{xx} = S_{xy} \tag{15-26b}$$

The solution to these equations is $\hat{\mu} = \bar{y}_{..}$ and $\hat{\beta} = S_{xy}/S_{xx}$, and the reduction in the total sum of squares due to fitting the reduced model is

$$R(\mu, \beta) = \hat{\mu} y_{..} + \hat{\beta} S_{xy}$$

$$= (\bar{y}_{..}) y_{..} + (S_{xy}/S_{xx}) S_{xy} \tag{15-27}$$

$$= y_{..}^2/an + (S_{xy})^2/S_{xx}$$

This sum of squares has 2 degrees of freedom.

We may find the appropriate sum of squares for testing $H_0: \tau_i = 0$ as

$$R(\tau \mid \mu, \beta) = R(\mu, \tau, \beta) - R(\mu, \beta)$$

$$= y_{..}^2/an + T_{yy} + (E_{xy})^2/E_{xx} - y_{..}^2/an - (S_{xy})^2/S_{xx} \tag{15-28}$$

$$= S_{yy} - (S_{xy})^2/S_{xx} - [E_{yy} - (E_{xy})^2/E_{xx}]$$

using $T_{yy} = S_{yy} - E_{yy}$. Note that $R(\tau \mid \mu, \beta)$ has $a + 1 - 2 = a - 1$ degrees of freedom, and is identical to the sum of squares given by $SS'_E - SS_E$ in Section 15-2. Thus the test statistic for $H_0: \tau_i = 0$ is

$$F_0 = \frac{R(\tau \mid \mu, \beta)/(a - 1)}{SS_E/[a(n - 1) - 1]} = \frac{(SS'_E - SS_E)/(a - 1)}{SS_E/[a(n - 1) - 1]} \tag{15-29}$$

which we gave previously as (15-16). Therefore, by using the general regression significance test we have justified the heuristic development of the analysis of covariance in Section 15-2.

15-4 OTHER COVARIANCE MODELS

The general procedure for the analysis of covariance can be embedded into any experimental design model. For example, consider the randomized complete block design with a single covariate

$$y_{ij} = \mu + \tau_i + \gamma_j + \beta(x_{ij} - \bar{x}_{..}) + \epsilon_{ij} \begin{cases} i = 1, 2, \ldots, a \\ j = 1, 2, \ldots, b \end{cases} \tag{15-30}$$

where y_{ij} is the observation on treatment i in block j, μ is an overall mean, τ_i is the effect of the ith treatment, γ_j is the effect of the jth block, β is a regression coefficient, x_{ij} is the ijth observation on the covariate, and ϵ_{ij} is the NID$(0, \sigma^2)$ random error component.

The computational details of the analysis of covariance for the randomized complete block design are displayed in Table 15-6. Definitions of the quantities in this table are

$$R_{xx} = \sum \frac{x_{.j}^2}{a} - \frac{x_{..}^2}{ab}, \quad R_{yy} = \sum \frac{y_{.j}^2}{a} - \frac{y_{..}^2}{ab}$$

$$R_{xy} = \sum \frac{x_{.j} y_{.j}}{a} - \frac{x_{..} y_{..}}{ab}$$

$$T_{xx} = \sum \frac{x_{i.}^2}{b} - \frac{x_{..}^2}{ab}, \quad T_{yy} = \sum \frac{y_{i.}^2}{b} - \frac{y_{..}^2}{ab}$$

$$T_{xy} = \sum \frac{x_{i.} y_{i.}}{b} - \frac{x_{..} y_{..}}{ab}$$

$$S_{xx} = \sum \sum x_{ij}^2 - \frac{x_{..}^2}{ab}, \quad S_{yy} = \sum \sum y_{ij}^2 - \frac{y_{..}^2}{ab}$$

$$S_{xy} = \sum \sum x_{ij} y_{ij} - \frac{x_{..} y_{..}}{ab}$$

$$E_{xx} = S_{xx} - R_{xx} - T_{xx}, \quad E_{yy} = S_{yy} - R_{yy} - T_{yy}$$

$$E_{xy} = S_{xy} - R_{xy} - T_{xy}$$

Notice that the procedure is similar to the usual analysis of variance, except for the adjustment afforded by regression. Comparing Tables 15-6 and 15-2, we see that in the randomized complete block design we have reduced the residuals E_{yy}, E_{xx}, and E_{xy} by a sum of squares or products due to blocks. Also, the adjusted treatments sum of squares is computed as the difference between the adjusted error sum of squares and an adjusted sum of squares for "treatments plus error," so that block effects may be eliminated. The regression coefficient β is estimated by

$$\hat{\beta} = \frac{E_{xy}}{E_{xx}} \tag{15-31}$$

and the appropriate statistic for testing H_0: $\beta = 0$ is

$$F_0 = \frac{(E_{xy})^2 / E_{xx}}{MS_E} \tag{15-32}$$

where the definition of MS_E is found in Table 15-6. Under the null hypothesis F_0 is distributed as $F_{1, ab-a-b}$, so H_0: $\beta = 0$ would be rejected if $F_0 > F_{\alpha, 1, ab-a-b}$.

The analysis of covariance may be extended to Latin squares, incomplete blocks, factorials, nested designs, and so on. It may also be used to compare the slopes of several regression lines. Finally, situations may arise in which a multiple covariance structure is necessary; that is, either y is related to x in a nonlinear fashion (such as

Table 15-6 Analysis of Covariance for the Randomized Complete Block Design

Source of Variation	Degrees of Freedom	Sums of Squares and Products			Adjusted for Regression			
		x	xy	y	y	Degrees of Freedom	Mean Square	F_0
Blocks	$b - 1$	R_{xx}	R_{xy}	R_{yy}	—	—	—	—
Treatments	$a - 1$	T_{xx}	T_{xy}	T_{yy}	—	—	—	—
Error	$(a - 1)(b - 1)$	E_{xx}	E_{xy}	E_{yy}	$SS_E = E_{yy} - (E_{xy})^2/E_{xx}$	$ab - a - b$	$MS_E = \dfrac{SS_E}{ab - a - b}$	—
Total	$ab - 1$	S_{xx}	S_{xy}	S_{yy}		$ab - 2$	—	
Treatments plus error	$b(a - 1)$	$S'_{xx} = T_{xx} + E_{xx}$	$S'_{xy} = T_{xy} + E_{xy}$	$S'_{yy} = T_{yy} + E_{yy}$	$SS'_E = S'_{yy} - (S'_{xy})^2/S'_{xx}$	$b(a - 1) - 1$	—	—
Adjusted treatments					$SS'_E - SS_E$	$a - 1$	$MS_A = \dfrac{SS'_E - SS_E}{a - 1}$	$\dfrac{MS_A}{MS_E}$

$y = \beta_1 x + \beta_2 x^2$), or there are two or more covariates. These and other related topics are discussed in Duncan (1974), Ostle (1963), Cochran (1957), and Federer (1957). The latter two references are found in the third volume of *Biometrics*, 1957, which is devoted almost exclusively to the analysis of covariance.

15-5 PROBLEMS

15-1. A soft drink distributor is studying the effectiveness of delivery methods. Three different types of hand trucks have been developed, and an experiment is performed in the company's methods engineering laboratory. The variable of interest is the delivery time in minutes (y); however, delivery time is also strongly related to the case volume delivered (x). Each hand truck is used four times and the data below obtained. Analyze this data and draw appropriate conclusions.

		Hand Truck Type			
1		2		3	
y	x	y	x	y	x
27	24	25	26	40	38
44	40	35	32	22	26
33	35	46	42	53	50
41	40	26	25	18	20

15-2. Compute the adjusted treatment means and the standard errors of the adjusted treatment means for the data in Problem 15-1.

15-3. The sum of squares and products for a one-way classification analysis of covariance are shown below. Complete the analysis and draw appropriate conclusions.

Source of Variation	Degrees of Freedom	Sums of Squares and Products		
		x	xy	y
Treatment	3	1500	1000	650
Error	12	6000	1200	550
Total	15	7500	2200	1200

15-4. Find the standard errors of the adjusted treatment means in Example 15-1.

15-5. Four different formulations of an industrial glue are being tested. The tensile strength of the glue is also related to the thickness. Five observations on strength (y) in pounds and thickness (x) in 0.01 inches are obtained for each formulation. The data are shown on page 385. Analyze this data and draw appropriate conclusions.

Glue Formulation							
1		2		3		4	
y	x	y	x	y	x	y	x
46.5	13	48.7	12	46.3	15	44.7	16
45.9	14	49.0	10	47.1	14	43.0	15
49.8	12	50.1	11	48.9	11	51.0	10
46.1	12	48.5	12	48.2	11	48.1	12
44.3	14	45.2	14	50.3	10	48.6	11

15-6. Compute the adjusted treatment means and their standard errors using the data in Problem 15-5.

15-7. An engineer is studying the effect of cutting speed on the rate of metal removal in a machining operation. However, the rate of metal removal is also related to the hardness of the test specimen. Five observations are taken at each cutting speed. The amount of metal removed (y) and the hardness of the specimen (x) are shown in the following table. Analyze the data using an analysis of covariance.

Cutting Speed (rpm)					
1000		1200		1400	
y	x	y	x	y	x
68	120	112	165	118	175
90	140	94	140	82	132
98	150	65	120	73	124
77	125	74	125	92	141
88	136	85	133	80	130

15-8. Show that in a one-way analysis of covariance with a single covariate a $100(1 - \alpha)$ percent confidence interval on the ith adjusted treatment mean is

$$\left\{ \bar{y}_{i.} - \hat{\beta}(\bar{x}_{i.} - \bar{x}_{..}) \pm t_{\alpha/2,a(n-1)-1} \left[MS_E \left(\frac{1}{n} + \frac{(\bar{x}_{i.} - \bar{x}_{..})^2}{E_{xx}} \right) \right]^{1/2} \right\}$$

Using this formula, calculate a 95 percent confidence interval on the adjusted mean of machine one in Example 15-1.

15-9. Show that in a one-way analysis of covariance with a single covariate the standard error of the difference between any two adjusted treatment means is

$$S_{adj.\bar{y}_{i.}-adj.\bar{y}_{j.}} = \left[MS_E \left(\frac{2}{n} + \frac{(\bar{x}_{i.} - \bar{x}_{j.})^2}{E_{xx}} \right) \right]^{1/2}$$

15-10. A randomized complete block design with one covariate gave the following table of corrected sums of squares and products. Complete the analysis of covariance and draw appropriate conclusions.

Source of Variation	Degrees of Freedom	Sums of Squares and Products		
		x	xy	y
Blocks	8	200	600	1200
Treatments	4	100	300	800
Error	32	600	700	1400

15-11. Write down the normal equations for a randomized complete block design with a single covariate. Solve the normal equations for the estimators of the model parameters.

15-12. Discuss how the operating characteristic curves for the analysis of variance can be used in the analysis of covariance.

BIBLIOGRAPHY

1. Addelman, S. (1961). "Irregular Fractions of the 2^n Factorial Experiments." *Technometrics*, Vol. 3, pp. 479–496.

2. Addelman, S. (1963). "Techniques for Constructing Fractional Replicate Plans." *Journal of the American Statistical Association*, Vol. 58, pp. 45–71.

3. Anderson, V. L. and R. A. McLean (1974). *Design of Experiments: A Realistic Approach*. Marcel Dekker Inc., New York.

4. Anscombe, F. J. (1960). "Rejection of Outliers." *Technometrics*, Vol. 2, pp. 123–147.

5. Anscombe, F. J. and J. W. Tukey (1963). "The Examination and Analysis of Residuals." *Technometrics*, Vol. 5, pp. 141–160.

6. Bancroft, T. A. (1968). *Topics in Intermediate Statistical Methods*. Iowa State University Press, Ames, Iowa.

7. Bartlett, M. S. (1947). "The Use of Transformations." *Biometrics*, Vol. 3, pp. 39–52.

8. Bennett, C. A. and N. L. Franklin (1954). *Statistical Analysis in Chemistry and the Chemical Industry*. Wiley, New York.

9. Bose, R. C. and T. Shimamoto (1952). "Classification and Analysis of Partially Balanced Incomplete Block Designs with Two Associate Classes." *Journal of the American Statistical Association*, Vol. 47, pp. 151–184.

10. Bose, R. C., W. H. Clatworthy, and S. S. Shrikhande (1954). *Tables of Partially Balanced Designs With Two Associate Classes*. Technical Bulletin No. 107, North Carolina Argicultural Experiment Station.

11. Bowker, A. H. and G. J. Lieberman (1972). *Engineering Statistics*. 2nd Edition. Prentice-Hall, Englewood Cliffs, N.J.

12. Box, G. E. P. and K. G. Wilson (1951). "On the Experimental Attainment of Optimum Conditions." *Journal of the Royal Statistical Society*, B, Vol. 13, pp. 1–45.

13. Box, G. E. P. (1954a). "Some Theorems on Quadratic Forms Applied in the Study of Analysis of Variance Problems: I. Effect of Inequality of Variance in the One-Way Classification." *Annals of Mathematical Statistics*, Vol. 25, pp. 290–302.

14. Box, G. E. P. (1954b). "Some Theorems on Quadratic Forms Applied in the Study of Analysis of Variance Problems: II. Effect of Inequality of Variance and of Correlation of Errors in the Two-Way Classification." *Annals of Mathematical Statistics*, Vol. 25, pp. 484–498.

15. Box, G. E. P. (1957). "Evolutionary Operation: A Method for Increasing Industrial Productivity." *Applied Statistics*, Vol. 6, pp. 81–101.

16. Box, G. E. P. and J. S. Hunter (1957). "Multifactor Experimental Designs for Exploring Response Surfaces." *Annals of Mathematical Statistics*, Vol. 28, pp. 195–242.

17. Box, G. E. P. and J. S. Hunter (1961a). "The 2^{k-p} Fractional Factorial Designs, Part I." *Technometrics*, Vol. 3, pp. 311–352.

18. Box, G. E. P. and J. S. Hunter (1961b). "The 2^{k-p} Fractional Factorial Designs, Part II." *Technometrics*, Vol. 3, pp. 449–458.

19. Box, G. E. P. and D. R. Cox (1964). "An Analysis of Transformations." *Journal of the Royal Statistical Society*, B, Vol. 26, pp. 211–243.

20. Box, G. E. P. and N. R. Draper (1969). *Evolutionary Operation*. Wiley, New York.

21. Carmer, S. G. and M. R. Swanson (1973). "Evaluation of Ten Pairwise Multiple Comparison Procedures by Monte Carlo Methods." *Journal of the American Statistical Association*, Vol. 68, No. 314, pp. 66–74.

22. Cochran, W. G. (1947). "Some Consequences When the Assumptions for the Analysis of Variance are not Satisfied." *Biometrics*, Vol. 3, pp. 22–38.

23. Cochran, W. G. (1957). "Analysis of Covariance: Its Nature and Uses." *Biometrics*, Vol. 13, No. 3, pp. 261–281.

24. Cochran, W. G. and G. M. Cox (1957). *Experimental Designs*. Second Edition. Wiley, New York.

25. Connor, W. S. and M. Zelen (1959). *Fractional Factorial Experimental Designs for Factors at Three Levels*. National Bureau of Standards, Washington, D.C. Applied Mathematics Series, No. 54.

26. Daniel, C. (1959). "Use of Half-Normal Plots in Interpreting Factorial Two Level Experiments." *Technometrics*, Vol. 1, pp. 311–342.

27. Davies, O. L. (1956). *Design and Analysis of Industrial Experiments*. Second Edition. Hafner Publishing Company, New York.

28. Dolby, J. L. (1963). "A Quick Method for Choosing a Transformation." *Technometrics*, Vol. 5, pp. 317–326.

29. Draper, N. R. and H. Smith (1966). *Applied Regression Analysis*. Wiley, New York.

30. Draper, N. R. and W. G. Hunter (1969). "Transformations: Some Examples Revisted." *Technometrics*, Vol. 11, pp. 23–40.

31. Duncan, A. J. (1974). *Quality Control and Industrial Statistics*. 4th Edition. Richard D. Irwin, Inc., Homewood, Ill.

32. Duncan, D. B. (1955). "Multiple Range and Multiple F Tests." *Biometrics*, Vol. 11, pp. 1–42.

33. Eisenhart, C. (1947). "The Assumptions Underlying the Analysis of Variance." *Biometrics*, Vol. 3, pp. 1–21.

34. Federer, W. T. (1957). "Variance and Covariance Analysis for Unbalanced Classifications." *Biometrics*, Vol. 13, No. 3, 333-362.

35. Fisher, R. A. and F. Yates (1953). *Statistical Tables for Biological, Agricultural, and Medical Research*. 4th Edition. Oliver and Boyd, Edinburgh.

36. Fisher, R. A. (1958). *Statistical Methods for Research Workers*. 13th Edition. Oliver and Boyd, Edinburgh.

37. Fisher, R. A. (1966). *The Design of Experiments*. 8th Edition. Hafner Publishing Company, New York.

38. Gaylor, D. W. and T. D. Hartwell (1969). "Expected Mean Squares for Nested Classifications." *Biometrics*, Vol. 25, pp. 427-430.

39. Gaylor, D. W. and F. N. Hopper (1969). "Estimating the Degrees of Freedom for Linear Combinations of Mean Squares by Satterthwaite's Formula." *Technometrics*, Vol. 11, No. 4, pp. 69-706.

40. Good, I. J. (1955). "The Interaction Algorithm and Practical Fourier Analysis." *Journal of the Royal Statistical Society*, B, Vol. 20, pp. 361-372.

41. Good, I. J. (1958). Addendum to "The Interaction Algorithm and Practical Fourier Analysis." *Journal of the Royal Statistical Society*, B, Vol. 22, pp. 372-375.

42. Graybill, F. A. (1961). *An Introduction to Linear Statistical Models*. Vol. 1. McGraw-Hill, New York.

43. Graybill, F. A. and D. L. Weeks (1959). "Combining Interblock and Intrablock Information in Balanced Incomplete Blocks." *Annals of Mathematical Statistics*, Vol. 30, pp. 799-805.

44. Herzberg, A. M. and D. R. Cox (1969). "Recent Work on the Design of Experiments: A Bibliography and a Review," *Journal of the Royal Statistical Society*, A, 132, pp. 29-67.

45. Hicks, C. R. (1973). *Fundamental Concepts in the Design of Experiments*. 2nd Edition. Holt, Rinehart and Winston, New York.

46. Hill, W. G. and W. G. Hunter (1966). "A Review of Response Surface Methodology: A Literature Survey." *Technometrics*, Vol. 8, pp. 571-590.

47. Hines, W. W. and D. C. Montgomery (1972). *Probability and Statistics in Engineering and Management Science*. The Ronald Press, New York.

48. Hocking, R. R. (1973). "A Discussion of the Two-Way Mixed Model." *The American Statistician*, Vol. 27, No. 4, pp. 148-152.

49. Hunter, J. S. (1966). "The Inverse Yates Algorithm." *Technometrics*, Vol. 8, pp. 177-183.

50. John. P. W. M. (1961). "The Three-Quarter Replicates of 2^4 and 2^5 Designs." *Biometrics*, Vol. 17, pp. 319-321.

51. John. P. W. M. (1962). "Three-Quarter Replicates of 2^n Designs." *Biometrics*, Vol. 18, pp. 171–184.

52. John. P. W. M. (1964). "Blocking a $3(2^{n-k})$ Designs." *Technometrics*, Vol. 6, pp. 371–376.

53. John. P. W. M. (1971). *Statistical Design and Analysis of Experiments*. The MacMillan Company, New York.

54. Kempthorne, O. (1952). *The Design and Analysis of Experiments*. Wiley, New York.

55. Keuls, M. (1952). "The Use of the Studentized Range in Connection with an Analysis of Variance." *Euphytica*, Vol. 1, pp. 112–122.

56. Margolin, B. H. (1967). "Systematic Methods of Analyzing $2^n 3^m$ Factorial Experiments with Applications." *Technometrics*, Vol. 9, pp. 245–260.

57. Margolin, B. H. (1969). "Results on Factorial Designs of Resolution IV for the 2^n and $2^n 3^m$ Series." *Technometrics*, Vol. 11, pp. 431–444.

58. Miller, R. G. (1966). *Simultaneous Statistical Inference*. McGraw-Hill, New York.

59. Myers, R. H. (1971). *Response Surface Methodology*. Allyn and Bacon, Inc., Boston.

60. Newman, D. (1939). "The Distribution of the Range in Samples from a Normal Population, Expressed in Terms of an Independent Estimate of Standard Deviation." *Biometrika*, Vol. 31, pp. 20–30.

61. Ostle, B. (1963). *Statistics in Research*. 2nd Edition. Iowa State Press, Ames, Iowa.

62. Pearson, E. S. and H. O. Hartley (1966). *Biometrika Tables for Statisticians*. Vol. 1, 3rd edition, Cambridge University Press, Cambridge.

63. Pearson, E. S. and H. O. Hartley (1972). *Biometrika Tables for Statisticians*. Vol. 2, Cambridge University Press, Cambridge.

64. Plackett, R. L. and J. P. Burman (1946). "The Design of Optimum Multifactorial Experiments." *Biometrika*, Vol. 33, pp. 305–325.

65. Quenouille, M. H. (1953). *The Design and Analysis of Experiments*. Charles Griffin and Company, London.

66. Quenouille, M. H. (1955). "Checks on the Calculation of Main Effects and Interactions in a 2^n Factorial Experiment." *Annals of Eugenics*, Vol. 19, pp. 151–152.

67. Rayner, A. A. (1967). "The Square Summing Check on the Main Effects and Interactions in a 2^n Experiment as Calculated by Yates' Algorithm." *Biometrics*, Vol. 23, pp. 571–573.

68. Satterthwaite, F. E. (1946). "An Approximate Distribution of Estimates of Variance Components." *Biometrics Bull.*, Vol. 2, pp. 110–112.

69. Scheffé, H. (1953). "A Method for Judging all Contrasts in the Analysis of Variance," *Biometrika*, Vol. 40, pp. 87–104.

70. Scheffé, H. (1956a). "A 'Mixed Model' for the Analysis of Variance." *Annals of Mathematical Statistics*, Vol. 27, pp. 23–36.

71. Scheffé, H. (1956b). "Alternative Models for the Analysis of Variance." *Annals of Mathematical Statistics*, Vol. 27, pp. 251–271.

72. Scheffé, H. (1959). *The Analysis of Variance*. Wiley, New York.

73. Searle, S. R. (1971a). *Linear Models*. Wiley, New York.

74. Searle, S. R. (1971b). "Topics in Variance Component Estimation." *Biometrics*, Vol. 27, pp. 1–76.

75. Searle, S. R. and R. F. Fawcett (1970). "Expected Mean Squares in Variance Component Models having Finite Populations." *Biometrics*, Vol. 26, pp. 243–254.

76. Smith, C. A. B. and H. O. Hartley (1948). "Construction of Youden Squares." *Journal of the Royal Statistical Society*, B, Vol. 10, pp. 262–264.

77. Smith, H. F. (1957). "Interpretations of Adjusted Treatment Means and Regressions in Analysis of Covariance." *Biometrics*, Vol. 13, No. 3, 282–308.

78. Tukey, J. W. (1949a). "One Degree of Freedom for Non-Additivity." *Biometrics*, Vol. 5, pp. 232–242.

79. Tukey, J. W. (1949b). "Comparing Individual Means in the Analysis of Variance." *Biometrics*, Vol. 5, pp. 99–114.

80. Tukey, J. W. (1951). "Quick and Dirty Methods in Statistics, Part II, Simple Analysis for Standard Designs." *Proceedings Fifth Annual Convention, American Society for Quality Control*, pp. 189–197.

81. Tukey, J. W. (1953). "The Problem of Multiple Comparisons." Unpublished Notes, Princeton University, pp. 396.

82. Wine, R. L. (1964). *Statistics for Scientists and Engineers*, Prentice-Hall, Inc., Englewood Cliffs, N.J.

83. Winer, B. J. (1971). *Statistical Principles in Experimental Design*. 2nd Edition. McGraw-Hill, New York.

84. Yates, F. (1937). *Design and Analysis of Factorial Experiments*. Tech. Comm. No. 35, Imperial Bureau of Soil Sciences, London.

85. Yates, F. (1940). "The Recovery of Interblock Information in Balanced Incomplete Block Designs." *Annals of Eugenics*, Vol. 10, pp. 317–325.

Appendix

I. Cumulative Standard Normal Distribution[a]

$$\Phi(z) = \int_{-\infty}^{z} \frac{1}{\sqrt{2\pi}} e^{-u^2/2} \, du$$

z	.00	.01	.02	.03	.04	z
.0	.500 00	.503 99	.507 98	.511 97	.515 95	.0
.1	.539 83	.543 79	.547 76	.551 72	.555 67	.1
.2	.579 26	.583 17	.587 06	.590 95	.594 83	.2
.3	.617 91	.621 72	.625 51	.629 30	.633 07	.3
.4	.655 42	.659 10	.662 76	.666 40	.670 03	.4
.5	.691 46	.694 97	.698 47	.701 94	.705 40	.5
.6	.725 75	.729 07	.732 37	.735 65	.738 91	.6
.7	.758 03	.761 15	.764 24	.767 30	.770 35	.7
.8	.788 14	.791 03	.793 89	.796 73	.799 54	.8
.9	.815 94	.818 59	.821 21	.823 81	.826 39	.9
1.0	.841 34	.843 75	.846 13	.848 49	.850 83	1.0
1.1	.864 33	.866 50	.868 64	.870 76	.872 85	1.1
1.2	.884 93	.886 86	.888 77	.890 65	.892 51	1.2
1.3	.903 20	.904 90	.906 58	.908 24	.909 88	1.3
1.4	.919 24	.920 73	.922 19	.923 64	.925 06	1.4
1.5	.933 19	.934 48	.935 74	.936 99	.938 22	1.5
1.6	.945 20	.946 30	.947 38	.948 45	.949 50	1.6
1.7	.955 43	.956 37	.957 28	.958 18	.959 07	1.7
1.8	.964 07	.964 85	.965 62	.966 37	.967 11	1.8
1.9	.971 28	.971 93	.972 57	.973 20	.973 81	1.9
2.0	.977 25	.977 78	.978 31	.978 82	.979 32	2.0
2.1	.982 14	.982 57	.983 00	.983 41	.938 82	2.1
2.2	.986 10	.986 45	.986 79	.987 13	.987 45	2.2
2.3	.989 28	.989 56	.989 83	.990 10	.990 36	2.3
2.4	.991 80	.992 02	.992 24	.992 45	.992 66	2.4
2.5	.993 79	.993 96	.994 13	.994 30	.994 46	2.5
2.6	.995 34	.995 47	.995 60	.995 73	.995 85	2.6
2.7	.996 53	.996 64	.996 74	.996 83	.996 93	2.7
2.8	.997 44	.997 52	.997 60	.997 67	.997 74	2.8
2.9	.998 13	.998 19	.998 25	.998 31	.998 36	2.9
3.0	.998 65	.998 69	.998 74	.998 78	.998 82	3.0
3.1	.999 03	.999 06	.999 10	.999 13	.999 16	3.1
3.2	.999 31	.999 34	.999 36	.999 38	.999 40	3.2
3.3	.999 52	.999 53	.999 55	.999 57	.999 58	3.3
3.4	.999 66	.999 68	.999 69	.999 70	.999 71	3.4
3.5	.999 77	.999 78	.999 78	.999 79	.999 80	3.5
3.6	.999 84	.999 85	.999 85	.999 86	.999 86	3.6
3.7	.999 89	.999 90	.999 90	.999 90	.999 91	3.7
3.8	.999 93	.999 93	.999 93	.999 94	.999 94	3.8
3.9	.999 95	.999 95	.999 96	.999 96	.999 96	3.9

[a]Reproduced with permission from *Probability and Statistics in Engineering and Management Science* by W. W. Hines and D. C. Montgomery, The Ronald Press, New York, 1972.

I. Cumulative Standard Normal Distribution (*continued*)

$$\Phi(z) = \int_{-\infty}^{z} \frac{1}{\sqrt{2\pi}} \, e^{-u^2/2} \, du$$

z	.05	.06	.07	.08	.09	z
.0	.519 94	.523 92	.527 90	.531 88	.535 86	.0
.1	.559 62	.563 56	.567 49	.571 42	.575 34	.1
.2	.598 71	.602 57	.606 42	.610 26	.614 09	.2
.3	.636 83	.640 58	.644 31	.648 03	.651 73	.3
.4	.673 64	.677 24	.680 82	.684 38	.687 93	.4
.5	.708 84	.712 26	.715 66	.719 04	.722 40	.5
.6	.742 15	.745 37	.748 57	.751 75	.754 90	.6
.7	.773 37	.776 37	.779 35	.782 30	.785 23	.7
.8	.802 34	.805 10	.807 85	.810 57	.813 27	.8
.9	.828 94	.831 47	.833 97	.836 46	.838 91	.9
1.0	.853 14	.855 43	.857 69	.859 93	.862 14	1.0
1.1	.874 93	.876 97	.879 00	.881 00	.882 97	1.1
1.2	.894 35	.896 16	.897 96	.899 73	.901 47	1.2
1.3	.911 49	.913 08	.914 65	.916 21	.917 73	1.3
1.4	.926 47	.927 85	.929 22	.930 56	.931 89	1.4
1.5	.939 43	.940 62	.941 79	.942 95	.944 08	1.5
1.6	.950 53	.951 54	.952 54	.953 52	.954 48	1.6
1.7	.959 94	.960 80	.961 64	.962 46	.963 27	1.7
1.8	.967 84	.968 56	.969 26	.969 95	.970 62	1.8
1.9	.974 41	.975 00	.975 58	.976 15	.976 70	1.9
2.0	.979 82	.980 30	.980 77	.981 24	.981 69	2.0
2.1	.984 22	.984 61	.985 00	.985 37	.985 74	2.1
2.2	.987 78	.988 09	.988 40	.988 70	.988 99	2.2
2.3	.990 61	.990 86	.991 11	.991 34	.991 58	2.3
2.4	.992 86	.993 05	.993 24	.993 43	.993 61	2.4
2.5	.994 61	.994 77	.994 92	.995 06	.995 20	2.5
2.6	.995 98	.996 09	.996 21	.996 32	.996 43	2.6
2.7	.997 02	.997 11	.997 20	.997 28	.997 36	2.7
2.8	.997 81	.997 88	.997 95	.998 01	.998 07	2.8
2.9	.998 41	.998 46	.998 51	.998 56	.998 61	2.9
3.0	.998 86	.998 89	.998 93	.998 97	.999 00	3.0
3.1	.999 18	.999 21	.999 24	.999 26	.999 29	3.1
3.2	.999 42	.999 44	.999 46	.999 48	.999 50	3.2
3.3	.999 60	.999 61	.999 62	.999 64	.999 65	3.3
3.4	.999 72	.999 73	.999 74	.999 75	.999 76	3.4
3.5	.999 81	.999 81	.999 82	.999 83	.999 83	3.5
3.6	.999 87	.999 87	.999 88	.999 88	.999 89	3.6
3.7	.999 91	.999 92	.999 92	.999 92	.999 92	3.7
3.8	.999 94	.999 94	.999 95	.999 95	.999 95	3.8
3.9	.999 96	.999 96	.999 96	.999 97	.999 97	3.9

II. Percentage Points of the t Distribution[a]

ν	.45	.40	.35	.30	.25	.125	.05	.025	.0125	.005	.0025
						α					
1	0.158	0.325	0.510	0.727	1.000	2.414	6.314	12.71	25.45	63.66	127.3
2	0.142	0.289	0.445	0.617	0.817	1.604	2.920	4.303	6.205	9.925	14.09
3	0.137	0.277	0.424	0.584	0.765	1.423	2.353	3.183	4.177	5.841	7.453
4	0.134	0.271	0.414	0.569	0.741	1.344	2.132	2.776	3.495	4.604	5.598
5	0.132	0.267	0.408	0.559	0.727	1.301	2.015	2.571	3.163	4.032	4.773
6	0.131	0.265	0.404	0.553	0.718	1.273	1.943	2.447	2.969	3.707	4.317
7	0.130	0.263	0.402	0.549	0.711	1.254	1.895	2.365	2.841	3.500	4.029
8	0.130	0.262	0.399	0.546	0.706	1.240	1.860	2.306	2.752	3.355	3.833
9	0.129	0.261	0.398	0.543	0.703	1.230	1.833	2.262	2.685	3.250	3.690
10	0.129	0.260	0.397	0.542	0.700	1.221	1.813	2.228	2.634	3.169	3.581
11	0.129	0.260	0.396	0.540	0.697	1.215	1.796	2.201	2.593	3.106	3.500
12	0.128	0.259	0.395	0.539	0.695	1.209	1.782	2.179	2.560	3.055	3.428
13	0.128	0.259	0.394	0.538	0.694	1.204	1.771	2.160	2.533	3.012	3.373
14	0.128	0.258	0.393	0.537	0.692	1.200	1.761	2.145	2.510	2.977	3.326
15	0.128	0.258	0.392	0.536	0.691	1.197	1.753	2.132	2.490	2.947	3.286
20	0.127	0.257	0.391	0.533	0.687	1.185	1.725	2.086	2.423	2.845	3.153
25	0.127	0.256	0.390	0.531	0.684	1.178	1.708	2.060	2.385	2.787	3.078
30	0.127	0.256	0.389	0.530	0.683	1.173	1.697	2.042	2.360	2.750	3.030
40	0.126	0.255	0.388	0.529	0.681	1.167	1.684	2.021	2.329	2.705	2.971
60	0.126	0.254	0.387	0.527	0.679	1.162	1.671	2.000	2.299	2.660	2.915
120	0.126	0.254	0.386	0.526	0.677	1.156	1.658	1.980	2.270	2.617	2.860
∞	0.126	0.253	0.385	0.524	0.674	1.150	1.645	1.960	2.241	2.576	2.807

ν = degrees of freedom.

[a]Reproduced with permission from *Probability and Statistics in Engineering and Management Science* by W. W. Hines and D. C. Montgomery, The Ronald Press, New York, 1972.

III. Percentage Points of the χ^2 Distribution[a]

ν	.995	.990	.975	.950	.500	.050	.025	.010	.005
1	0.00+	0.00+	0.00+	0.00+	0.45	3.84	5.02	6.63	7.88
2	0.01	0.02	0.05	0.10	1.39	5.99	7.38	9.21	10.60
3	0.07	0.11	0.22	0.35	2.37	7.81	9.35	11.34	12.84
4	0.21	0.30	0.48	0.71	3.36	9.49	11.14	13.28	14.86
5	0.41	0.55	0.83	1.15	4.35	11.07	12.38	15.09	16.75
6	0.68	0.87	1.24	1.64	5.35	12.59	14.45	16.81	18.55
7	0.99	1.24	1.69	2.17	6.35	14.07	16.01	18.48	20.28
8	1.34	1.65	2.18	2.73	7.34	15.51	17.53	20.09	21.96
9	1.73	2.09	2.70	3.33	8.34	16.92	19.02	21.67	23.59
10	2.16	2.56	3.25	3.94	9.34	18.31	20.48	23.21	25.19
11	2.60	3.05	3.82	4.57	10.34	19.68	21.92	24.72	26.76
12	3.07	3.57	4.40	5.23	11.34	21.03	23.34	26.22	28.30
13	3.57	4.11	5.01	5.89	12.34	22.36	24.74	27.69	29.82
14	4.07	4.66	5.63	6.57	13.34	23.68	26.12	29.14	31.32
15	4.60	5.23	6.27	7.26	14.34	25.00	27.49	30.58	32.80
16	5.14	5.81	6.91	7.96	15.34	26.30	28.85	32.00	34.27
17	5.70	6.41	7.56	8.67	16.34	27.59	30.19	33.41	35.72
18	6.26	7.01	8.23	9.39	17.34	28.87	31.53	34.81	37.16
19	6.84	7.63	8.91	10.12	18.34	30.14	32.85	36.19	38.58
20	7.43	8.26	9.59	10.85	19.34	31.41	34.17	37.57	40.00
25	10.52	11.52	13.12	14.61	24.34	37.65	40.65	44.31	46.93
30	13.79	14.95	16.79	18.49	29.34	43.77	46.98	50.89	53.67
40	20.71	22.16	24.43	26.51	39.34	55.76	59.34	63.69	66.77
50	27.99	29.71	32.36	34.76	49.33	67.50	71.42	76.15	79.49
60	35.53	37.48	40.48	43.19	59.33	79.08	83.30	88.38	91.95
70	43.28	45.44	48.76	51.74	69.33	90.53	95.02	100.42	104.22
80	51.17	53.54	57.15	60.39	79.33	101.88	106.63	112.33	116.32
90	59.20	61.75	65.65	69.13	89.33	113.14	118.14	124.12	128.30
100	67.33	70.06	74.22	77.93	99.33	124.34	129.56	135.81	140.17

ν = degrees of freedom

[a]Adapted with permission from *Biometrika Tables For Statisticians*, Vol. 1, 3rd Edition by E. S. Pearson and H. O. Hartley, Cambridge University Press, Cambridge, 1966.

IV. Percentage Points of the F Distribution[a]

$$F_{.25,\ \nu_1,\ \nu_2}$$

Degrees of freedom for the numerator (ν_1)

ν_2	1	2	3	4	5	6	7	8	9	10	12	15	20	24	30	40	60	120	∞
1	5.83	7.50	8.20	8.58	8.82	8.98	9.10	9.19	9.26	9.32	9.41	9.49	9.58	9.63	9.67	9.71	9.76	9.80	9.85
2	2.57	3.00	3.15	3.23	3.28	3.31	3.34	3.35	3.37	3.38	3.39	3.41	3.43	3.43	3.44	3.45	3.46	3.47	3.48
3	2.02	2.28	2.36	2.39	2.41	2.42	2.43	2.44	2.44	2.44	2.45	2.46	2.46	2.46	2.47	2.47	2.47	2.47	2.47
4	1.81	2.00	2.05	2.06	2.07	2.08	2.08	2.08	2.08	2.08	2.08	2.08	2.08	2.08	2.08	2.08	2.08	2.08	2.08
5	1.69	1.85	1.88	1.89	1.89	1.89	1.89	1.89	1.89	1.89	1.89	1.89	1.88	1.88	1.88	1.88	1.87	1.87	1.87
6	1.62	1.76	1.78	1.79	1.79	1.78	1.78	1.78	1.77	1.77	1.77	1.76	1.76	1.75	1.75	1.75	1.74	1.74	1.74
7	1.57	1.70	1.72	1.72	1.71	1.71	1.70	1.70	1.70	1.69	1.68	1.68	1.67	1.67	1.66	1.66	1.65	1.65	1.65
8	1.54	1.66	1.67	1.66	1.66	1.65	1.64	1.64	1.63	1.63	1.62	1.62	1.61	1.60	1.60	1.59	1.59	1.58	1.58
9	1.51	1.62	1.63	1.63	1.62	1.61	1.60	1.60	1.59	1.59	1.58	1.57	1.56	1.56	1.55	1.54	1.54	1.53	1.53
10	1.49	1.60	1.60	1.59	1.59	1.58	1.57	1.56	1.56	1.55	1.54	1.53	1.52	1.52	1.51	1.51	1.50	1.49	1.48
11	1.47	1.58	1.58	1.57	1.56	1.55	1.54	1.53	1.53	1.52	1.51	1.50	1.49	1.49	1.48	1.47	1.47	1.46	1.45
12	1.46	1.56	1.56	1.55	1.54	1.53	1.52	1.51	1.51	1.50	1.49	1.48	1.47	1.46	1.45	1.45	1.44	1.43	1.42
13	1.45	1.55	1.55	1.53	1.52	1.51	1.50	1.49	1.49	1.48	1.47	1.46	1.45	1.44	1.43	1.42	1.42	1.41	1.40
14	1.44	1.53	1.53	1.52	1.51	1.50	1.49	1.48	1.47	1.46	1.45	1.44	1.43	1.42	1.41	1.41	1.40	1.39	1.38
15	1.43	1.52	1.52	1.51	1.49	1.48	1.47	1.46	1.46	1.45	1.44	1.43	1.41	1.41	1.40	1.39	1.38	1.37	1.36
16	1.42	1.51	1.51	1.50	1.48	1.47	1.46	1.45	1.44	1.44	1.43	1.41	1.40	1.39	1.38	1.37	1.36	1.35	1.34
17	1.42	1.51	1.50	1.49	1.47	1.46	1.45	1.44	1.43	1.43	1.41	1.40	1.39	1.38	1.37	1.36	1.35	1.34	1.33
18	1.41	1.50	1.49	1.48	1.46	1.45	1.44	1.43	1.42	1.42	1.40	1.39	1.38	1.37	1.36	1.35	1.34	1.33	1.32
19	1.41	1.49	1.49	1.47	1.46	1.44	1.43	1.42	1.41	1.41	1.40	1.38	1.37	1.36	1.35	1.34	1.33	1.32	1.30
20	1.40	1.49	1.48	1.47	1.45	1.44	1.43	1.42	1.41	1.40	1.39	1.37	1.36	1.35	1.34	1.33	1.31	1.31	1.29
21	1.40	1.48	1.48	1.46	1.44	1.43	1.42	1.41	1.40	1.39	1.38	1.37	1.35	1.34	1.33	1.32	1.31	1.30	1.28
22	1.40	1.48	1.47	1.45	1.44	1.42	1.41	1.40	1.39	1.39	1.37	1.36	1.34	1.33	1.32	1.31	1.30	1.29	1.28
23	1.39	1.47	1.47	1.45	1.43	1.42	1.41	1.40	1.39	1.38	1.37	1.35	1.34	1.33	1.32	1.31	1.30	1.28	1.27
24	1.39	1.47	1.46	1.44	1.43	1.41	1.40	1.39	1.38	1.38	1.36	1.35	1.33	1.32	1.31	1.30	1.29	1.28	1.26
25	1.39	1.47	1.46	1.44	1.42	1.41	1.40	1.39	1.38	1.37	1.36	1.34	1.33	1.32	1.31	1.29	1.28	1.27	1.25
26	1.38	1.46	1.45	1.44	1.42	1.41	1.39	1.38	1.37	1.37	1.35	1.34	1.32	1.31	1.30	1.29	1.28	1.26	1.25
27	1.38	1.46	1.45	1.43	1.42	1.40	1.39	1.38	1.37	1.36	1.35	1.33	1.32	1.31	1.30	1.28	1.27	1.26	1.24
28	1.38	1.46	1.45	1.43	1.41	1.40	1.39	1.38	1.37	1.36	1.34	1.33	1.31	1.30	1.29	1.28	1.27	1.25	1.24
29	1.38	1.45	1.45	1.43	1.41	1.40	1.38	1.37	1.36	1.35	1.34	1.32	1.31	1.30	1.29	1.27	1.26	1.25	1.23
30	1.38	1.45	1.44	1.42	1.41	1.39	1.38	1.37	1.36	1.35	1.34	1.32	1.30	1.29	1.28	1.27	1.26	1.24	1.23
40	1.36	1.44	1.42	1.40	1.39	1.37	1.36	1.35	1.34	1.33	1.31	1.30	1.28	1.26	1.25	1.24	1.22	1.21	1.19
60	1.35	1.42	1.41	1.38	1.37	1.35	1.33	1.32	1.31	1.30	1.29	1.27	1.25	1.24	1.22	1.21	1.19	1.17	1.15
120	1.34	1.40	1.39	1.37	1.35	1.33	1.31	1.30	1.29	1.28	1.26	1.24	1.22	1.21	1.19	1.18	1.16	1.13	1.10
∞	1.32	1.39	1.37	1.35	1.33	1.31	1.29	1.28	1.27	1.25	1.24	1.22	1.19	1.18	1.16	1.14	1.12	1.08	1.00

Degrees of freedom for the denominator (ν_2)

[a] Adapted with permission from Biometrika Tables For Statisticians, Vol. 1, 3rd Edition, by E. S. Pearson and H. O. Hartley, Cambridge University Press, Cambridge, 1966.

IV. Percentage Points of the F Distribution (continued)

$$F_{.10,\,\nu_1,\,\nu_2}$$

Degrees of freedom for the numerator (ν_1)

ν_2	1	2	3	4	5	6	7	8	9	10	12	15	20	24	30	40	60	120	∞
1	39.86	49.50	53.59	55.83	57.24	58.20	58.91	59.44	59.86	60.19	60.71	61.22	61.74	62.00	62.26	62.53	62.79	63.06	63.33
2	8.53	9.00	9.16	9.24	9.29	9.33	9.35	9.37	9.38	9.39	9.41	9.42	9.44	9.45	9.46	9.47	9.47	9.48	9.49
3	5.54	5.46	5.39	5.34	5.31	5.28	5.27	5.25	5.24	5.23	5.22	5.20	5.18	5.18	5.17	5.16	5.15	5.14	5.13
4	4.54	4.32	4.19	4.11	4.05	4.01	3.98	3.95	3.94	3.92	3.90	3.87	3.84	3.83	3.82	3.80	3.79	3.78	3.76
5	4.06	3.78	3.62	3.52	3.45	3.40	3.37	3.34	3.32	3.30	3.27	3.24	3.21	3.19	3.17	3.16	3.14	3.12	3.10
6	3.78	3.46	3.29	3.18	3.11	3.05	3.01	2.98	2.96	2.94	2.90	2.87	2.84	2.82	2.80	2.78	2.76	2.74	2.72
7	3.59	3.26	3.07	2.96	2.88	2.83	2.78	2.75	2.72	2.70	2.67	2.63	2.59	2.58	2.56	2.54	2.51	2.49	2.47
8	3.46	3.11	2.92	2.81	2.73	2.67	2.62	2.59	2.56	2.54	2.50	2.46	2.42	2.40	2.38	2.36	2.34	2.32	2.29
9	3.36	3.01	2.81	2.69	2.61	2.55	2.51	2.47	2.44	2.42	2.38	2.34	2.30	2.28	2.25	2.23	2.21	2.18	2.16
10	3.29	2.92	2.73	2.61	2.52	2.46	2.41	2.38	2.35	2.32	2.28	2.24	2.20	2.18	2.16	2.13	2.11	2.08	2.06
11	3.23	2.86	2.66	2.54	2.45	2.39	2.34	2.30	2.27	2.25	2.21	2.17	2.12	2.10	2.08	2.05	2.03	2.00	1.97
12	3.18	2.81	2.61	2.48	2.39	2.33	2.28	2.24	2.21	2.19	2.15	2.10	2.06	2.04	2.01	1.99	1.96	1.93	1.90
13	3.14	2.76	2.56	2.43	2.35	2.28	2.23	2.20	2.16	2.14	2.10	2.05	2.01	1.98	1.96	1.93	1.90	1.88	1.85
14	3.10	2.73	2.52	2.39	2.31	2.24	2.19	2.15	2.12	2.10	2.05	2.01	1.96	1.94	1.91	1.89	1.86	1.83	1.80
15	3.07	2.70	2.49	2.36	2.27	2.21	2.16	2.12	2.09	2.06	2.02	1.97	1.92	1.90	1.87	1.85	1.82	1.79	1.76
16	3.05	2.67	2.46	2.33	2.24	2.18	2.13	2.09	2.06	2.03	1.99	1.94	1.89	1.87	1.84	1.81	1.78	1.75	1.72
17	3.03	2.64	2.44	2.31	2.22	2.15	2.10	2.06	2.03	2.00	1.96	1.91	1.86	1.84	1.81	1.78	1.75	1.72	1.69
18	3.01	2.62	2.42	2.29	2.20	2.13	2.08	2.04	2.00	1.98	1.93	1.89	1.84	1.81	1.78	1.75	1.72	1.69	1.66
19	2.99	2.61	2.40	2.27	2.18	2.11	2.06	2.02	1.98	1.96	1.91	1.86	1.81	1.79	1.76	1.73	1.70	1.67	1.63
20	2.97	2.59	2.38	2.25	2.16	2.09	2.04	2.00	1.96	1.94	1.89	1.84	1.79	1.77	1.74	1.71	1.68	1.64	1.61
21	2.96	2.57	2.36	2.23	2.14	2.08	2.02	1.98	1.95	1.92	1.87	1.83	1.78	1.75	1.72	1.69	1.66	1.62	1.59
22	2.95	2.56	2.35	2.22	2.13	2.06	2.01	1.97	1.93	1.90	1.86	1.81	1.76	1.73	1.70	1.67	1.64	1.60	1.57
23	2.94	2.55	2.34	2.21	2.11	2.05	1.99	1.95	1.92	1.89	1.84	1.80	1.74	1.72	1.69	1.66	1.62	1.59	1.55
24	2.93	2.54	2.33	2.19	2.10	2.04	1.98	1.94	1.91	1.88	1.83	1.78	1.73	1.70	1.67	1.64	1.61	1.57	1.53
25	2.92	2.53	2.32	2.18	2.09	2.02	1.97	1.93	1.89	1.87	1.82	1.77	1.72	1.69	1.66	1.63	1.59	1.56	1.52
26	2.91	2.52	2.31	2.17	2.08	2.01	1.96	1.92	1.88	1.86	1.81	1.76	1.71	1.68	1.65	1.61	1.58	1.54	1.50
27	2.90	2.51	2.30	2.17	2.07	2.00	1.95	1.91	1.87	1.85	1.80	1.75	1.70	1.67	1.64	1.60	1.57	1.53	1.49
28	2.89	2.50	2.29	2.16	2.06	2.00	1.94	1.90	1.87	1.84	1.79	1.74	1.69	1.66	1.63	1.59	1.56	1.52	1.48
29	2.89	2.50	2.28	2.15	2.06	1.99	1.93	1.89	1.86	1.83	1.78	1.73	1.68	1.65	1.62	1.58	1.55	1.51	1.47
30	2.88	2.49	2.28	2.14	2.03	1.98	1.93	1.88	1.85	1.82	1.77	1.72	1.67	1.64	1.61	1.57	1.54	1.50	1.46
40	2.84	2.44	2.23	2.09	2.00	1.93	1.87	1.83	1.79	1.76	1.71	1.66	1.61	1.57	1.54	1.51	1.47	1.42	1.38
60	2.79	2.39	2.18	2.04	1.95	1.87	1.82	1.77	1.74	1.71	1.66	1.60	1.54	1.51	1.48	1.44	1.40	1.35	1.29
120	2.75	2.35	2.13	1.99	1.90	1.82	1.77	1.72	1.68	1.65	1.60	1.55	1.48	1.45	1.41	1.37	1.32	1.26	1.19
∞	2.71	2.30	2.08	1.94	1.85	1.77	1.72	1.67	1.63	1.60	1.55	1.49	1.42	1.38	1.34	1.30	1.24	1.17	1.00

Degrees of freedom for the denominator (ν_2)

IV. Percentage Points of the F Distribution (continued)

$$F_{.05}, \nu_1, \nu_2$$

Degrees of freedom for the numerator (ν_1)

ν_2 \ ν_1	1	2	3	4	5	6	7	8	9	10	12	15	20	24	30	40	60	120	∞
1	161.4	199.5	215.7	224.6	230.2	234.0	236.8	238.9	240.5	241.9	243.9	245.9	248.0	249.1	250.1	251.1	252.2	253.3	254.3
2	18.51	19.00	19.16	19.25	19.30	19.33	19.35	19.37	19.38	19.40	19.41	19.43	19.45	19.45	19.46	19.47	19.48	19.49	19.50
3	10.13	9.55	9.28	9.12	9.01	8.94	8.89	8.85	8.81	8.79	8.74	8.70	8.66	8.64	8.62	8.59	8.57	8.55	8.53
4	7.71	6.94	6.59	6.39	6.26	6.16	6.09	6.04	6.00	5.96	5.91	5.86	5.80	5.77	5.75	5.72	5.69	5.66	5.63
5	6.61	5.79	5.41	5.19	5.05	4.95	4.88	4.82	4.77	4.74	4.68	4.62	4.56	4.53	4.50	4.46	4.43	4.40	4.36
6	5.99	5.14	4.76	4.53	4.39	4.28	4.21	4.15	4.10	4.06	4.00	3.94	3.87	3.84	3.81	3.77	3.74	3.70	3.67
7	5.59	4.74	4.35	4.12	3.97	3.87	3.79	3.73	3.68	3.64	3.57	3.51	3.44	3.41	3.38	3.34	3.30	3.27	3.23
8	5.32	4.46	4.07	3.84	3.69	3.58	3.50	3.44	3.39	3.35	3.28	3.22	3.15	3.12	3.08	3.04	3.01	2.97	2.93
9	5.12	4.26	3.86	3.63	3.48	3.37	3.29	3.23	3.18	3.14	3.07	3.01	2.94	2.90	2.86	2.83	2.79	2.75	2.71
10	4.96	4.10	3.71	3.48	3.33	3.22	3.14	3.07	3.02	2.98	2.91	2.85	2.77	2.74	2.70	2.66	2.62	2.58	2.54
11	4.84	3.98	3.59	3.36	3.20	3.09	3.01	2.95	2.90	2.85	2.79	2.72	2.65	2.61	2.57	2.53	2.49	2.45	2.40
12	4.75	3.89	3.49	3.26	3.11	3.00	2.91	2.85	2.80	2.75	2.69	2.62	2.54	2.51	2.47	2.43	2.38	2.34	2.30
13	4.67	3.81	3.41	3.18	3.03	2.92	2.83	2.77	2.71	2.67	2.60	2.53	2.46	2.42	2.38	2.34	2.30	2.25	2.21
14	4.60	3.74	3.34	3.11	2.96	2.85	2.76	2.70	2.65	2.60	2.53	2.46	2.39	2.35	2.31	2.27	2.22	2.18	2.13
15	4.54	3.68	3.29	3.06	2.90	2.79	2.71	2.64	2.59	2.54	2.48	2.40	2.33	2.29	2.25	2.20	2.16	2.11	2.07
16	4.49	3.63	3.24	3.01	2.85	2.74	2.66	2.59	2.54	2.49	2.42	2.35	2.28	2.24	2.19	2.15	2.11	2.06	2.01
17	4.45	3.59	3.20	2.96	2.81	2.70	2.61	2.55	2.49	2.45	2.38	2.31	2.23	2.19	2.15	2.10	2.06	2.01	1.96
18	4.41	3.55	3.16	2.93	2.77	2.66	2.58	2.51	2.46	2.41	2.34	2.27	2.19	2.15	2.11	2.06	2.02	1.97	1.92
19	4.38	3.52	3.13	2.90	2.74	2.63	2.54	2.48	2.42	2.38	2.31	2.23	2.16	2.11	2.07	2.03	1.98	1.93	1.88
20	4.35	3.49	3.10	2.87	2.71	2.60	2.51	2.45	2.39	2.35	2.28	2.20	2.12	2.08	2.04	1.99	1.95	1.90	1.84
21	4.32	3.47	3.07	2.84	2.68	2.57	2.49	2.42	2.37	2.32	2.25	2.18	2.10	2.05	2.01	1.96	1.92	1.87	1.81
22	4.30	3.44	3.05	2.82	2.66	2.55	2.46	2.40	2.34	2.30	2.23	2.15	2.07	2.03	1.98	1.94	1.89	1.84	1.78
23	4.28	3.42	3.03	2.80	2.64	2.53	2.44	2.37	2.32	2.27	2.20	2.13	2.05	2.01	1.96	1.91	1.86	1.81	1.76
24	4.26	3.40	3.01	2.78	2.62	2.51	2.42	2.36	2.30	2.25	2.18	2.11	2.03	1.98	1.94	1.89	1.84	1.79	1.73
25	4.24	3.39	2.99	2.76	2.60	2.49	2.40	2.34	2.28	2.24	2.16	2.09	2.01	1.96	1.92	1.87	1.82	1.77	1.71
26	4.23	3.37	2.98	2.74	2.59	2.47	2.39	2.32	2.27	2.22	2.15	2.07	1.99	1.95	1.90	1.85	1.80	1.75	1.69
27	4.21	3.35	2.96	2.73	2.57	2.46	2.37	2.31	2.25	2.20	2.13	2.06	1.97	1.93	1.88	1.84	1.79	1.73	1.67
28	4.20	3.34	2.95	2.71	2.56	2.45	2.36	2.29	2.24	2.19	2.12	2.04	1.96	1.91	1.87	1.82	1.77	1.71	1.65
29	4.18	3.33	2.93	2.70	2.55	2.43	2.35	2.28	2.22	2.18	2.10	2.03	1.94	1.90	1.85	1.81	1.75	1.70	1.64
30	4.17	3.32	2.92	2.69	2.53	2.42	2.33	2.27	2.21	2.16	2.09	2.01	1.93	1.89	1.84	1.79	1.74	1.68	1.62
40	4.08	3.23	2.84	2.61	2.45	2.34	2.25	2.18	2.12	2.08	2.00	1.92	1.84	1.79	1.74	1.69	1.64	1.58	1.51
60	4.00	3.15	2.76	2.53	2.37	2.25	2.17	2.10	2.04	1.99	1.92	1.84	1.75	1.70	1.65	1.59	1.53	1.47	1.39
120	3.92	3.07	2.68	2.45	2.29	2.17	2.09	2.02	1.96	1.91	1.83	1.75	1.66	1.61	1.55	1.50	1.43	1.35	1.25
∞	3.84	3.00	2.60	2.37	2.21	2.10	2.01	1.94	1.88	1.83	1.75	1.67	1.57	1.52	1.46	1.39	1.32	1.22	1.00

Degrees of freedom for the denominator (ν_2)

IV. Percentage Points of the F Distribution (continued)

$$F.025, \nu_1, \nu_2$$

Degrees of freedom for the numerator (ν_1)

ν_2	1	2	3	4	5	6	7	8	9	10	12	15	20	24	30	40	60	120	∞
1	647.8	799.5	864.2	899.6	921.8	937.1	948.2	956.7	963.3	968.6	976.7	984.9	993.1	997.2	1001	1006	1010	1014	1018
2	38.51	39.00	39.17	39.25	39.30	39.33	39.36	39.37	39.39	39.40	39.41	39.43	39.45	39.46	39.46	39.47	39.48	39.49	39.50
3	17.44	16.04	15.44	15.10	14.88	14.73	14.62	14.54	14.47	14.42	14.34	14.25	14.17	14.12	14.08	14.04	13.99	13.95	13.90
4	12.22	10.65	9.98	9.60	9.36	9.20	9.07	8.98	8.90	8.84	8.75	8.66	8.56	8.51	8.46	8.41	8.36	8.31	8.26
5	10.01	8.43	7.76	7.39	7.15	6.98	6.85	6.76	6.68	6.62	6.52	6.43	6.33	6.28	6.23	6.18	6.12	6.07	6.02
6	8.81	7.26	6.60	6.23	5.99	5.82	5.70	5.60	5.52	5.46	5.37	5.27	5.17	5.12	5.07	5.01	4.96	4.90	4.85
7	8.07	6.54	5.89	5.52	5.29	5.12	4.99	4.90	4.82	4.76	4.67	4.57	4.47	4.42	4.36	4.31	4.25	4.20	4.14
8	7.57	6.06	5.42	5.05	4.82	4.65	4.53	4.43	4.36	4.30	4.20	4.10	4.00	3.95	3.89	3.84	3.78	3.73	3.67
9	7.21	5.71	5.08	4.72	4.48	4.32	4.20	4.10	4.03	3.96	3.87	3.77	3.67	3.61	3.56	3.51	3.45	3.39	3.33
10	6.94	5.46	4.83	4.47	4.24	4.07	3.95	3.85	3.78	3.72	3.62	3.52	3.42	3.37	3.31	3.26	3.20	3.14	3.08
11	6.72	5.26	4.63	4.28	4.04	3.88	3.76	3.66	3.59	3.53	3.43	3.33	3.23	3.17	3.12	3.06	3.00	2.94	2.88
12	6.55	5.10	4.47	4.12	3.89	3.73	3.61	3.51	3.44	3.37	3.28	3.18	3.07	3.02	2.96	2.91	2.85	2.79	2.72
13	6.41	4.97	4.35	4.00	3.77	3.60	3.48	3.39	3.31	3.25	3.15	3.05	2.95	2.89	2.84	2.78	2.72	2.66	2.60
14	6.30	4.86	4.24	3.89	3.66	3.50	3.38	3.29	3.21	3.15	3.05	2.95	2.84	2.79	2.73	2.67	2.61	2.55	2.49
15	6.20	4.77	4.15	3.80	3.58	3.41	3.29	3.20	3.12	3.06	2.96	2.86	2.76	2.70	2.64	2.59	2.52	2.46	2.40
16	6.12	4.69	4.08	3.73	3.50	3.34	3.22	3.12	3.05	2.99	2.89	2.79	2.68	2.63	2.57	2.51	2.45	2.38	2.32
17	6.04	4.62	4.01	3.66	3.44	3.28	3.16	3.06	2.98	2.92	2.82	2.72	2.62	2.56	2.50	2.44	2.38	2.32	2.25
18	5.98	4.56	3.95	3.61	3.38	3.22	3.10	3.01	2.93	2.87	2.77	2.67	2.56	2.50	2.44	2.38	2.32	2.26	2.19
19	5.92	4.51	3.90	3.56	3.33	3.17	3.05	2.96	2.88	2.82	2.72	2.62	2.51	2.45	2.39	2.33	2.27	2.20	2.13
20	5.87	4.46	3.86	3.51	3.29	3.13	3.01	2.91	2.84	2.77	2.68	2.57	2.46	2.41	2.35	2.29	2.22	2.16	2.09
21	5.83	4.42	3.82	3.48	3.25	3.09	2.97	2.87	2.80	2.73	2.64	2.53	2.42	2.37	2.31	2.25	2.18	2.11	2.04
22	5.79	4.38	3.78	3.44	3.22	3.05	2.93	2.84	2.76	2.70	2.60	2.50	2.39	2.33	2.27	2.21	2.14	2.08	2.00
23	5.75	4.35	3.75	3.41	3.18	3.02	2.90	2.81	2.73	2.67	2.57	2.47	2.36	2.30	2.24	2.18	2.11	2.04	1.97
24	5.72	4.32	3.72	3.38	3.15	2.99	2.87	2.78	2.70	2.64	2.54	2.44	2.33	2.27	2.21	2.15	2.08	2.01	1.94
25	5.69	4.29	3.69	3.35	3.13	2.97	2.85	2.75	2.68	2.61	2.51	2.41	2.30	2.24	2.18	2.12	2.05	1.98	1.91
26	5.66	4.27	3.67	3.33	3.10	2.94	2.82	2.73	2.65	2.59	2.49	2.39	2.28	2.22	2.16	2.09	2.03	1.95	1.88
27	5.63	4.24	3.65	3.31	3.08	2.92	2.80	2.71	2.63	2.57	2.47	2.36	2.25	2.19	2.13	2.07	2.00	1.93	1.85
28	5.61	4.22	3.63	3.29	3.06	2.90	2.78	2.69	2.61	2.55	2.45	2.34	2.23	2.17	2.11	2.05	1.98	1.91	1.83
29	5.59	4.20	3.61	3.27	3.04	2.88	2.76	2.67	2.59	2.53	2.43	2.32	2.21	2.15	2.09	2.03	1.96	1.89	1.81
30	5.57	4.18	3.59	3.25	3.03	2.87	2.75	2.65	2.57	2.51	2.41	2.31	2.20	2.14	2.07	2.01	1.94	1.87	1.79
40	5.42	4.05	3.46	3.13	2.90	2.74	2.62	2.53	2.45	2.39	2.29	2.18	2.07	2.01	1.94	1.88	1.80	1.72	1.64
60	5.29	3.93	3.34	3.01	2.79	2.63	2.51	2.41	2.33	2.27	2.17	2.06	1.94	1.88	1.82	1.74	1.67	1.58	1.48
120	5.15	3.80	3.23	2.89	2.67	2.52	2.39	2.30	2.22	2.16	2.05	1.94	1.82	1.76	1.69	1.61	1.53	1.43	1.31
∞	5.02	3.69	3.12	2.79	2.57	2.41	2.29	2.19	2.11	2.05	1.94	1.83	1.71	1.64	1.57	1.48	1.39	1.27	1.00

Degrees of freedom for the denominator (ν_2)

IV. Percentage Points of the F Distribution (continued)

$$F_{.01},\ \nu_1,\ \nu_2$$

Degrees of freedom for the numerator (ν_1)

ν_2	1	2	3	4	5	6	7	8	9	10	12	15	20	24	30	40	60	120	∞
1	4052	4999.5	5403	5625	5764	5859	5928	5982	6022	6056	6106	6157	6209	6235	6261	6287	6313	6339	6366
2	98.50	99.00	99.17	99.25	99.30	99.33	99.36	99.37	99.39	99.40	99.42	99.43	99.45	99.45	99.47	99.47	99.48	99.49	99.50
3	34.12	30.82	29.46	28.71	28.24	27.91	27.67	27.49	27.35	27.23	27.05	26.87	26.69	26.60	26.50	26.41	26.32	26.22	26.13
4	21.20	18.00	16.69	15.98	15.52	15.21	14.98	14.80	14.66	14.55	14.37	14.20	14.02	13.93	13.84	13.75	13.65	13.56	13.46
5	16.26	13.27	12.06	11.39	10.97	10.67	10.46	10.29	10.16	10.05	9.89	9.72	9.55	9.47	9.38	9.29	9.20	9.11	9.02
6	13.75	10.92	9.78	9.15	8.75	8.47	8.26	8.10	7.98	7.87	7.72	7.56	7.40	7.31	7.23	7.14	7.06	6.97	6.88
7	12.25	9.55	8.45	7.85	7.46	7.19	6.99	6.84	6.72	6.62	6.47	6.31	6.16	6.07	5.99	5.91	5.82	5.74	5.65
8	11.26	8.65	7.59	7.01	6.63	6.37	6.18	6.03	5.91	5.81	5.67	5.52	5.36	5.28	5.20	5.12	5.03	4.95	4.86
9	10.56	8.02	6.99	6.42	6.06	5.80	5.61	5.47	5.35	5.26	5.11	4.95	4.81	4.73	4.65	4.57	4.48	4.40	4.31
10	10.04	7.56	6.55	5.99	5.64	5.39	5.20	5.05	4.94	4.85	4.71	4.55	4.41	4.33	4.25	4.17	4.08	4.00	3.91
11	9.65	7.21	6.22	5.67	5.32	5.07	4.89	4.74	4.63	4.54	4.40	4.25	4.10	4.02	3.94	3.86	3.78	3.69	3.60
12	9.33	6.93	5.95	5.41	5.06	4.82	4.64	4.50	4.39	4.30	4.16	4.01	3.86	3.78	3.70	3.62	3.54	3.45	3.36
13	9.07	6.70	5.74	5.21	4.86	4.62	4.44	4.30	4.19	4.10	3.96	3.82	3.66	3.59	3.51	3.43	3.34	3.25	3.17
14	8.86	6.51	5.56	5.04	4.69	4.46	4.28	4.14	4.03	3.94	3.80	3.66	3.51	3.43	3.35	3.27	3.18	3.09	3.00
15	8.68	6.36	5.42	4.89	4.56	4.32	4.14	4.00	3.89	3.80	3.67	3.52	3.37	3.29	3.21	3.13	3.05	2.96	2.87
16	8.53	6.23	5.29	4.77	4.44	4.20	4.03	3.89	3.78	3.69	3.55	3.41	3.26	3.18	3.10	3.02	2.93	2.84	2.75
17	8.40	6.11	5.18	4.67	4.34	4.10	3.93	3.79	3.68	3.59	3.46	3.31	3.16	3.08	3.00	2.92	2.83	2.75	2.65
18	8.29	6.01	5.09	4.58	4.25	4.01	3.84	3.71	3.60	3.51	3.37	3.23	3.08	3.00	2.92	2.84	2.75	2.66	2.57
19	8.18	5.93	5.01	4.50	4.17	3.94	3.77	3.63	3.52	3.43	3.30	3.15	3.00	2.92	2.84	2.76	2.67	2.58	2.49
20	8.10	5.85	4.94	4.43	4.10	3.87	3.70	3.56	3.46	3.37	3.23	3.09	2.94	2.86	2.78	2.69	2.61	2.52	2.42
21	8.02	5.78	4.87	4.37	4.04	3.81	3.64	3.51	3.40	3.31	3.17	3.03	2.88	2.80	2.72	2.64	2.55	2.46	2.36
22	7.95	5.72	4.82	4.31	3.99	3.76	3.59	3.45	3.35	3.26	3.12	2.98	2.83	2.75	2.67	2.58	2.50	2.40	2.31
23	7.88	5.66	4.76	4.26	3.94	3.71	3.54	3.41	3.30	3.21	3.07	2.93	2.78	2.70	2.62	2.54	2.45	2.35	2.26
24	7.82	5.61	4.72	4.22	3.90	3.67	3.50	3.36	3.26	3.17	3.03	2.89	2.74	2.66	2.58	2.49	2.40	2.31	2.21
25	7.77	5.57	4.68	4.18	3.85	3.63	3.46	3.32	3.22	3.13	2.99	2.85	2.70	2.62	2.54	2.45	2.36	2.27	2.17
26	7.72	5.53	4.64	4.14	3.82	3.59	3.42	3.29	3.18	3.09	2.96	2.81	2.66	2.58	2.50	2.42	2.33	2.23	2.13
27	7.68	5.49	4.60	4.11	3.78	3.56	3.39	3.26	3.15	3.06	2.93	2.78	2.63	2.55	2.47	2.38	2.29	2.20	2.10
28	7.64	5.45	4.57	4.07	3.75	3.53	3.36	3.23	3.12	3.03	2.90	2.75	2.60	2.52	2.44	2.35	2.26	2.17	2.06
29	7.60	5.42	4.54	4.04	3.73	3.50	3.33	3.20	3.09	3.00	2.87	2.73	2.57	2.49	2.41	2.33	2.23	2.14	2.03
30	7.56	5.39	4.51	4.02	3.70	3.47	3.30	3.17	3.07	2.98	2.84	2.70	2.55	2.47	2.39	2.30	2.21	2.11	2.01
40	7.31	5.18	4.31	3.83	3.51	3.29	3.12	2.99	2.89	2.80	2.66	2.52	2.37	2.29	2.20	2.11	2.02	1.92	1.80
60	7.08	4.98	4.13	3.65	3.34	3.12	2.95	2.82	2.72	2.63	2.50	2.35	2.20	2.12	2.03	1.94	1.84	1.73	1.60
120	6.85	4.79	3.95	3.48	3.17	2.96	2.79	2.66	2.56	2.47	2.34	2.19	2.03	1.95	1.86	1.76	1.66	1.53	1.38
∞	6.63	4.61	3.78	3.32	3.02	2.80	2.64	2.51	2.41	2.32	2.18	2.04	1.88	1.79	1.70	1.59	1.47	1.32	1.00

Degrees of freedom for the denominator (ν_2)

V. Operating Characteristic Curves for the Fixed Effects Model Analysis of Variance[a]

ν_1 = numerator degrees of freedom, ν_2 = denominator degrees of freedom

[a]Adapted with permission from *Biometrika Tables for Statisticians*, Vol. 2, by E. S. Pearson and H. O. Hartley, Cambridge University Press, Cambridge, 1972.

V. Operating Characteristic Curves for the Fixed Effects Model Analysis of Variance
(*continued*)

V. Operating Characteristic Curves for the Fixed Effects Model Analysis of Variance
(*continued*)

V. Operating Characteristic Curves for the Fixed Effects Model Analysis of Variance
(*continued*)

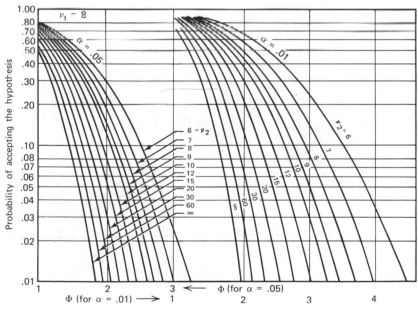

VI. Operating Characteristic Curves for the Random Effects Model Analysis
of Variance[a]

[a]Reproduced with permission from *Engineering Statistics*, 2nd Edition, by A. H. Bowker and
G. J. Lieberman, Prentice-Hall, Inc., Englewood Cliffs, N.J., 1972.

VI. Operating Characteristic Curves for the Random Effects Model Analysis of Variance (*continued*)

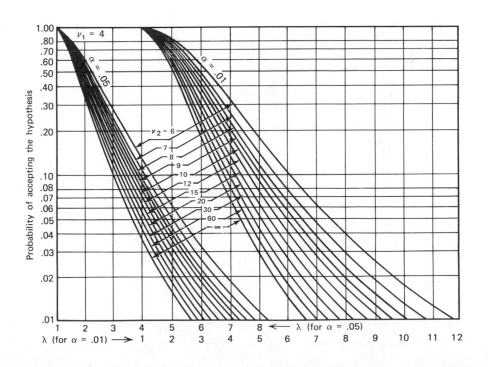

VI. Operating Characteristic Curves for the Random Effects Model Analysis of Variance (*continued*)

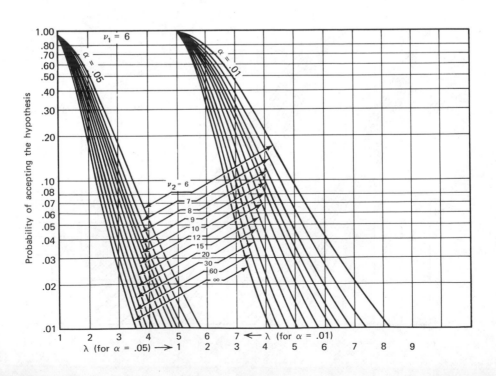

VI. Operating Characteristic Curves for the Random Effects Model Analysis
of Variance (*continued*)

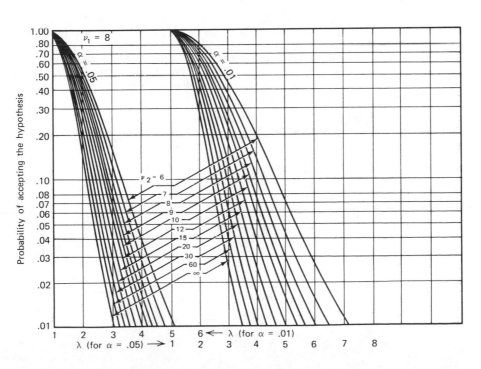

VII. Significant Ranges for Duncan's Multiple Range Test[a]

$$r_{.01}(p, f)$$

f	p												
	2	3	4	5	6	7	8	9	10	20	50	100
1	90.0	90.0	90.0	90.0	90.0	90.0	90.0	90.0	90.0	90.0	90.0	90.0
2	14.0	14.0	14.0	14.0	14.0	14.0	14.0	14.0	14.0	14.0	14.0	14.0
3	8.26	8.5	8.6	8.7	8.8	8.9	8.9	9.0	9.0	9.3	9.3	9.3
4	6.51	6.8	6.9	7.0	7.1	7.1	7.2	7.2	7.3	7.5	7.5	7.5
5	5.70	5.96	6.11	6.18	6.26	6.33	6.40	6.44	6.5	6.8	6.8	6.8
6	5.24	5.51	5.65	5.73	5.81	5.88	5.95	6.00	6.0	6.3	6.3	6.3
7	4.95	5.22	5.37	5.45	5.53	5.61	5.69	5.73	5.8	6.0	6.0	6.0
8	4.74	5.00	5.14	5.23	5.32	5.40	5.47	5.51	5.5	5.8	5.8	5.8
9	4.60	4.86	4.99	5.08	5.17	5.25	5.32	5.36	5.4	5.7	5.7	5.7
10	4.48	4.73	4.88	4.96	5.06	5.13	5.20	5.24	5.28	5.55	5.55	5.55
11	4.39	4.63	4.77	4.86	4.94	5.01	5.06	5.12	5.15	5.39	5.39	5.39
12	4.32	4.55	4.68	4.76	4.84	4.92	4.96	5.02	5.07	5.26	5.26	5.26
13	4.26	4.48	4.62	4.69	4.74	4.84	4.88	4.94	4.98	5.15	5.15	5.15
14	4.21	4.42	4.55	4.63	4.70	4.78	4.83	4.87	4.91	5.07	5.07	5.07
15	4.17	4.37	4.50	4.58	4.64	4.72	4.77	4.81	4.84	5.00	5.00	5.00
16	4.13	4.34	4.45	4.54	4.60	4.67	4.72	4.76	4.79	4.94	4.94	4.94
17	4.10	4.30	4.41	4.50	4.56	4.63	4.68	4.73	4.75	4.89	4.89	4.89
18	4.07	4.27	4.38	4.46	4.53	4.59	4.64	4.68	4.71	4.85	4.85	4.85
19	4.05	4.24	4.35	4.43	4.50	4.56	4.61	4.64	4.67	4.82	4.82	4.82
20	4.02	4.22	4.33	4.40	4.47	4.53	4.58	4.61	4.65	4.79	4.79	4.79
30	3.89	4.06	4.16	4.22	4.32	4.36	4.41	4.45	4.48	4.65	4.71	4.71
40	3.82	3.99	4.10	4.17	4.24	4.30	4.34	4.37	4.41	4.59	4.69	4.69
60	3.76	3.92	4.03	4.12	4.17	4.23	4.27	4.31	4.34	4.53	4.66	4.66
100	3.71	3.86	3.98	4.06	4.11	4.17	4.21	4.25	4.29	4.48	4.64	4.65
∞	3.64	3.80	3.90	3.98	4.04	4.09	4.14	4.17	4.20	4.41	4.60	4.68

f = degrees of freedom.

[a]Reproduced with permission from "Multiple Range and Multiple F Tests," by D. B. Duncan, *Biometrics*, Vol. 1, No. 1, pp. 1–42, 1955.

VII. Significant Ranges for Duncan's Multiple Range Test (*continued*)

$$r_{.05}(p, f)$$

						p							
f	2	3	4	5	6	7	8	9	10	20	50	100
1	18.0	18.0	18.0	18.0	18.0	18.0	18.0	18.0	18.0	18.0	18.0	18.0
2	6.09	6.09	6.09	6.09	6.09	6.09	6.09	6.09	6.09	6.09	6.09	6.09
3	4.50	4.50	4.50	4.50	4.50	4.50	4.50	4.50	4.50	4.50	4.50	4.50
4	3.93	4.01	4.02	4.02	4.02	4.02	4.02	4.02	4.02	4.02	4.02	4.02
5	3.64	3.74	3.79	3.83	3.83	3.83	3.83	3.83	3.83	3.83	3.83	3.83
6	3.46	3.58	3.64	3.68	3.68	3.68	3.68	3.68	3.68	3.68	3.68	3.68
7	3.35	3.47	3.54	3.58	3.60	3.61	3.61	3.61	3.61	3.61	3.61	3.61
8	3.26	3.39	3.47	3.52	3.55	3.56	3.56	3.56	3.56	3.56	3.56	3.56
9	3.20	3.34	3.41	3.47	3.50	3.52	3.52	3.52	3.52	3.52	3.52	3.52
10	3.15	3.30	3.37	3.43	3.46	3.47	3.47	3.47	3.47	3.48	3.48	3.48
11	3.11	3.27	3.35	3.39	3.43	3.44	3.45	3.46	3.46	3.48	3.48	3.48
12	3.08	3.23	3.33	3.36	3.40	3.42	3.44	3.44	3.46	3.48	3.48	3.48
13	3.06	3.21	3.30	3.35	3.38	3.41	3.42	3.44	3.45	3.47	3.47	3.47
14	3.03	3.18	3.27	3.33	3.37	3.39	3.41	3.42	3.44	3.47	3.47	3.47
15	3.01	3.16	3.25	3.31	3.36	3.38	3.40	3.42	3.43	3.47	3.47	3.47
16	3.00	3.15	3.23	3.30	3.34	3.37	3.39	3.41	3.43	3.47	3.47	3.47
17	2.98	3.13	3.22	3.28	3.33	3.36	3.38	3.40	3.42	3.47	3.47	3.47
18	2.97	3.12	3.21	3.27	3.32	3.35	3.37	3.39	3.41	3.47	3.47	3.47
19	2.96	3.11	3.19	3.26	3.31	3.35	3.37	3.39	3.41	3.47	3.47	3.47
20	2.95	3.10	3.18	3.25	3.30	3.34	3.36	3.38	3.40	3.47	3.47	3.47
30	2.89	3.04	3.12	3.20	3.25	3.29	3.32	3.35	3.37	3.47	3.47	3.47
40	2.86	3.01	3.10	3.17	3.22	3.27	3.30	3.33	3.35	3.47	3.47	3.47
60	2.83	2.98	3.08	3.14	3.20	3.24	3.28	3.31	3.33	3.47	3.48	3.48
100	2.80	2.95	3.05	3.12	3.18	3.22	3.26	3.29	3.32	3.47	3.53	3.53
∞	2.77	2.92	3.02	3.09	3.15	3.19	3.23	3.26	3.29	3.47	3.61	3.67

f = degrees of freedom.

VIII. Coefficients of Orthogonal Polynomials[a]

SEE P. 55, 148

n = 3

X_j	P_1	P_2
1	-1	1
2	0	-2
3	1	1
$\sum_{j=1}^{n} \{P_i(x_j)\}^2$	2	6
λ	1	3

n = 4

X_j	P_1	P_2	P_3
1	-3	1	-1
2	-1	-1	3
3	1	-1	-3
4	3	1	1
$\sum_{j=1}^{n} \{P_i(x_j)\}^2$	20	4	20
λ	2	1	$\frac{10}{3}$

n = 5

X_j	P_1	P_2	P_3	P_4
1	-2	2	-1	1
2	-1	-1	2	-4
3	0	-2	0	6
4	1	-1	-2	-4
5	2	2	1	1
$\sum_{j=1}^{n} \{P_i(x_j)\}^2$	10	14	10	70
λ	1	1	$\frac{5}{6}$	$\frac{35}{12}$

n = 6

X_j	P_1	P_2	P_3	P_4	P_5
1	-5	5	-5	1	-1
2	-3	-1	7	-3	5
3	-1	-4	4	2	-10
4	1	-4	-4	2	10
5	3	-1	-7	-3	-5
6	5	5	5	1	1
$\sum_{j=1}^{n} \{P_i(x_j)\}^2$	70	84	180	28	252
λ	2	$\frac{3}{2}$	$\frac{5}{3}$	$\frac{7}{12}$	$\frac{21}{10}$

n = 7

X_j	P_1	P_2	P_3	P_4	P_5	P_6
1	-3	5	-1	3	-1	1
2	-2	0	1	-7	4	-6
3	-1	-3	1	1	-5	15
4	0	-4	0	6	0	-20
5	1	-3	-1	1	5	15
6	2	0	-1	-7	-4	-6
7	3	5	1	3	1	1
$\sum_{j=1}^{n} \{P_i(x_j)\}^2$	28	84	6	154	84	924
λ	1	1	$\frac{1}{6}$	$\frac{7}{12}$	$\frac{7}{20}$	$\frac{77}{60}$

n = 8

X_j	P_1	P_2	P_3	P_4	P_5	P_6
1	-7	7	-7	7	-7	1
2	-5	1	5	-13	23	-5
3	-3	-3	7	-3	-17	9
4	-1	-5	3	9	-15	-5
5	1	-5	-3	9	15	-5
6	3	-3	-7	-3	17	9
7	5	1	-5	-13	-23	-5
8	7	7	7	7	7	1
$\sum_{j=1}^{n} \{P_i(x_j)\}^2$	168	168	264	616	2,184	264
λ	2	1	$\frac{2}{3}$	$\frac{7}{12}$	$\frac{7}{10}$	$\frac{11}{60}$

n = 9

X_j	P_1	P_2	P_3	P_4	P_5	P_6
1	-4	28	-14	14	-4	4
2	-3	7	7	-21	11	-17
3	-2	-8	13	-11	-4	22
4	-1	-17	9	9	-9	1
5	0	-20	0	18	0	-20
6	1	-17	-9	9	9	1
7	2	-8	-13	-11	4	22
8	3	7	-7	-21	-11	-17
9	4	28	14	14	4	4
$\sum_{j=1}^{n} \{P_i(x_j)\}^2$	60	2,772	990	2,002	468	1,980
λ	1	3	$\frac{5}{6}$	$\frac{7}{12}$	$\frac{3}{20}$	$\frac{11}{60}$

n = 10

X_j	P_1	P_2	P_3	P_4	P_5	P_6
1	-9	6	-42	18	-6	3
2	-7	2	14	-22	14	-11
3	-5	-1	35	-17	-1	10
4	-3	-3	31	3	-11	6
5	-1	-4	12	18	-6	-8
6	1	-4	-12	18	6	-8
7	3	-3	-31	3	11	6
8	5	-1	-35	-17	1	10
9	7	2	-14	-22	-14	-11
10	9	6	42	18	6	3
$\sum_{j=1}^{n} \{P_i(x_j)\}^2$	330	132	8,580	2,860	780	660
λ	2	$\frac{1}{2}$	$\frac{5}{3}$	$\frac{5}{12}$	$\frac{1}{10}$	$\frac{11}{240}$

[a]Adapted with permission from *Biometrika Tables For Statisticians*, Vol. 1, 3rd Edition by E. S. Pearson and H. O. Hartley, Cambridge University Press, Cambridge, 1966.

IX. Random Numbers[a]

10480	15011	01536	02011	81647	91646	69179	14194	62590
22368	46573	25595	85393	30995	89198	27982	53402	93965
24130	48360	22527	97265	76393	64809	15179	24830	49340
42167	93093	06243	61680	07856	16376	39440	53537	71341
37570	39975	81837	16656	06121	91782	60468	81305	49684
77921	06907	11008	42751	27756	53498	18602	70659	90655
99562	72905	56420	69994	98872	31016	71194	18738	44013
96301	91977	05463	07972	18876	20922	94595	56869	69014
89579	14342	63661	10281	17453	18103	57740	84378	25331
85475	36857	53342	53988	53060	59533	38867	62300	08158
28918	69578	88231	33276	70997	79936	56865	05859	90106
63553	40961	48235	03427	49626	69445	18663	72695	52180
09429	93969	52636	92737	88974	33488	36320	17617	30015
10365	61129	87529	85689	48237	52267	67689	93394	01511
07119	97336	71048	08178	77233	13916	47564	81056	97735
51085	12765	51821	51259	77452	16308	60756	92144	49442
02368	21382	52404	60268	89368	19885	55322	44819	01188
01011	54092	33362	94904	31273	04146	18594	29852	71585
52162	53916	46369	58586	23216	14513	83149	98736	23495
07056	97628	33787	09998	42698	06691	76988	13602	51851
48663	91245	85828	14346	09172	30168	90229	04734	59193
54164	58492	22421	74103	47070	25306	76468	26384	58151
32639	32363	05597	24200	13363	38005	94342	28728	35806
29334	27001	87637	87308	58731	00256	45834	15398	46557
02488	33062	28834	07351	19731	92420	60952	61280	50001
81525	72295	04839	96423	24878	82651	66566	14778	76797
29676	20591	68086	26432	46901	20849	89768	81536	86645
00742	57392	39064	66432	84673	40027	32832	61362	98947
05366	04213	25669	26422	44407	44048	37937	63904	45766
91921	26418	64117	94305	26766	25940	39972	22209	71500
00582	04711	87917	77341	42206	35126	74087	99547	81817
00725	69884	62797	56170	86324	88072	76222	36086	84637
69011	65795	95876	55293	18988	27354	26575	08625	40801
25976	57948	29888	88604	67917	48708	18912	82271	65424
09763	83473	73577	12908	30883	18317	28290	35797	05998
91567	42595	27958	30134	04024	86385	29880	99730	55536
17955	56349	90999	49127	20044	59931	06115	20542	18059
46503	18584	18845	49618	02304	51038	20655	58727	28168
92157	89634	94824	78171	84610	82834	09922	25417	44137
14577	62765	35605	81263	39667	47358	56873	56307	61607

IX. Random Numbers[a] (*continued*)

98427	07523	33362	64270	01638	92477	66969	98420	04880
34914	63976	88720	82765	34476	17032	87589	40836	32427
70060	28277	39475	46473	23219	53416	94970	25832	69975
53976	54914	06990	67245	68350	82948	11398	42878	80287
76072	29515	40980	07391	58745	25774	22987	80059	39911
90725	52210	83974	29992	65831	38857	50490	83765	55657
64364	67412	33339	31926	14883	24413	59744	92351	97473
08962	00358	31662	25388	61642	34072	81249	35648	56891
95012	68379	93526	70765	10592	04542	76463	54328	02349
15664	10493	20492	38391	91132	21999	59516	81652	27195

[a]Reproduced with permission from *Probability and Statistics in Engineering and Management Science* by W. W. Hines and D. C. Montgomery, The Ronald Press, New York, 1972.

Index